FIELD GUIDE TO THE WILD FLOWERS of the
CANARY ISLANDS

FIELD GUIDE TO THE WILD FLOWERS of the
CANARY ISLANDS

Chris Thorogood
Mark Carine
J. Alfredo Reyes-Betancort

Kew Publishing
Royal Botanic Gardens, Kew

© The Board of Trustees of the Royal Botanic Gardens, Kew 2024
Text and illustrations © the authors
Photographs © the authors, unless otherwise stated

The authors have asserted their rights to be identified as the authors of this work in accordance with the Copyright, Designs and Patents Act 1988.

All rights reserved. No part of this publication may be reproduced, stored in a retrieval system, or transmitted, in any form, or by any means, electronic, mechanical, photocopying, recording or otherwise, without the written permission of the publisher unless in accordance with the provisions of the Copyright Designs and Patents Act 1988.

Great care has been taken to maintain the accuracy of the information contained in this work. However, neither the publisher, the editors nor the authors can be held responsible for any consequences arising from use of the information contained herein. The views expressed in this work are those of the individual authors and do not necessarily reflect those of the publisher or of the Board of Trustees of the Royal Botanic Gardens, Kew.

First published in 2024 by
Royal Botanic Gardens, Kew
Richmond, Surrey, TW9 3AB, UK
www.kew.org

ISBN 978 1 84246 822 7
eISBN 978 1 84246 824 1

Distributed on behalf of the Royal Botanic Gardens, Kew in North America by the University of Chicago Press, 1427 East 60th Street, Chicago, IL 60637, USA.

British Library Cataloguing in Publication Data
A catalogue record for this book is available from the British Library

Copy-editing: James Kingsland
Design and page layout: Christine Beard
Proofreading: Sharon Whitehead
Production management: Georgina Hills

COVER ILLUSTRATIONS
Front: *Euphorbia handiensis*
Back, clockwise from top left: *Malva acerifolia*; *Lotus pyranthus*; *Ferula lancerotensis*; *Canarina canariensis*; *Aeonium subplanum*; *Dracaena tamaranae*
FRONTISPIECE: *Dracunculus canariensis*
CONTENTS PAGE: *Euphorbia canariensis*

Printed and bound in Great Britain by Gomer Press

For information or to purchase all Kew titles please visit
shop.kew.org/kewbooksonline or email publishing@kew.org

Kew's mission is to understand and protect plants and fungi, for the wellbeing of people and the future of all life on Earth.

Kew receives approximately one third of its running costs from Government through the Department for Environment, Food and Rural Affairs (Defra). All other funding needed to support Kew's vital work comes from members, foundations, donors and commercial activities including book sales.

Acknowledgements

Our sincere thanks to the students who contributed significantly to the working manuscript, checking data and distributions for the plants included in this guide: Isaac Terrington, Catherine Walter, George MacKay and Jonas Nohturfft. The project would not have been possible without their help. Thank you to those who provided additional photographs, especially Rachael Graham, Oliver White and Amy Jackson. Thanks to all those who supported us during the many years spent in the field, especially Arnoldo Santos Guerra on the western islands, and Matías Hernández González on Lanzarote and La Graciosa.

Contents

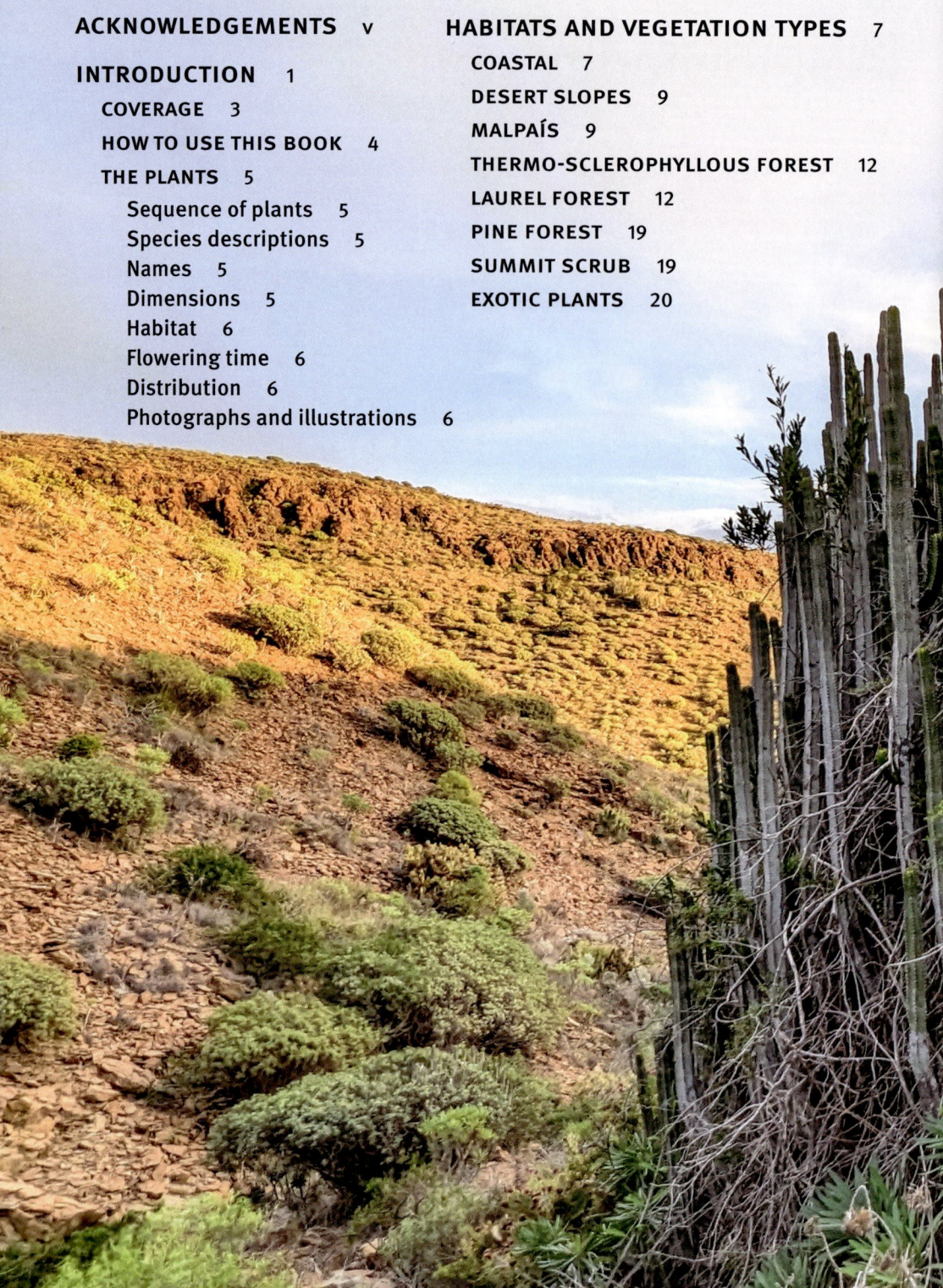

ACKNOWLEDGEMENTS v

INTRODUCTION 1
   COVERAGE 3
   HOW TO USE THIS BOOK 4
   THE PLANTS 5
      Sequence of plants 5
      Species descriptions 5
      Names 5
      Dimensions 5
      Habitat 6
      Flowering time 6
      Distribution 6
      Photographs and illustrations 6

HABITATS AND VEGETATION TYPES 7
   COASTAL 7
   DESERT SLOPES 9
   MALPAÍS 9
   THERMO-SCLEROPHYLLOUS FOREST 12
   LAUREL FOREST 12
   PINE FOREST 19
   SUMMIT SCRUB 19
   EXOTIC PLANTS 20

THE EASTERN ISLANDS 21
    LANZAROTE 21
    LA GRACIOSA 22
    FUERTEVENTURA 24

THE WESTERN ISLANDS 27
    GRAN CANARIA 27
    TENERIFE 30
    LA GOMERA 34
    LA PALMA 36
    EL HIERRO 38

DIVERSITY AND EVOLUTION 40

SPECIES DESCRIPTIONS 43

GLOSSARY OF TERMS 395

INDEX OF SPANISH NAMES 405

INDEX OF ENGLISH NAMES 411

INDEX OF SCIENTIFIC NAMES 417

Ceropegia fusca, Gran Canaria

Introduction

The Canary Islands archipelago comprises eight major islands, and their associated islets, located in the Atlantic off the northwest coast of Africa. The main islands, from largest to smallest in area, are: Tenerife, Fuerteventura, Gran Canaria, Lanzarote, La Palma, La Gomera, El Hierro and La Graciosa. The archipelago includes many smaller islands and islets, including Alegranza, Isla de Lobos, Montaña Clara, Roque del Oeste and Roque del Este. The islands' varying climates, topography and geology have driven the evolution of distinct vegetation zones including thermo-sclerophyllous (hard-leaved) forest, laurel forest, *fayal-brezal*, pine forest and high-montane broom, as well as arid and semiarid lowland desert scrubs, dunes and malpaís communities. Each has its own distinct community of plants.

The Canary Islands fall within the wider region called Macaronesia, which also includes the Azores, Madeira, Selvagens and Cabo Verde, although the validity of this collection of archipelagos as a biogeographic unit is questioned by some authors. These five volcanic archipelagos are also considered part of the wider Mediterranean Basin Biodiversity Hotspot (an area of high biodiversity under threat). The main Canary Islands originated as submarine seamount volcanoes on the floor of the Atlantic Ocean. In this guide, we divide the eight islands into two groups for ease: the Eastern Canary Islands (Fuerteventura, Lanzarote and Graciosa) and the Western Canary Islands (Gran Canaria, Tenerife, La Gomera, La Palma and El Hierro), together with their smaller islets. There is a progression in age from the oldest and most eroded easternmost islands, to the steeper, more rugged and generally more humid westernmost islands. The western islands consist of mountain peaks that rise directly from a deep ocean floor; the two easternmost islands (once a single island) rise from a separate submarine plateau. Volcanic and subsequent erosional processes have had a profound effect on topography across the archipelago, from the island-building strato-volcanoes of Teide on Tenerife, to the island-eroding landslips of El Hierro.

Map of the Canary Islands, courtesy of NASA Worldview.

The general climate of the islands is Mediterranean, characterised by scarce autumn and winter rainfall and summer drought. However, there is variation in the climate both across the archipelago, with the eastern islands notably drier, and also within islands where the topography and prevailing weather patterns result in considerable climatic variation over short distances. This means that areas of temperate rainforest and of desert-like conditions occur in close proximity.

The flora of the Canary Islands is rich, comprising almost 700 endemic taxa (species and subspecies), over a land area of 7,492 km², accounting collectively for over half of the total native flora. Many of these native species are restricted in their distribution; across the islands, several areas have been identified that have a particularly high concentration of local endemics. The flora has been shaped by the Canary Islands' long history of human occupation dating from the first inhabitants, the Guanches – pre-European occupants in the first millennium BCE. Today this rich flora is highly vulnerable to environmental change: the islands are home to 2.2 million residents and host over 15 million international tourists per year, on a relatively small landmass. Meanwhile, over 30% of the endemic flora is included on the Canary Islands Red List, and is of conservation concern. In order to conserve the unique and threatened flora, vegetation, landscapes and culture of the Canary Islands, an extensive network of protected areas has been established that covers approximately 40% of the archipelago's total land area.

This book's mission is to be the most comprehensive single-volume field guide to the wild flowers of the Canary Islands, giving the reader the best chance of finding and identifying a given plant in a rich and complex flora.

The Canary Island endemic *Canarina canariensis* in the forests of Los Tilos de Moya in Barranco de la Virgen, Gran Canaria.

COVERAGE

It would be impossible to depict in a single, easy-to-use volume all the thousands of species, native and introduced, that occur in the Canary Islands. Coverage is therefore balanced by island, habitat and vegetation type, and classification (taxonomy).

Digitalis canariensis

Island coverage: species encountered across all islands are covered, for example the widespread tabaiba dulce (*Euphorbia balsamifera*). Conspicuous plants, such as *Digitalis canariensis*, which occurs on three islands, are included because they are likely to be spotted regardless of their distribution. Meanwhile, many single-island endemics are also covered. For example, *Euphorbia handiensis* can only be encountered in a restricted area of Fuerteventura; nevertheless, it is an important constituent of the flora.

La Graciosa, seen from the north of Lanzarote.

Habitat and vegetation type: coverage spans all the major habitats on the islands, including plant communities found in thermophilous, laurel and pine forests, as well as in montane, arid, coastal habitats and malpaís. A few species may be found in more than one habitat type; for example, where coastal dunes merge into jables (sand flats) and malpaís inland, or in transition zones along altitudinal gradients.

Echium wildpretii in Tenerife.

Classification (taxonomy): all major flowering plant families are included, although some families are represented in greater detail than others. This is because some families are of wider interest or more conspicuous, and better suited to a field guide. For example, the Boraginaceae, and specifically the echiums (genus *Echium*) are given greater coverage because they are an iconic example of a Canary Islands radiation. Grasses and sedges are given relatively less coverage because they are less conspicuous or not unique to the archipelago. Ferns are included, as are some commonly planted ornamentals and naturalised exotics; while the latter are not native, they form a significant part of the flora, furthermore it may not be obvious whether a plant is native or not.

HOW TO USE THIS BOOK

This book is a visual guide in which the plants are sequenced by family, in order of evolutionary relatedness. Readers with more advanced knowledge will head straight to the relevant plant family and search for the species in question. Beginners are likely to scan the photographs and use the associated descriptions and illustrations to confirm identification, cross-checking descriptions for closely related species. Key distinguishing characteristics are highlighted to assist this process. The less experienced enthusiast will typically follow the sequence below to identify a plant:

1. Identify the relevant section of the book by checking if the plant belongs to one of the major plant families.

2. Scan the photographs for that family to find a species similar to the plant in question.

3. For larger or more complex groups, find the relevant subgroup heading A, B, C, etc., which summarises the salient features.

4. Use the descriptions and illustrations associated with the photographs to confirm identification.

5. Verify identification by cross-checking descriptions for closely related species, noting carefully the key distinguishing characteristics highlighted.

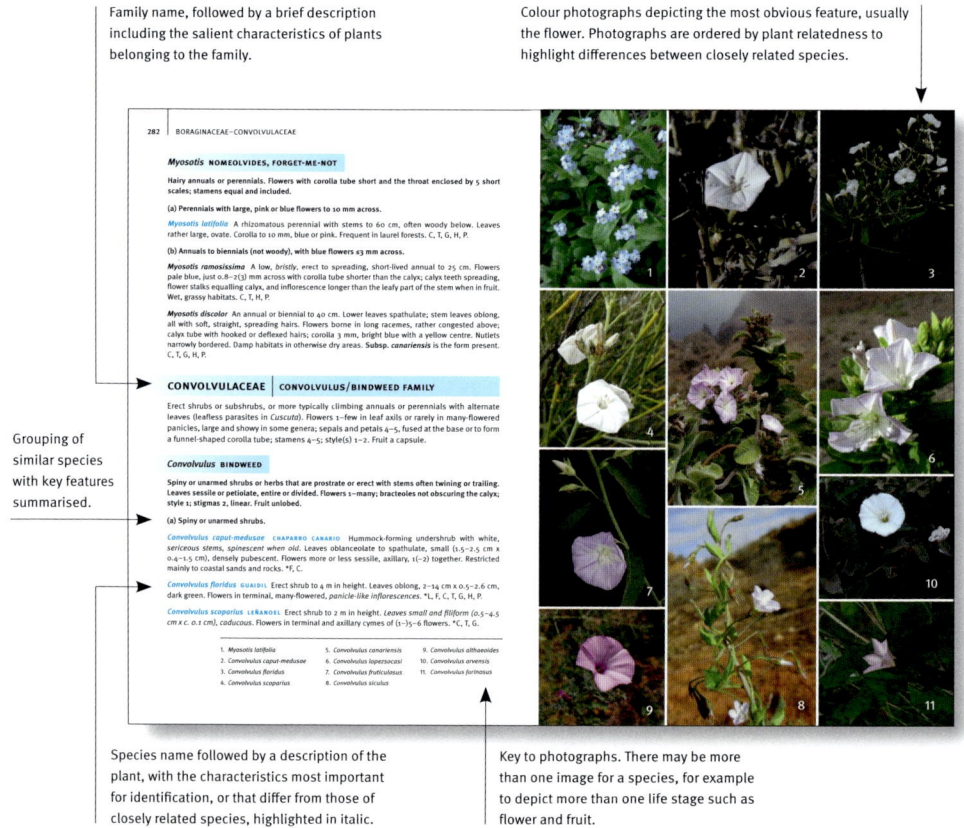

Family name, followed by a brief description including the salient characteristics of plants belonging to the family.

Colour photographs depicting the most obvious feature, usually the flower. Photographs are ordered by plant relatedness to highlight differences between closely related species.

Grouping of similar species with key features summarised.

Species name followed by a description of the plant, with the characteristics most important for identification, or that differ from those of closely related species, highlighted in italic.

Key to photographs. There may be more than one image for a species, for example to depict more than one life stage such as flower and fruit.

THE PLANTS

Sequence of plants

All plants in this book are assigned to a hierarchy that begins with five groups: ferns (spore-producing plants), gymnosperms (non-flowering seed plants), basal angiosperms (a set of ancient lineages of flowering plants), monocots (a group including bulbs, orchids and grasses) and eudicots (representing the majority of plants). Within each of these groups, families are in turn made up of genera and species. The relatedness of families on which this sequence is based follows the APG IV (Angiosperm Phylogeny Group IV) system, which is the globally accepted classification for flowering plants.

The assignment of species to a family in this book will sometimes differ from that of traditional floras, in which plants were classified according to their morphology (appearance), rather than their relatedness, an obvious example being the Canary Island foxglove (*Digitalis canariensis*), which was formerly in the family Scrophulariaceae but is now assigned to the family Plantaginaceae. The sequence of families may also deviate subtly from texts that followed earlier versions of the APG system, which has been updated four times, but is now relatively stable.

Species descriptions

Similar plants within a large or complex family have been grouped by important shared features under headings (a), (b), (c), etc. These replace traditional keys, for which there is not space in this book and which can lead to incorrect identification unless all of the species in a given family are included. The descriptions include dimensions, habitat and distribution, with the most important features highlighted in italics. Not all descriptions are written in exactly the same way. For example, it is useful to highlight a particular set of characteristics (such as leaf shape) in one genus but not in another. The use of technical terms is unavoidable to differentiate species reliably, so a list of definitions is included in the Glossary.

Names

Each plant is assigned a Spanish name and an English name (if it has one), and a scientific name, which is made up of two parts: the first is the genus to which the species belongs; the second is the specific epithet. Some are further split into subspecies, which differ subtly but are not considered distinct enough to be treated as separate at the species rank. Plant species names are frequently revised as more information becomes available about their relatedness. This can lead to confusion, especially where botanists do not agree on the status of a species, subspecies or variety. For these reasons, plant names may vary from one guide to another as well as over time. To avoid confusion, commonly used synonyms are included for species that are widely known by another name, indicated by '(syn.)'. Subspecies are included where they are geographically important or particularly distinct. While it is conventional to include all described subspecies for a plant, it is beyond the scope of this guide to do so, particularly for variable species in which numerous subspecies exist.

Dimensions

Dimensions are based on the authors' observations and information from scientific papers. However, they vary under different environmental conditions and with plant age, so should be viewed as an approximate guide unless emphasised specifically. In some cases, maximum and minimum dimensions outside of the typical ranges are also given in parentheses.

Habitat

The habitat is described because this is often useful for identification. For example, a plant typically found on sea cliffs is unlikely to be encountered in a humid laurel forest. Where several similar species that share a habitat are described together, the habitat is only described for the first, to avoid repetition.

Flowering time

The precise flowering time is not given in the descriptions because this varies from year to year, with each island and with elevation. Many species flower in February, March and April; for those that are not spring-flowering, the approximate flowering time is highlighted in the description.

Distribution

It would be impossible to include a map for every species, and there is no other precise yet simple means of indicating distribution. The distributions given are based on regional floras and databases, and on the authors' observations. It is possible that a species may be encountered outside the distribution given in this guide. Islands are indicated with abbreviations (see map in Introduction). Canary Islands endemics (plants that occur nowhere else), are denoted with an asterisk, e.g. a plant with '*L, F' listed as its distribution, is endemic to Lanzarote and Fuerteventura.

Photographs and illustrations

Photographs have been selected to reflect coverage based on geography, habitat, taxonomy and classification. Names of plants for which photographs are included are indicated in blue bold. The majority of photographs depict the flower, which is usually the most useful for identification. Black-and-white illustrations are provided to complement the descriptions and to illustrate the characteristics that are most useful for identification.

Aeonium tabulaeforme, Tenerife

Habitats and vegetation types

The Canary Islands archipelago supports a wide diversity of habitats, given its small land mass. Among the richest of these are the vestiges of ancient humid forest that cover much of the western islands – especially those at the margins and on cliffs. Further east, drier habitats, including malpaís, desertic slopes and coastal habitats, host a different flora, with strong links to continental Africa. Here a brief guide is given to the main types of habitat, some of which intergrade in transition zones. Some of the notable or more prominent species are highlighted from each habitat.

COASTAL

Extensive sand dune systems extend for miles along the coasts of some of the Canary Islands, interrupted in places where ancient black lava fields flow into the sea. The flora here shares links with Madeira, and has North African elements. It is well developed in the eastern islands and extends to the southern coast of Tenerife, west to La Gomera.

Vast unspoiled stretches of sand in the Corralejo Natural Park form a veritable Sahara Desert in the northeast corner of Fuerteventura, and indeed, the park shares many plant species with the Sahara region. Vegetation here is dominated by shrubs such as *Traganum moquinii*, *Atriplex halimus*, *Tetraena fontanesii* and *Ononis hesperia*, interrupted by flexuous stands of *Euphorbia paralias*, and patchy carpets of various *Suaeda* spp. among the rocks. *Salsola divaricata*, an endemic of the Canary Islands, is common on most islands. Other common species include patch-forming *Mesembryanthemum crystallinum* and *M. nodiflorum*, and *Heliotropium bacciferum*.

Locally in Lanzarote and Fuerteventura, the thick, pineapple-yellow spikes of *Cistanche phelypaea* pierce the sand and can be seen from some distance along the roadsides. This leafless parasite steals food from the roots of desert shrubs; its fruiting spikes, blackened by the sun, persist for many months after flowering. A more widespread parasitic plant is

Cystanche phelypaea

8 HABITATS AND VEGETATION TYPES

Coastal plants: A. *Traganum moquinii*, **B.** *Mesembryanthemum crystallinum*, **C.** *Heliotropium bacciferum*,
D. *Astydamia latifolia*, **E.** *Polygonum maritimum*, **F.** *Atriplex halimus*, **G.** *Tetraena fontanesii*, **H.** *Euphorbia paralias*.

dodder (*Cuscuta approximata*) which produces tangled balls of wiry yellow stems which often completely engulf their host plants.

Lagoons and saltmarshes are dominated by verdant-looking saltworts such as *Arthrocaulon macrostachyum* and *Sarcocornia perennis*, and ravine mouths in saline areas are vegetated with thickets of *Tamarix canariensis*, *Atriplex glauca* subsp. *ifniensis*, *Lycium intricatum*, *Caroxylon* spp. and *Suaeda* spp.

DESERT SLOPES

Much of the drier eastern Canary Islands is desert. The barren, honey-coloured interior slopes are dominated by just a few common species of desert shrub. Among the most abundant are those in the family Amaranthaceae. Mediterranean saltwort (*Caroxylon vermiculatum*) is a ubiquitous, tough, drought-tolerant species that produces conspicuous, often bright red, papery-winged fruits in late summer. Aulaga (*Launaea arborescens*) is also abundant on dry slopes and studs the landscape with its intricate balls of blue-grey, virtually leafless stems. Rocky screes and red and black volcanic slopes are commonly blotched with thickets of *Lycium intricatum* (family Solanaceae), known locally as espino. In the spring, its sun-bleached, robustly spiny stems sprout leaves and purple flowers, followed by prominent red, ovoid berries. On unstable or degraded slopes, sparse thickets of shrub tobacco (*Nicotiana glauca*) bear pendulous clusters of yellow flowers. An interesting spring-flowering species on dry, stony hillsides on Fuerteventura is *Pallenis hierichuntica*, a so-called 'resurrection plant' that only blooms after winter rains.

MALPAÍS

Malpaís (meaning 'bad land') is a barren, rocky landscape derived from lava fields. It is rough and difficult to traverse, and forms great swathes across many of the islands. Malpaís is particularly rich in lichen species, which give a distinct greenish tint to the landscape. Volcanic rocks are often heavily festooned and encrusted with pale grey-green lichens belonging to the genera *Stereocaulon*, *Seirophora* and *Ramalina*, as well as occasional patches of *Cladonia foliacea*, *Squamarina cartilaginea* and *Toninia tristis*. Tracksides where water collects are a good hunting ground for spring-flowering annuals such as *Calendula arvensis* and *Reichardia tingitana*. Bulbs such as *Asphodelus tenuifolius* and the attractive, brown-flowered *Dipcadi serotinum* are conspicuous in winter and spring.

Common shrubs of desert slopes include species in the family Amaranthaceae (top inset, left to right: *Suaeda vera*, *Suaeda ifniensis*, *Caroxylon vermiculatum* in flower; *Caroxylon vermiculatum* in leaf).

Older malpaís landscapes are often dominated by *Euphorbia* scrub vegetation. This vegetation has an ancient, dry-tropical origin and occurs on all the islands. *Euphorbia balsamifera* typically predominates, and more locally *Euphorbia canariensis* (absent across much of the eastern islands, for example Lanzarote). Other common plants include *Caroxylon vermiculatum*, *Launaea*

Plants of the malpaís: **A.** *Caroxylon vermiculatum*, **B.** *Carrichtera annua*, **C.** *Rumex vesicarius*, **D.** *Asteriscus intermedius*, **E.** *Reichardia tingitana*, **F.** *Vicia benghalensis*, **G.** *Tetraena gaetula*, **H.** *Frankenia pulverulenta*, **I.** *Bassia tomentosa*, **J.** *Cenchrus ciliaris*.

Plants of the malpaís: A. *Launaea arborescens*, **B.** *Fagonia cretica*, **C.** *Patellifolia patellaris*, **D.** *Kleinia neriifolia*, **E.** *Asphodelus tenuifolius*, **F.** *Lycium intricatum*, **G.** *Aizoon canariense*.

arborescens, *Helianthemum canariense*, *Kleinia neriifolia* and *Lycium intricatum*. *Rumex lunaria* is locally dominant in some areas, for example on the volcanic slopes near Máguez in Lanzarote where the species was introduced as a forage crop.

THERMO-SCLEROPHYLLOUS FOREST

Warm forests on the western islands are of palaeo-Mediterranean origin and have a distinctly Mediterranean feel. Little of the original vegetation survives; most is restricted to vestiges. These forests often intergrade with stands of Canary Island pine (*Pinus canariensis*) (see below), and are dominated by smaller sclerophyllous trees and shrubs such as juniper (*Juniperus canariensis*), olive (*Olea cerasiformis*) and mastics (*Pistacia atlantica* and *P. lentiscus*). Mixed juniper and pine woods form a thick green mantle over cliffs and outcrops of Tenerife and El Hierro in particular. In degraded plant communities, *Hypericum canariense* is common on Gran Canaria, Tenerife, Gomera and La Palma; *Rhamnus crenulata* is also common in Tenerife (both species are also frequent in laurel forests). Grazed, burnt or stripped areas of thermophilic forests feature swathes of open scrub dominated by white broom (*Retama rhodorhizoides*), various echiums (*Echium*) and rockroses (*Cistus*). Bare cliffs and outcrops in the forests often harbour an interesting assemblage of rock-dwelling succulents in the family Crassulaceae, such as *Aeonium*, *Aichryson*, *Greenovia* and *Monanthes*. Damper valleys and springs are home to rare native stands of Canary Island palm (*Phoenix canariensis*). These wild palm groves often include date palms (*P. dactylifera*), together with their hybrids, and are best preserved on La Gomera, Gran Canaria, and locally in Fuerteventura.

LAUREL FOREST

Humid cloud forests cover the trade-wind-facing slopes at higher elevations of the western islands (and locally in Gran Canaria), with residual elements on the summit of southern Fuerteventura. These mist-filled forests are a relic of a Tertiary flora dominated by perennial broadleaf laurifolious trees, and are some of the most diverse vegetation on the archipelago. The dominance of the various species in this rich assemblage of plants varies with altitude. At altitudes below 800 m, plants adapted to drier conditions are frequent, for example *Arbutus*, *Persea barbujana*, *Visnea* and *Picconia*. Above 800 m, conditions are more humid, and *Laurus novocanariensis*, *Prunus lusitanica* subsp. *hixa*, *Ilex perado* subsp. *platyphylla*, *I. canariensis* and *Persea indica* are common to sub-dominant. This vegetation is particularly well preserved in La Gomera and La Palma where, over much of the interior landscape, damp, mossy forests form a green mantle to which clouds cling. Shady ravine forests are dominated by *Ocotea foetens* and *Persea indica*, with an exuberant undergrowth of ferns including *Davallia canariensis*, *Dryopteris oligodonta*, *Asplenium onopteris*, *Diplazium caudatum* and *Woodwardia radicans*. At altitudes above 1200 m, conditions are cooler and drier; at such elevations, cold-tolerant evergreen forests become dominated by *Myrica faya* and *Erica canariensis*.

A unique form of vegetation within these forests, called *fayal-brezal*, arose as a result of laurel deforestation. It is dominated by tree heaths (*Erica canariensis*) and broad-leaved shrubs such as *Myrica faya*. *Fayal-brezal* occurs on all western islands where laurel forests have been cut down, and includes species such as *Daphne gnidium*, *Hypericum grandifolium* and *Ilex canariensis*.

Canary Island willow (*Salix canariensis*) is locally common near springs and running water in the laurel forests; it is typically associated with *Myrica faya* and thickets of the endemic bramble (*Rubus bollei*) in the best-preserved stands.

HABITATS AND VEGETATION TYPES | 13

Some common ferns of the Canary Islands: (clockwise from top left) *Davallia canariensis*, *Dryopteris oligodonta*, *Asplenium onopteris*, *Pteridium aquilinum*, *Polypodium macaronesicum*.

Plants of the laurel forests with needle-like or opposite leaves: A. *Viburnum rugosum* (leaf rough); **B.** *Picconia excelsa* (leaves smooth); **C.** *Erica platycodon* (leaves ± horizontal); **D.** *Erica canariensis* (leaves ± erect).

Plants of the laurel forests with domatia (including galls and glands): **A.** *Persea barbujana* (note galls); **B.** *Laurus novocanariensis* (note glands at each vein axis); **C.** *Rhamnus glandulosa* (note glands along midrib); **D.** *Ocotea foetens* (note 2 prominent glands at leaf base).

Plants of the laurel forests with leaves with prominent secondary veins (A), reddish colouration (B) or variably pubescent (C,D): **A.** *Pleiomeris canariensis* (note prominent secondary veins); **B.** *Heberdenia excelsa* (leaves with minute red dots); **C.** *Salix canariensis* (hairy on veins beneath); **D.** *Visnea mocanera* (hairy on veins beneath).

Plants of the laurel forests with spiny, serrate or toothed margins: A. *Prunus lusitanica* subsp. *hixa* (note long-acuminate tip); **B.** *Ilex perado* subsp. *platyphylla* (note apical spine; margins variably spiny); **C.** *Myrica faya* (note irregular serrate margin); **D.** *Arbutus canariensis* (note conspicuous veins).

Plants of the laurel forests with entire leaves: A. *Persea indica* (note long petiole); **B.** *Ilex canariensis* (note prominent fungal spots); **C.** *Euphorbia mellifera*; **D.** *Sideroxylon canariense* (note long petiole).

PINE FOREST

At higher altitudes and beyond the reach of cloud-bearing trade winds, hillslopes become drier. Here, acidic, sandy or rocky substrates are dominated by Canary Island pine (*Pinus canariensis*), which forms extensive bands above the humid laurel and thermo-sclerophyllous forests, between 900 and 2,250 m. Winter frosts and occasional snowfall occur at this elevation, while in summer, the forests can be ravaged by fires. These pine forests comprise the most extensive forest vegetation on the islands, and cover large parts of the western islands, especially Tenerife. The forests have long been exploited for timber, and reforestation has extended their cover locally. The undergrowth is poor compared with the humid laurel forests; it typically comprises various shrubs and woody perennials such as *Chamaecytisus proliferus*, *Adenocarpus viscosus*, *Cistus symphytifolius* and *Sideritis* spp. White-flowered *Cistus monspeliensis* is also common on rocky outcrops and roadsides in the forest. At low altitude where pine forests intergrade with evergreen laurel forests, conditions become more humid, and the vegetation is more species-rich. At their highest elevations, pine forests give way to an open summit scrub (see below).

SUMMIT SCRUB

Summit scrub vegetation is dominated by brooms (*Adenocarpus* and *Spartocytisus*) on Tenerife and La Palma. These plant communities have been impacted by fire and grazing historically and are now protected elements of the islands' flora. Various rare shrubs perch on the island summits overlooking the pine forests, and occasionally extend to lower altitudes in ravines. *Spartocytisus supranubius* grows on Tenerife, above the pine woods. Other frequent species in this habitat include *Argyranthemum tenerifae*, *Arrhenatherum calderae*, *Nepeta teydea*, *Pterocephalus lasiospermus* and *Scrophularia glabrata*.

Arguably the most impressive species in this plant community is *Echium wildpretii*, which forms forests of red spires in early summer that extend as far as the eye can see beneath the crowning summit of Mount Teide.

Pinus canariensis forests in the hills of Tenerife.

At altitudes between 2,400 and 3,500 m in Las Cañadas and the slopes of Pico del Teide, a rather species-poor community occurs with *Viola cheiranthifolia* interspersed with domes of white daisy-flowered *Argyranthemum tenerifae*.

EXOTIC PLANTS

Many of the exotic plants in the Canary Islands are remnants of introductions from other arid climates associated with traditional farming activities following European conquest in the late 15th century. In recent decades, with urban development and tourism, the number of alien species has increased. Prickly pears (*Opuntia*), native to the Americas, were once widely planted as boundary hedges, for their edible fruits and as the host of the cochineal insect (an important industry in the 19th century). Remnant populations of former cultivation are scattered across the islands. Century plants (*Agave*) are also a prominent feature of the landscape; plantations of *A. fourcroydes* and *A. sisalana* were planted in early 20th century for their tough fibres, and persist in conspicuous spiky swathes in some areas. *Agave americana* is also common on hillsides, where it grows in rows along the brows of the hills. All three species produce magnificent, towering yellow inflorescences in early summer. The related *A. attenuata* is commonly planted on roadsides and in towns as an unusual ornamental for its succulent, blue-leaf rosettes and cernuous, pale yellow flowering stems, the tips of which drape along the ground. *Aloe vera* is becoming an increasingly important cash crop on the eastern islands for use in the cosmetics industry. Fields of their squat, prickly red leaves topped with slender yellow flowering scapes that ripple in the breeze have become a familiar feature of the landscape and have even become a tourist attraction in their own right.

Agave attenuata naturalised in northwest Lanzarote.

Some of the islands' introductions are considered to be invasive, and can have a profound impact on native plant communities. *Arundo donax*, an Asiatic species that is common throughout the tropics and subtropics, locally forms vast swathes, for example around reservoirs and the water courses of ravines. *Cenchrus setaceus* is an African species introduced in the first half of the 20th century that is now widespread across the Canary Islands (especially in Tenerife, Gran Canaria and La Palma). Meanwhile, the creeping succulent *Mesembryanthemum crystallinum* (native to South Africa) is now common on dunes, roadsides and bare ground across the islands, as is the related species *Malephora crocea* more locally, which has showier yellow, pink or orange flowers.

The Eastern Islands

Lanzarote, La Graciosa and Fuerteventura make up the easternmost Canary Islands. The flora of this sub-province comprises Mediterranean, North African and Macaronesian elements, as well as more recent introductions. Together, they exhibit lower levels of endemic species richness than the western islands due to their comparatively low habitat diversity, and greater human impact. Lying close to the Saharan belt, just 100 km from the coast of Africa, these volcanic 'desert islands' have a warm and windy climate that is classified as 'oceanic-desertic/ xeric inframediterranean'. Only 60 mm of rain falls per year around some coastal areas, rising to about 200–250 mm in higher parts, and there is virtually no rainfall at all during the summer months.

The natural climax vegetation across much of these islands is *Euphorbia* scrub, which is typically dominated by succulent euphorbias such as *Euphorbia balsamifera* and *E. regis-jubae*. Pristine succulent thicket vegetation is now largely restricted to natural parks and isolated areas such as ravines and steep slopes, due to land degradation caused by human activity across much of the islands. Nevertheless, the eastern islands harbour a bountiful flora and are best visited in late January and March.

LANZAROTE

Lanzarote has an area of 845,9 km² and rises to 671 m above sea level. Owing to its accessibility, and popularity as a tourist destination, Lanzarote is one of the best known of the Canary Islands. The island comprises a landscape of jagged volcanic peaks, extensive lava beds, lava springs surrounding the Montañas del Fuego, and white, sandy coastal plains.

It has over 700 flowering plant species and many endemics shared with the neighbouring island of Fuerteventura. The richest vegetation on Lanzarote is to be found in the Famara massif in the north, which is a geologically old formation of high cliffs and ridges that are relatively inaccessible to man and goat, and humid due to sea mists. Most of the island's endemic plants are concentrated in this mountainous region; indeed, 12 taxa can be found nowhere else. Colonies of *Asteriscus schultzii*, a small, shrubby daisy relative, can be found below the cliffs; this species is confined to Lanzarote and Fuerteventura (and neighbouring Morocco). The rocky ravines and crevices of the cliffs support populations of *Aichryson tortuosum* and two species of *Aeonium*: pink-flowered *A. lancerottense* and yellow-flowered *A. balsamiferum*. On the higher peaks, *Limonium bourgeaui* occurs along with *Echium decaisnei* subsp. *purpuriense* and *Argyranthemum maderense*. To the north of Haría – a town set between the hills – the peaks of Peñas del Chache (671 m) are home to the spectacular giant fennel (*Ferula lancerotensis*) and stands of broad-leaved *Lavandula pinnata*. The salt marsh of Salinas del Río by the Famara cliffs supports populations of *Arthrocaulon* and various *Suaeda* species. Meanwhile to the south of Famara is a stretch of sandy plains that support various coastal species. The Elvira Sánchez ravine originates from the eastern slopes of the massif and is a tributary to the Valle de Haría. Numerous interesting species occur here, including the striking *Argyranthemum maderense*, succulents such as *Aichryson tortuosum* and *Sedum nudum* subsp. *lancerottense*, wild thyme (*Thymus origanoides*) and the related *Sideritis pumila*.

There are two significant dune systems on the main island: Mancha Vagal, and Jable Grande and Jable Chico east of Órzola. Mancha Vagal supports stands of the desert shrub *Traganum moquinii*, interspersed with clumps of *Euphorbia paralias*, *Tetraena* and *Polycarpaea nivea*. Jable Grande and Jable Chico host a diverse assemblage of coastal species, including the impressive yellow parasitic plant *Cistanche phelypaea*. Sheltered ravines on the island also harbour an interesting flora, such as Tenegüime, where a large population of pink-flowered *Campylanthus salsoloides* can be found. Extensive sheets of malpaís on the island support a rich assortment of lichens, and the island endemic *Aeonium lancerottense*, which grows in fissures in the rock.

LA GRACIOSA

La Graciosa is a small island about 27 km², lying about 2 km off the northernmost point of Lanzarote. This wind-beaten island is arid and dominated by saltworts (*Suaeda ifniensis* and *Caroxylon* spp.). Of interest in the coastal flora here are the uncommon purple broomrape (*Phelipanche gratiosa*) as well as the rare sea lavender, *Limonium papillatum*, which is restricted to the eastern Canary Islands. The island is home to the archipelago's sole surviving populations of *Cynomorium coccineum* which grows on the wilder stretches of dunes, and requires written permission to visit. The island is best visited after late winter rainfall, typically in February.

Above: the dunes of La Graciosa (left), are the only remaining habitat in the Canary Islands for the rare parasitic plant *Cynomorium coccineum* (right). Opposite: the plants of Lanzarote: **A.** *Helichrysum monogynum*, **B.** *Periploca laevigata* (in fruit), **C.** *Cuscuta approximata*, **D.** *Ferula lancerotensis*, **E.** *Scilla latifolia*, **F.** *Helichrysum gossypinum*.

FUERTEVENTURA

Fuerteventura is the oldest of the Canary Islands, shaped over the course of 21 million years. The volcanic island's rolling plains of orange dust, studded with white, lonely windmills and jagged, wind-blasted coasts, may seem devoid of life at first glance. In fact, its 1,650 km² harbours a varied flora for its size. A network of undulating hills, rocky and sandy plains with deep depressions, and extensive dune systems form the northern seaboard. The Corralejo Natural Park in the northeast is famous for its sandy beaches. Here, vast white dune systems are flanked by barracks of ragged, sand-blasted *Traganum moquinii* – a plant characteristic of the Sahara. Low, sheltered dune slacks mark the eastern boundary of the park. These are characterised by a short, heath-like sward that runs parallel to the roadside and hosts plentiful annuals in the spring.

White dunes interspersed with sheets of black, volcanic rock between Majanicho and El Cotillo to the west are a stronghold for the rare, purple-flowered sea lavender, *Limonium papillatum*, along with little grey patches of *Polycarpaea nivea* and large clumps of exuberant, glossy-leaved *Astydamia latifolia*. The Isla de Lobos ('Lobos Island' which translates as 'Wolves Island'), named after its once plentiful sea wolves or Mediterranean monk seals (*Monachus monachus*), is an islet of just 5 km² to the northeast of Corralejo, accessible only by boat. A renowned Special Protection Area for birds, the islet also has a varied succulent thicket and coastal flora, and is home to stands of endemic sea lavender (*Limonium bollei*) which grows in the wet, saline depressions not far from the paths.

To the south and west of Corralejo, dunes and sandy plains give way to undulating, knobbly grey-green encrusted lava fields. Here, sheltered rocky walls and gullies host a spiny thicket of arching *Asparagus pastorianus* stems, succulent *Euphorbia* bushes, the short, muscular trunks of *Kleinia neriifolia*, and on the hillsides, rangy shrub tobacco trees. To the southwest of the village of Lajares lies the Barranco del Jable – an interesting geological labyrinth of barrancos (ravines), flats and intriguing rocky formations. Here on the seasonally rain-soaked jables (sandflats), grows the little, rare eastern Canary Island crocus-like endemic *Colchicum psammophilum*, which blooms for a short period in winter or early spring. Robust clumps of the rare *Asteriscus schultzii* also occur locally on rocky ledges and in crevices.

One of the richest floristic areas of central Fuerteventura is the protected Parque Rural de Betancuria. Baked dry in summer, the Betancuria Massif's concertinaed hillsides are verdant by late winter in most years and harbour the island's few permanent springs. Exposed crags and outcrops beside the footpaths are a habitat for carpets of the curious, square-stemmed succulent *Apteranthes burchardii*, which sends out little white-felted, starfish-like flowers in the winter months, followed by twinned, horn-like fruiting pods in spring. Higher slopes here are studded by the conspicuous flowering spikes of asphodel (*Asphodelus ramosus*), robust, spiny, grey rosettes of *Cynara cardunculus*, as well as cushion-like balls of yellow-flowered *Asteriscus sericeus*.

Montaña Cardón is the most spectacular of the mountainous massifs on the southern mainland, and is the island's richest floristic area, besides Jandía, with an estimated 18 endemic plants. At 691 m, the mountain's steep and imposing, virtually inaccessible rocky walls are incised with narrow valleys and damp crevices, and have remained largely shielded from human impact for millennia. The mountain's eastern slopes, cut by cavernous ravines, are scarcely accessible, and are home to Fuerteventura's only remnant population of the imposing, candelabra-like succulent *Euphorbia canariensis* outside of Jandía.

The plants and landscapes of Fuerteventura: **A.** *Pulicaria canariensis*, **B.** *Apteranthes burchardii*, **C.** *Apteranthes burchardii* (in fruit), **D.** Montaña Cardón, **E.** the summit cliffs of the Jandía peninsula, **F.** the dunes of Corralejo.

The Jandía peninsula, once an isolated volcanic island, is now joined to the mainland by a narrow isthmus of golden sand. Despite development for tourism along the southern coast, most of the remote, rocky peninsula remains unspoiled and is Fuerteventura's most botanically rich region, with many endemic plant taxa. A prominent volcanic ridge with sharp, abruptly precipitous peaks runs parallel to the remote and wind-swept Cofete plain of sand to the north; meanwhile a series of dry hills, dissected by deep valleys, runs more gently to the rockier southern coastline. Steep and inaccessible, mist-soaked cliffs and ledges have been sheltered from human impact and grazing, and are home to fascinating summit-scrub vegetation with plants that are rare or absent elsewhere on the island. Rare point endemics on the summit include shrubs such as *Argyranthemum winteri* and *Echium handiense*, which grow side-by-side with more abundant species such as the chickweed-like *Rhodalsine platyphylla*. Small perennial herbs, such as *Reichardia famarae* and *Polycarpaea nivea*, perch precariously on the damp, gravelly shelves. One or two of the remote, rocky valleys to the south of the long, jagged peak are home to Fuerteventura's most celebrated endemic plant: the cactoid *Euphorbia handiensis*. The yellow-flowered *Pulicaria burchardii*, shared with the west coast of the Sahara, occurs only on Jandía in the Canary Islands.

Euphorbia handiensis, endemic to the Jandía peninsula of Fuerteventura.

The Western Islands

The western islands – Gran Canaria, Tenerife, La Palma, La Gomera and El Hierro – consist of mountain peaks rising directly from the deep ocean floor. These islands have a particularly complex geology characterised by recent periods of significant volcanic activity on Tenerife, El Hierro and La Palma, and more localised activity giving rise to malpaís. Colossal landslips have also contributed to changes in the islands' topography which, followed by erosion, have given rise to deep valleys and imposing cliffs.

The western islands are steeper and more rugged than the eastern islands, and have a Mediterranean climate subject to humid trade-winds, which form banks of cloud on the windward slopes. They support vegetation types that are scarce or absent on the eastern islands, for example humid laurel and pine forest vegetation, and a distinct montane summit vegetation above the treeline on the highest islands of Tenerife and La Palma.

GRAN CANARIA

Gran Canaria is a circular island crowned by the Pico de Las Nieves, from which a radial series of valleys fan out to the sea. Some of these valleys harbour extensive *Pistacia lentiscus* woods, especially in the northwest. The deepest valleys lie in the south and west of the island. *Euphorbia aphylla* scrub dominates the northern rocky coasts and cliffs, which are blasted by trade-winds. Laurel forests were once extensive on the island, but only a small area remains. The most pristine lie in the valleys of Los Tilos de Moya and Barranco de la Virgen. Willows (*Salix canariensis*) line the mossy, boulder-strewn watercourse of the valley in Los Tilos de Moya, and the purple-flowered *Convolvulus canariensis* is common in the area. Rare species that occur here include *Digitalis chalcantha* and *Sideritis discolor*.

Cruz de Tejeda rises to 1,600 m and has an interesting flora. The golden-flowered succulent, *Aeonium simsii*, is common in the area; pine woods intergrade with a summit scrub dominated by *Genista microphylla*, *Sideritis dasygnaphala* and *Erysimum albescens*. *Gonospermum ptarmiciflorum* grows locally, notable for its conspicuous white hairy coating (indumentum). To the west of Cruz de Tejeda, the rare *Hypericum coadunatum* occurs on the wet cliffs along the roads. To the east, down to Caldera de Bandama, species on the rocky hillsides include *Parolinia glabriuscula*, *Polycarpaea filifolia*, *Adiantum reniforme* and *Camptoloma canariense*.

To the south, white pyramids of *Echium decaisnei* abound on the roadsides. In the Barranco de Guayadeque, many Canarian rarities and endemics occur, including *Lavatera acerifolia*, *Salvia canariensis*, *Periploca laevigata*, as well as Gran Canarian endemics such as *Descurainia preauxiana*, *Kickxia pendula* and *Helianthemum tholiforme*. *Kunkeliella canariensis* and *Parolinia platypetala* grow exclusively in this ravine.

In the Barranco de Fataga, great palms (*Phoenix canariensis*) form groves around the bed, framed by sheer rocky walls. In the slopes of this great valley, endemics include: *Convolvulus glandulosus*, *Limonium preauxii* and *Teline rosmarinifolia*. The black echium (*Echium onosmifolium*) is common at higher elevations here. The peculiar, wine-coloured flowers of *Ceropegia fusca* can be found on the dry cliffs of the ravine.

The plants and landscapes of Gran Canaria. Opposite: **A.** *Canarina canariensis*, **B.** *Pericallis webbii*, **C.** *Sonchus acaulis*, **D.** Los Tiles de Moya and Barranco de la Virgen, **E.** the mountainous centre of the island. This page: **A.** *Ceropegia fusca*, **B.** *Euphorbia canariensis*, **C.** the arid barranco south of Fataga, in the south of the island.

The south coast is dominated by *Euphorbia balsamifera* scrub. Rangy, white-flowered stems of *Convolvulus scoparius* are frequent in the valley of Arguineguín, which also harbours the island's few remaining individuals of the Gran Canarian dragon tree (*Dracaena tamaranae*). Meanwhile, the dry, sandy slopes around Arinaga in the southeast of Gran Canaria are the habitat of the rare shrubby bindweed *Convolvulus caput-medusae*, which otherwise grows locally only on Fuerteventura, as well as the attractive *Atractylis preauxiana*, which is shared with Tenerife, and the local endemic *Lotus arinagensis*.

In the west, on rocky walls along the winding roadsides leading to the deep valley of La Aldea, are found endemics such as *Prenanthes pendula*, *Dendriopoterium pulidoi* and *Crambe scoparia*. The Tamadaba Massif in the northwest of the island is dominated by stands of Canary Island pine (*Pinus canariensis*). Pink-flowered *Cistus ocreatus* and *Micromeria pineolens* are common in the scrubby undergrowth, while damper areas are a stronghold for the fern *Ophioglossum lusitanicum*. Rusty-orange-flowered *Digitalis isabelliana* abounds on the forested cliffs here. Rocky outcrops in the northeast of the island host an interesting succulent flora characterised by *Aeonium aureum* and *Aeonium simsii*, as well as the critically endangered *Pericallis hadrosoma*.

TENERIFE

Tenerife is a teardrop-shaped island with a torn edge, 80 km long. A furrowed spinal ridge dissects the island from end to end with gullies fanning out to the coast, dominated in the centre by the gargantuan Mount Teide (3,715 m) and its vast crater, Las Cañadas. This relatively young, eroded volcano straddles the most ancient parts of the island: the Roque del Conde volcano, followed in order of age by Teno and then Anaga.

The north of the island drops steeply to the sea; by contrast, the southern edge has broad coastal plains. Teno and Anaga form the island's western and eastern promontories, respectively. Their forest-crested slopes and valleys have a rich flora. Zones of vegetation are associated with the different climatic zones in the island. On the northern slopes, *Euphorbia* scrub predominates at lower altitudes, and at higher elevations is replaced progressively by juniper woodland (although now heavily impacted by anthropic change), laurel forest, pine forest and summit scrub.

Las Cañadas del Teide comprises an ancient volcanic crater, from the centre of which projects the younger great cone of Pico del Teide, which is often snow-covered during the colder months. Here, endemic plant species abound. Perhaps the most iconic is the red-flowered *Echium wildpretii* which grows along the cliff walls and rocky ledges to the east of the summit, often in large numbers visible from the roadside. On closer inspection, the flowers are typically teeming with pollinating bees (both native and introduced, the latter possibly of concern). Early in the season, they are bird-pollinated. The much smaller, blue-flowered species, *E. auberianum*, grows locally on bare slopes around Las Arenas Negras alongside mounds of pink-flowered *Pterocephalus lasiospermus*. Rarities include the pink-flowered

---

The plants and landscapes of Tenerife. Opposite: **A.** *Aeonium tabulaeforme*, **B.** the montane forests of the Sierra de Anaga, **C.** remote sea cliffs on the north of the island. Following page: **A.** *Dracunculus canariensis*, **B.** the ancient specimen of *Juniperus cedrus* near Las Cañadas del Teide, **C.** *Aeonium aureum*, **D.** the montane pine forests of Arafo, **E.** *Malva acerifolia*, **F.** an extensive tract of humid laurel forest.

Cistus osbeckiifolius and an ancient specimen of *Juniperus cedrus*, which is surrounded by a protective fence. Other plants of note in the area include: *Tolpis webbii*, *Erysimum scoparium*, *Argyranthemum tenerifae*, *Descurainia bourgaeana*, *Adenocarpus viscosus*, *Pterocephalus lasiospermus*, *Scrophularia glabrata*, *Nepeta teydea* and *Andryala pinnatifida* subsp. *teydensis*. Towards the highest reaches of Teide, the endemic violet *Viola cheiranthifolia* grows virtually in isolation.

The Sierra de Anaga is worth visiting in summer. The mountains here rise to almost 1,000 m and stretch from the town of La Laguna to the island's northernmost, finger-like promontory. Dense, humid laurel forests in the area are dominated by an emerald-green canopy of *Laurus novocanariensis*, *Prunus lusitanica* subsp. *hixa*, *Heberdenia excelsa*, *Persea indica*, *P. barbujana*, *Ilex perado* subsp. *platyphylla* and *I. canariensis*, interspersed with the chocolate-coloured trunks of *Erica canariensis*. It is easy to see some of the Canary Islands' finest botanical treasure in these forests: candelabras of flame-like *Digitalis canariensis* are common in the thickets of the deepest forests, along with delicate sprays of white-flowered *Crambe strigosa*. Loose scarps under the canopy host stands of *Dracunculus canariensis*, which produces beautiful white, reflexed spathes in late spring. *Bencomia caudata* is an endemic shrub frequently seen on the forested cliffs that has pendulous strings of fruits; meanwhile, little succulents such as *Aichryson* spp. are common among the shady boulders and rocky ledges. An impressive species on the loose coastal cliffs of the northern slopes is the endemic *Echium simplex*, which in spring sends up dense, tail-like spires of white flowers.

The Montañas de Teno are botanically rich. A well-preserved laurel forest can be seen at Monte del Agua, where *Visnea mocanera* and *Arbutus canariensis* are especially common. Sheltered cliffs and slopes are thickly forested in *Laurus novocanariensis*, *Heberdenia excelsa*, *Pleiomeris canariensis* and *Picconia excelsa*, providing a refuge for laurel forest pigeons. To the west, the ancient basalt sea cliffs abound in rarities and endemics including *Genista salsoloides*, *Limonium spectabile* and *Argyranthemum coronopifolium* (a species shared with the Anaga Massif). They are also a refuge for the endemic giant lizard *Gallotia intermedia*. Fissures in the rock harbour little cushions of succulent *Monanthes pallens* and *M. polyphylla*, and the curious disks of *Aeonium tabulaeforme*, which also perch along the crest of the cliffs. Large stands of *Aeonium pseudourbicum* lean out of the boulders along with blue-green mounds of *Bupleurum salicifolium* subsp. *aciphyllum*. Pencil-like stems of *Ceropegia dichotoma* with their yellow, claw-shaped flowers grow around the coastal cliffs and slopes near Punta de Teno. Pink-flowered *Parolinia intermedia* is another endemic shrub found in the area.

The Laderas de Güímar on the island's east-central coast is home to an exceptional array of rare flora. Buttressed rocky walls drop steeply into a canyon that bakes dry in the summer months. *Ruta pinnata* and *Salvia canariensis* are common around the mouth of the ravine. Steep, dry scarps host inconspicuous sprays of broom-like *Thesium retamoides*. *Crambe arborea* var. *indivisa* and the iconic pink-flowered *Malva acerifolia* sprout from the bare grey rock. On the coast, the Malpaís de Güímar is a refuge for well-preserved coastal succulent scrub dominated by *Euphorbia canariensis* and *E. balsamifera*.

To the south and west, deep ravines (such as Tamadaya, Barranco del Río and Barranco del Infierno) are worth examination and harbour some interesting endemics, for instance, *Euphorbia atropurpurea*, *Pericallis lanata*, *Bethencourtia palmensis*, *Sideritis infernalis* and *Anagyris latifolia*.

LA GOMERA

La Gomera is a circular island, roughly 24 km across at its maximum width, rising to about 1,450 m at its central peak, from which some 40 ravines that are pleated into sheer cliffs fall steeply to the sea. Volcanic activity has occurred within historic times on all of the Canary Islands, with the exception of La Gomera, where erosion has been an important driver of the island's topography.

The island shares vegetative characteristics with northern Tenerife, and lacks extensive dunes or plains. The largest valley is the Barranco de la Villa, which runs from the centre to the capital, San Sebastian, on the east coast. Around the roads, *Aeonium decorum* and *A. viscatum* are common on the cliffs and rocky walls; to the north *A. castello-paivae* is frequent; far rarer is *A. rubrolineatum*, which occurs on just a few rocky cliffs mainly to the south and west of the island, many of them inaccessible. Another prominent succulent is *Aeonium diplocyclum*, which forms little pale, cabbage-like balls on the rock in its summer form.

The northern slopes of La Gomera are home to the best-conserved remnants of thermo-sclerophyllous forests in the Canary Islands. They are dominated by the tree *Juniperus canariensis*, while the under-layer is composed of endemics such as *Brachypodium arbuscula* and *Lotus emeroides*. *Euphorbia bravoana*, *Cheirolophus ghomerythus* and *Convolvulus subauriculatus* occur on the cliffs here.

Traquitic and phonolitic outcrops are a characteristic feature of La Gomera landscapes and harbour interesting endemics such as *Bethencourtia hermosae* and *B. rupicola*, as well as *Cistus chinamadensis* subsp. *gomerae*.

The forests of El Cedro at the centre of La Gomera are among the best-preserved laurel forests on the Canary Islands, and support large, ancient trees: these are considered the core of the Garajonay National Park. The wet, mossy woods are dominated by leathery-leaved trees, including *Laurus novocanariensis*, *Myrica faya*, *Ilex canariensis* and *Persea indica*, while willows (*Salix canariensis*) colonise the watercourses. Ferns are abundant and blanket the banks and tracksides of the forest.

The steep, dry barrancos west and south of Chipude are largely inaccessible on foot. Rocky slopes here are home to rarities, including the arborescent sea lavender, *Limonium dendroides*, the white-flowered *Cheirolophus sarataensis*, and felted *Sideritis nutans*. At lower elevations of the Barranco de Argaga, the only known population of *Parolinia schizogynoides* can be found, while nearby, the sparse, pinkish-white-flowered *Echium triste* grows along the coast.

Plants and landscapes of La Gomera: A. dry barranco in the northwest of the island, **B.** the forests of El Cedro on the centre of the island, **C.** *Aeonium diplocyclum* (summer habit), **D.** *Aeonium rubrolineatum*, **E.** *Aeonium castello-paivae*.

LA PALMA

La Palma is 45 km long and roughly the outline of South America. It is a mountainous island with a prominent central crater, La Caldera de Taburiente. The outer rim of the crater forms the island's peak, Roque de los Muchachos, at 2,483 m. The north of the island is wrinkled with valleys that fan out to the coastal, cliffy edge. The south of the island slopes more gently towards Fuencaliente. Notably, in 2021, a violent volcanic eruption occurred on the slopes of the Cumbre Vieja rift that saw vast volumes of lava pour out over the west of the island – the greatest ever recorded there.

The cloud-cloaked, jagged edge of the La Caldera de Taburiente, which is a designated national park, has a subalpine flora similar to that of the slopes of Teide on Tenerife. It includes pine forests and specimens of the endangered *Juniperus cedrus*, as well as some point endemics, such as the rare blue-flowered *Echium gentianoides* and the pink-flowered *Echium perezii*. The latter was only recently recognised as a species distinct from *E. wildpretii*, which is restricted to the high mountains of Tenerife. Also found here are clumps of violet-flowered *Viola palmensis*, a white form of the yellow daisy-flowered *Tolpis calderae*, and the rare Canary Island representative of the genus *Genista* (*G. benehoavensis*). Flowing out of the Caldera, to the west, is the Barranco de las Angustias where the conspicuous, red-flowered *Aeonium nobile* sprouts from the dry rocky shelves.

In the northeast of the island, Cubo de la Galga is a damp, mossy forest, dominated by old specimens of *Ocotea foetens*, *Myrica faya* and *Erica canariensis*. Accessible tracks pass a long, forested cliff where bright green, rose-like *Aeonium canariense* subsp. *christii* (syn. *A. palmense*) sprouts from the rocky scrub. *Echium pininana*, a common garden plant in much of the world, is an endemic rarity that grows wild in the heart of these forests, forming spectacular spires that rise from the dense vegetation. The smaller, shrubby species *E. webbii* occurs further north, near the barranco of Los Tilos. Here shady woods and wet gorges support large stands of giant ferns, *Woodwardia radicans*, which drape the rock, and yellow-flowered *Sonchus palmensis*, leaning out of the scrub. The softly forested, steep slopes around Barlovento are largely inaccessible and remain wild and unspoiled.

The northeast coast harbours interesting populations of *Tolpis santosii*, *Sonchus bornmuelleri* and the sea lavender *Limonium imbricatum*. On the north-facing cliffs above the sea, white-flowered *Echium bethencourtii* and *Cheirolophus sventenii* are conspicuous. To the west, warmer slopes that are sheltered from the trade winds are dominated by white-flowered *Retama rhodorhizoides*.

Dragon trees (*Dracaena draco*) on the slopes overlooking the sea near Las Tricias in the northwest have a distinct, branched and upright habit. This is apparently the consequence of a long history of artificial selection for plants that yield plentiful leaves for animal fodder.

The south of the island is of more recent geological origin and features volcanic structures including cones, cinders and lava fields that are still in the process of plant colonization. Here *Rumex lunaria* and *Echium brevirame* breathe life onto the bare, nascent land. Roque de Teneguía, a traquitic-phonolitic outcrop exposed among the young basaltic surroundings, is a refuge for the endemic, pink-flowered *Cheirolophus junonianus*, which is a relative of knapweed.

Finally, the ravines near Santa Cruz de La Palma are a mainstay for interesting endemics, including *Cheirolophus santos-abreui*, *Aichryson brevipetalum*, *Aeonium goochiae* and *Teline splendens*.

Top: Caldera de Taburiente, La Palma (PHOTOGRAPH BY OLIVER WHITE); below: *Echium perezii* in La Palma (PHOTOGRAPH BY RACHAEL GRAHAM).

EL HIERRO

El Hierro, 33 km long, is the youngest and one of the smallest Canary Islands, but still reaches an altitude of over 1500 m at its highest elevation. Geologically, the island has experienced significant growth and destructive events over the course of its 1.12 million year history, with major landslips removing much of its volume.

A central ridge concertinas into rocky folds that form high cliffs on all sides. The cliffs facing to the northwest coast, El Golfo, are at high altitudes, densely blanketed with laurel forest. Many of the island's endemics occur here, including *Bencomia sphaerocarpa*, *Crambe feulleei*, *Adenocarpus ombriosus*, *Cistus chinamadensis* subsp. *ombriosus* on rocky slopes, and stands of lemon yellow-flowered *Teline stenopetala* subsp. *microphylla* on the cliffs. In late spring, *Aeonium diplocyclum* prints yellow along the rocky walls on the road down to Frontera.

Dry evergreen forests dominated by *Visnea mocanera* remain relatively intact on the island; by contrast in the north, the humid constituents of the forest have been almost eradicated, and are replaced by patches of *fayal-brezal* vegetation and animal fodder crops.

Euphorbia scrub (*E. lamarckii*) is a common form of vegetation across much of the island, meanwhile juniper woods, interspersed with pine, cover the cliffs and slopes of Riscos de Las Playas and El Julan. In the Riscos de Las Playas, pink-flowered *Cistus asper* and *Orchis canariensis* can be found in spring.

From February, the white broom scrub (*Retama rhodorhizoides*) is a spectacle, especially in the east of the island, where it substitutes the ancient forests of *Juniperus canariensis*. *Sonchus hierrensis* forms splashes of yellow across the island.

At higher elevations of the western corner of the island, unusual forests of *Juniperus canariensis* occur in which specimens have acquired strange wind-beaten forms over the millennia. On the approach to Mirador de Bascos, bright-flowered *Argyranthemum hierrense*, *Erysimum virescens*, and *Pericallis murrayi* (here treated as conspecific with *P. steetzii*), are all easily-spotted, framed by an extended view of El Golfo which shifts between the clouds. Slopes and sheltered cliffs above the village of Sabinosa host endemics including *Silene sabinosae, Tolpis proustii* or *Cheirolophus duranii*.

On the south of the island, lava flows dominate the landscape. Here *Aeonium valverdense* is a common volcanic pioneer species, and tufts of *Schizogyne sericea* are a conspicuous feature of the black, basaltic sand and crevices.

Opposite. Above: Las Playas, El Hierro; below: Jinama, El Hierro (PHOTOGRAPHS BY OLIVER WHITE). Inset (above): the endemic *Argyranthemum hierrense*; inset (below): an ancient specimen of *Juniperus canariensis*.

Diversity and Evolution

Volcanic, oceanic archipelagos have been described as 'natural laboratories' for studying evolution, and they provide some of the most spectacular examples of evolutionary radiations. The Canary Islands are no exception. Within the archipelago are many species-rich groups that are the result of speciation following a single colonisation from the continent. In some cases, colonisation has been accompanied by the evolution of a woody habit: anyone familiar with foxgloves (*Digitalis*) could not fail to be impressed by the Canary Islands' endemic representatives of the genus that have become robust, woody shrubs. 'Island woodiness' is perhaps the most conspicuous aspect of island floras, comparable to flightlessness in island birds, and it has evolved in many different Canary Island groups. Recent research suggests that the absence of herbivores and adaptation to drought may best explain the phenomenon.

Notable examples of Canary Island evolutionary radiations include the *Aeonium* alliance (comprising the genera *Aeonium*, *Aichryson* and *Monanthes*), the Macaronesian endemic genus *Argyranthemum* and the endemic *Echium* and *Sonchus* lineages. While the Canary Islands are the centre for diversification of these groups, further dispersal from the Canaries to Madeira, the Azores, the Cape Verde Islands and, in some cases, even to north Africa, has given rise to species that are endemic to those areas, too.

DNA-based analyses have helped to provide new insights into the evolutionary processes that generated the diversity of species observed within Canary Island groups. The results suggest that speciation was often very recent, as might be anticipated in an archipelago where at least some of the islands are young. It is also clear from those analyses that geographical isolation has been an important driver of diversification. With the exception of Lanzarote and Fuerteventura, which merged in the last glacial maximum to form a single landmass ('Mahan'), the Canary Islands have never been physically connected and the ocean between them has served as a very effective barrier to dispersal. It is remarkable that La Gomera and Tenerife, separated by just 28 km of ocean, have more than 30 and 80 single-island endemic species, respectively. Over time, plant populations isolated on different islands have diverged and given rise to new species. Geographical isolation and subsequent divergence have also occurred at a smaller scale within islands, where giant landslips have fragmented populations and as a result of rugged and highly dissected terrain.

Shifts between habitat zones have also been an important evolutionary driver. Of the four giant monocarpic rosette species of *Echium*, two occur in the laurel forest (*E. simplex* on Tenerife and *E. pininana* on La Palma) and two in the subalpine zone (*E. wildpretii* on Tenerife and *E. perezii* on La Palma). Restricted to the summits of Tenerife and La Palma, the subalpine zone is a particularly harsh environment. There is very little precipitation and average temperatures are generally low, although high levels of solar radiation during the day lead to significant diurnal temperature variation. This is a challenging place for plants to survive and contrasts markedly with the wetter, warmer conditions found in the laurel forest. DNA analyses have shown that there was one shift to the subalpine zone within this group, giving rise to the *E. wildpretii*/*E. perezii* lineage that is restricted to this habitat. Similar shifts between habitats have occurred in many groups. New techniques promise insights into the rapid genetic changes that underpin such habitat shifts.

Hybrids between species within the same group can make identification challenging for the field botanist but in an evolutionary context, hybridization can generate individuals with novel characteristics and can even lead to the evolution of new species. Hybridization generates a unique combination of genes from the two parental species. Hybrids may be morphologically distinct from their parents and may have different characteristics. They may, for example, be able to flourish in a different habitat to the parents. DNA analyses of Canary Island groups suggest that hybridization has contributed to the evolution of the flora. A particularly well-studied example can be found in *Argyranthemum* in the Anaga peninsula where *A. frutescens* occurs in arid low elevation habitats and *A. broussonetii* occurs in the laurisilva on the upper slopes. Also found in the region, at intermediate elevations, are *A. sundingii* and *A. lemsii*. Genetic data indicate that those two species each arose independently, through hybridization between *A. frutescens* and *A. broussonetii*, the hybrid species in each case able to thrive in habitats that neither of the parental species are adapted to.

The evolution of Canary Island plant groups has often been rapid. The evolutionary history of each group is different but it is through a combination of geographical isolation, ecological shifts and hybridization that the spectacular endemic diversity we see in the archipelago has evolved.

References

del Arco Aguilar, M.J., González-González, R., Garzón-Machado, V. et al. (2010). Actual and potential natural vegetation on the Canary Islands and its conservation status. *Biodivers Conserv.* 19, 3089–3140. https://doi.org/10.1007/s10531-010-9881-2

del Arco Aguilar, M.J. & Rodríguez Delgado, O. (2018). *Vegetation of the Canary Islands*. Plant and Vegetation Vol. 16. Springer. https://doi.org/10.1007/978-3-319-77255-4

Species descriptions

PTERIDOPHYTES (FERNS)

Spore plants without flowers or seeds. Spores borne in sporangia grouped in sori.

EQUISETACEAE | HORSETAIL FAMILY

Stems with reduced leaves borne in whorls. Spores produced in cone-like strobili at the tips of the fertile stems.

Equisetum ramosissimum COLA DE CABALLO, BRANCHED HORSETAIL A horsetail with bristly, ascending, slender, hollow stems 20 cm–1.5 m. Sterile stems with 8–20 grooves and leaves reduced to sheath-like whorls. Fertile stems similar but with apical, club-shaped strobili. Damp habitats; local. C, T, G, P.

OPHIOGLOSSACEAE | ADDER'S-TONGUE FAMILY

Ferns, typically with fleshy, solitary, entire fronds.

Ophioglossum polyphyllum LENGUA DE SERPIENTE A small fern to 5–10 cm with sterile, fleshy, leafy fronds, and a spike-like structure bearing two rows of spore-producing sporangia. Very local on sand dunes and other naturally bare habitats. L, F, C, T, H, P.

PTERIDACEAE | RIBBON FERN FAMILY

Rhizomes scaly, branched, with leaves borne at the tips, spirally coiled when young. Sori various.

Adiantum

Rhizomes short. Fronds 1–3-pinnate with wedge- or kidney-shaped segments. Sori marginal.

Adiantum capillus-veneris CULANTRILLO DE POZO, MAIDENHAIR FERN A small fern, 5–50 cm, with slender, creeping rhizomes. Fronds pale green, wedge-shaped with fine, brownish-black petioles; sori borne under the margins. Damp rocks and walls. L, F, C, T, G, H, P. *A. reniforme* TOSTONERA has kidney-shaped fronds. Var. *reniforme* is found throughout; the endemic var. *pusillum* is absent from the eastern islands. Rocky habitats. L, F, C, T, G, H, P.

Cheilanthes

Rock-dwelling ferns, often dormant during dry periods. Fronds 2-pinnate. Sori marginal.

Cheilanthes maderensis A small fern with reddish petioles and hairless, rather crinkled fronds with oval, *irregular lobes* (pinnules). Sori borne in irregular rows along the lobe margins. Rocky crevices. L, F, C, T, G, H, P. *C. pulchella* DORADILLA FINA has linear–lanceolate pinnules with a continuous false indusium (sorus covering). Rocky habitats. *C, T, G, H, P.

1. *Equisetum ramosissimum*
2. *Equisetum ramosissimum* (habit)
3. *Ophioglossum polyphyllum*
4. *Adiantum capillus-veneris*
5. *Adiantum reniforme*
6. *Cheilanthes maderensis*
7. *Cheilanthes pulchella*

Notholaena

Rock-dwelling ferns. Fronds possess *farina* (floury coating). Sori marginal.

Notholaena marantae subsp. *subcordata* DORADILLA CANELA A fern 10–35 cm with powdery, bipinnatisect fronds with lanceolate lobes (pinnules) and *undersides conspicuously brown-hairy*. Rocky crevices. C, T, G, H, P.

Cosentinia

Rock-dwelling ferns with white-woolly, *thread-like hairs*. Sori marginal. (One species.)

Cosentinia vellea DORADILLA VELLUDA A small, *white-woolly* fern with narrowly lanceolate, 2-pinnately-lobed fronds, curled at the apex; underside pale, ageing bronze. Rocky crevices. L, F, C, T, G, H, P.

DENNSTAEDTIACEAE

Rhizomes spreading, deeply buried with leaves borne singly, spirally coiled when young, with strong petioles. Leaves *strong-smelling when crushed*.

Pteridium aquilinum HELECHERA, BRACKEN A thicket-forming fern with *branched fronds* to 2 m, with *broadly triangular segments*; sori (if present), *borne partly concealed in rows along the down-turned leaf margins*. Disturbed forest fringe habitats. L, C, T, G, H, P.

ASPLENIACEAE | SPLEENWORT FAMILY

Most species small. Rhizomes short, scaly, with leaves borne in tufts, spirally coiled when young. Sori often *elongated*.

Asplenium aureum (syn. *Ceterach aureum*) DORADILLA MEDICINAL A small fern with pale-margined, *pinnatifid-lobed* fronds, the undersides densely covered in oblong-linear, outwardly splayed, narrow rows of sori. **Var.** *aureum* has oblong-elliptic lobes (pinnae); **var.** *parvifolium* has semicircular to oblong-oval pinnae. Damp rocks and malpaís. L, F, C, T, G, H, P. *A. marinum* CULANTRILLO MARINO, SEA SPLEENWORT has prominently veined, yellowish-green pinnate fronds with conspicuous brown sori on the undersides. Coastal rocks and cliffs. L, F, C, T, G, H, P.

Asplenium hemionitis HIERBA CANDIL, PATAGALLO A small fern with leathery, *ivy-shaped* leaves with attenuate lobes and sori borne in slender, parallel stripes. Common in rocky crevices and malpaís. L, F, C, T, G, H, P.

Asplenium onopteris CULANTRILLO NEGRO A fern 15–50 cm with reddish-brown petioles and pinnate fronds, *triangular* in outline, narrowly tapered into long points; segments with acute teeth; lower leaflets somewhat arched upwards. Common in damp woods and on rocks. L, F, C, T, G, H, P.

---

1. *Notholaena marantae* (showing sporangia on underside)
2. *Notholaena marantae* subsp. *subcordata*
3. *Cosentinia vellea*
4. *Pteridium aquilinum*
5. *Asplenium aureum*
6. *Asplenium hemionitis*
7. *Asplenium hemionitis* (sporangia)
8. *Asplenium marinum*
9. *Asplenium onopteris*
10. *Asplenium onopteris* (sori)

BLECHNACEAE | HARD FERN FAMILY

Rhizomes short to long, scaly. Fronds leathery, borne in tufts at the ends of rhizome branches, spirally coiled when young. Sori *linear and continuous along the midribs*.

Blechnum

Leaves 1-pinnate, fertile or sterile.

Blechnum spicant HELECHO PEINE, DEER FERN A fern 10–75 cm with leathery, many-lobed, pinnate fronds; sterile fronds spreading, fertile fronds sub-erect with narrower lobes. *Sori borne in continuous rows* along the undersides of the lobes. Damp rocks and woods. T, G.

Woodwardia

Leaves 2-pinnate, of one sort.

Woodwardia radicans PÍJARA, PIRGUA, CHAIN FERN A *large* fern 2–3 m with arching or pendulous, bipinnate fronds; fronds dark green and leathery. Sori borne in *broken lines along the midribs*. Damp forests and shaded cliffs. C, T, G, H, P.

DRYOPTERIDACEAE | BUCKER-FERN FAMILY

Rhizomes short, scaly, with leaves borne in tufts at terminal branches. *Sori spot-like*.

Dryopteris guanchica HELECHO PENCO DENTADO A fern with *3-pinnate* fronds, with asymmetrical basal leaflets (pinnae), and lobes (pinnules) with *upturned teeth along the margins*. Sori borne in spots. Damp woods and thickets. T, G, H. *D. oligodonta* HELECHO MACHO is similar, but pinnules not petiolate and lack prominent marginal teeth. C, T, G, H, P.

DAVALLIACEAE | HARE'S-FOOT FERN FAMILY

Long, densely silky rhizomes. Leaves 3(–4)-pinnate.

Davallia canariensis BATATILLA, COCHINITA, HARE'S-FOOT FERN A fern to 40 cm with thick, fibrous rhizomes and intricately lobed 3(–4)-pinnate, rather leathery fronds with yellowish sori borne *under the tips of* the lobes. Rocky habitats and woods. L, F, C, T, G, H, P.

POLYPODIACEAE | POLYPODY FAMILY

Rhizomes extended with leaves borne singly, spirally coiled when young. Sori prominent, spot-like, not sunken.

Polypodium macaronesicum (syn. *P. cambricum* subsp. *macaronesicum*) POLIPODIO DEL PAÍS A fern 15–50 cm with pale green, papery fronds, triangular or broadly oval in outline, sometimes with slightly toothed margins; sori oval, yellow, ageing orange, borne in neat parallel rows. Various habitats, especially woods and cliffs. L, F, C, T, G, H, P.

1. *Blechnum spicant*
2. *Blechnum spicant* (sporangia)
3. *Woodwardia radicans* ('bulbil')
4. *Woodwardia radicans*
5. *Woodwardia radicans* (sporangia)
6. *Dryopteris guanchica* (sporangia)
7. *Dryopteris guanchica*

OVERLEAF:

1. *Dryopteris oligodonta*
2. *Dryopteris oligodonta* (sporangia)
3. *Dryopteris oligodonta* (2)
4. *Davallia canariensis*
5. *Davallia canariensis* (sori)
6. *Davallia canariensis* (rhizomes)
7. *Polypodium macaronesicum*
8. *Polypodium macaronesicum* (sori)
9. *Polypodium macaronesicum* (2)

GYMNOSPERMS

Seed plants without true flowers, in which ovules are 'naked' rather than enclosed within an ovary.

PINACEAE | PINE FAMILY

Evergreen or deciduous trees with *needle-like leaves borne spirally* and vegetative buds with brown scales. Male cones with copious pollen; female cones borne on the same tree and more substantial, with woody scales, taking years to mature.

Pinus PINUS, PINE

Evergreen trees with stiff, often long, needle-like leaves borne in clusters of 2, 3 or 5 on very small shoots.

(a) Needles in pairs.

Pinus halepensis PINO CARRASCO, ALEPPO PINE A tree to 20 m with an *irregular, rounded crown* and twisted branches; branches and twigs *greyish*, bark becoming reddish and fissured; buds not sticky. Leaves needle-like, borne in pairs, 60 mm–12(15) cm long and *narrow*, just 0.7–1 mm wide, straight or twisted. Female cones egg-shaped, 60 mm-12 cm long, borne on *recurved* stalks; seeds winged. Planted and naturalised. L, F, C, T, G, H. *P. pinaster* PINO MARÍTIMO, MARITIME PINE is a taller, pyramidal tree to 30 m with brown bark. Leaves needle-like, borne in pairs, 10–25 cm long. Female cones egg-shaped, 80 mm–22 cm, scarcely stalked. Seeds winged. Planted and naturalised. C, T.

(b) Needles normally in threes.

Pinus canariensis PINO CANARIO, CANARY ISLAND PINE A *large* tree to 30(60) m with ashy, reddish-brown bark. Leaves needle-like, borne *in threes*, *long*, to 30 cm. Female cones virtually stalkless, oblong to sub-cylindrical and *large*, to 20 cm; seeds winged. Forms extensive stands and widely planted at higher elevations. *L (planted), F (planted), C, T, G, H, P. *P. radiata* PINO DE MONTEREY, MONTEREY PINE has a conical to rounded crown with spreading branches and smaller female cones, to 7–14 cm. Widely planted and naturalised on western islands. L, C, T, G, H, P.

CUPRESSACEAE | JUNIPER FAMILY

Evergreen trees or shrubs with resin, leaves opposite or in whorls, scale-like or needle-like, usually in groups of 3; vegetative buds without bud-scales. Female cones with woody or succulent scales.

Juniperus JUNIPER

Evergreen shrubs with twigs spreading in 3 dimensions; leaves needle-like or opposite, scale-like when mature, borne on spreading (not flat) branches. Typically dioecious; female cones berry-like, with *fused, succulent* scales. Seeds not winged.

Juniperus canariensis (syn. *J. turbinata* subsp. *canariensis*) SABINA, CANARY ISLAND JUNIPER A shrub or small tree with oval–diamond-shaped scale-like leaves appressed to the stems. Female cones 9–12 mm with 3 or fewer seeds, reddish-brown when ripe. C, T, G, H, P. *J. cedrus* CEDRO has *needle-like leaves borne in threes and female cones 15 mm. *C (rare – only Montaña del Cedro), T, G, P.

1. *Pinus canariensis*
2. *Pinus canariensis* (female cone)
3. *Juniperus canariensis*
4. *Juniperus canariensis* (female cones)
5. *Juniperus cedrus*
6. *Araucaria heterophylla*
7. *Ephedra nebrodensis* (male cones)
8. *Ephedra nebrodensis*
9. *Ephedra fragilis*

Cupressus CYPRESS

Evergreen trees with twigs spreading in 3 dimensions and opposite, scale-like leaves when mature. Female cones woody, separating into scales; seeds winged.

Cupressus sempervirens CIPRÉS, CYPRESS A tree to 30 m with either a conical or slender, columnar canopy. *Leaves dark green*, scale-like, just 0.5–1 mm long and closely overlapping. *Female cones spherical, 25–40 mm across* and yellowish. Native to the Mediterranean; planted. T, G, H. ***C. macrocarpa*** CIPRÉS DE MONTEREY, MONTEREY CYPRESS has dark green leaves 1–2 mm long and female cones 20–35 mm long, and an intense citrus aroma when rubbed. Native to the USA; planted and naturalised. C, T, G, P.

ARAUCARIACEAE | MONKEY-PUZZLE FAMILY

Very large, evergreen trees with columnar trunks and needle-like or flattened leaves borne spirally; vegetative buds without bud-scales. Male cones large, catkin-like and drooping. Seeds not winged.

Araucaria

Evergreen trees native to Australasia and South America; leaves broad and with many veins or awl-shaped and incurved.

Araucaria heterophylla ARAUCARIA, NORFOLK ISLAND PINE A tall, erect tree to 60 m, with horizontal spreading branches arranged somewhat symmetrically in whorls around the trunk; leaves awl-shaped and incurved. Female cones squat and spherical, with woody, spirally arranged scales and wingless seeds. Native to Norfolk Island in the Pacific, and widely planted. C.

EPHEDRACEAE | JOINT PINE FAMILY

Small shrubs with rush- or broom-like stems, small opposite or whorled leaves, often reduced to scales. Dioecious; male cones in small axillary clusters; female cones with several pairs of bracts. 'Fruit' (in fact, a female cone) berry-like with 1–2 seeds.

Ephedra

Broom-like shrubs with branched stems and inconspicuous, scale-like leaves. Male cones with several 2–3-celled stamens.

(a) Stems not brittle.

Ephedra nebrodensis TEPOPOTE FINO A broom-like shrub to 1 m with *slender* branches to 0.7 mm thick, *difficult to pull apart*. Leaves membranous, 3 mm. Male cones borne in stalkless clusters 2–4 mm across; female cones yellowish to reddish. Rocky slopes at high altitude. T, P.

(b) Stems easily broken at the nodes.

Ephedra fragilis TEPOPOTE FRÁGIL A dense, broom-like shrub to 3 m with *thick branches, easy to break at the nodes*. Leaves to 2 mm, soon-falling. Cones oblong–ellipsoidal, 7–9 x 2.5–3 mm, reddish, with seeds concealed completely. Rocky crags and crevices. L, C, T, G, H, P.

BASAL ANGIOSPERMS AND MAGNOLIIDS

Ancient flowering plant families including trees, shrubs, herbs and vines.

ARISTOLOCHIACEAE | BIRTHWORT FAMILY

Perennial herbs and woody climbers with creeping rhizomes. Leaves simple, alternate, untoothed, with heart-shaped bases. Flowers very distinctive, usually zygomorphic; stamens 6–12 in 1–2 whorls; ovary inferior; styles 6. Fruit a capsule.

Aristolochia

Herbs with stems >10 cm. *Flowers zygomorphic and tubular*, swollen at the base (pitcher-like); stamens 6. Root and micro-morphology are important but difficult to observe in the field.

Aristolochia paucinervis FLOR DE PATO A rhizomatous, creeping herb with alternate heart-shaped leaves. Flowers 30–50 mm, greenish-yellow, solitary, tubular, inflated at the base and expanded apically with a 5-nerved limb, borne in the leaf axils. Fruit a capsule. A Mediterranean native; naturalised. C, T.

LAURACEAE | LAUREL FAMILY

Aromatic, evergreen shrubs and trees with simple, alternate leaves. Flowers small and cosexual, regular, with 8–12 stamens; style 1. Fruit a 1-seeded berry.

Laurus BAY LAUREL

Aromatic trees and shrubs, mainly tropical. Leaves with shining oil glands.

Laurus novocanariensis LAUREL, LORO A tree to 20 m. Leaves oval–lanceolate, dark green with *small glands in the axils of the lateral veins*. Flowers in stalked umbels. Fruits ovoid, 10–15 mm, black when ripe. A characteristic species of laurel forests and fayal-brezal. C, T, G, H, P.

Persea

Trees and shrubs with fleshy, 1-seeded fruits.

Persea indica VIÑÁTIGO A tree to 20 m. Leaves aromatic, large, up to 20 cm, broadly lanceolate, light green, *turning reddish*; *glands absent*. Flowers inconspicuous; petals 6, greenish-white. Fruit ellipsoid, 20 mm, bluish-black when ripe. Laurel forests. C, T, G, H, P.

Persea barbujana (syn. *Apollonias barbujana*) BARBUSANO, CANARY LAUREL A tree to 25 m. Leaves 60–90 mm, oval, glossy, dark green with somewhat downturned margins; glands absent; *leaves frequently with prominent galls*. Flowers with 6 greenish-white petals, fragrant, borne in axillary and terminal panicles. Fruits ovoid, elongated, bluish-blackish when ripe. Laurel forests. **Subsp.** *barbujana* occurs on all the islands except Lanzarote and Fuerteventura; **subsp.** *ceballosi* BARBUSANO NEGRO is an endemic of La Gomera. C, T, G, H, P.

Persea americana AGUACATE, AVOCADO A tree to 15 m tall with large, oval–elliptic leaves, leathery and pointed at the apex. Flowers small, white, with 5 petals. Fruit an avocado pear. Native to Mexico and Central America; frequently planted in gardens in the region. Widely planted.

Ocotea

Trees and shrubs with simple leaves. Fruit a berry with a *cup-like cupule*.

Ocotea foetens TIL, TILO A broad-crowned tree to 30 m; bark with lenticels. Leaves 10–12 cm, broadly lanceolate, *glossy* dark green, with raised *glandular lumps only at the base or in the basal half*. Flowers greenish or yellowish-white, borne in summer. Fruit 30 mm, black when ripe, *shaped like an acorn*. Humid laurel forests. C, T, G, H, P.

1. *Laurus novocanariensis* (habit)
2. *Laurus novocanariensis*
3. *Persea indica*
4. *Persea barbujana* (habit)
5. *Persea barbujana* (flowers)
6. *Persea barbujana* (note the prominent galls)
7. *Persea americana*
8. *Ocotea foetens*
9. *Ocotea foetens* (fruits)

MONOCOTS

A major group of plants, typically with a single seed leaf, leaves with parallel (not net) veins, and flower parts in multiples of 3. Mostly herbaceous and with underground storage organs such as bulbs and corms.

ARACEAE | ARUM FAMILY (including former Lemnaceae)

Perennial, usually hairless, tuberous herbs with leaves all basal and stalked. Individual flowers tiny, borne in a compact spike (spadix), enfolded in a large, often leafy bract (spathe). Fruit a berry.

Arum

Tuberous, usually spring-flowering perennials with arrow-shaped, untoothed leaves. Flowers unisexual, borne at the base of a spadix, which is enveloped by the spathe. Fruit a red berry, produced generally after the leaves have withered.

Arum italicum subsp. *canariense* OREJA DE BURRO, ITALIAN ARUM A perennial with leaves 15–35 cm long, arrow-shaped with pointed lobes; mid to dark green with variable paler marbling. Visible portion of the spadix narrowly club-shaped and pale yellow, ¼–½ the length of the spathe limb; spathe to 40 cm long. Berries red, borne in dense spikes. This subspecies has purplish spathe interior and stalks. Damp hedgerows and ditches; probably confused with *Arisarum* in the east. L(?), F(?), C, T, G, H, P.

Arisarum

Small, tuberous perennials with heart-shaped leaves. Inflorescences like those of *Arum* but smaller and the spathe fused over most of its length, thus tube-like. Fruit green, several-seeded.

Arisarum simorrhinum CANDIL, CACHIMBA A winter-flowering perennial with all basal leaves 20 mm–17 cm long, broadly heart-shaped with rounded lobes. Spadix greenish or brown, swollen at the tip; slightly protruding from the *hooded spathe* which is whitish, striped purplish-brown. Berries greenish to blackish. Malpaís and rocky woods. L, F, C, T.

Zantedeschia ARUM LILY

Rhizomatous herbaceous perennials with long-stalked, broadly oval leaves with heart-shaped bases. Spadix without an appendix. Fruit a yellow berry, several-seeded. Native to southern Africa.

Zantedeschia aethiopica CALA, OREJA DE BURRO, ARUM LILY A hairless perennial with rather fleshy, arrowhead-shaped leaves 10–45 x 10–25 cm, borne on long stalks to 75 cm. Flower stalks somewhat exceeding the leaves; *spadix yellow*, to 15 cm long, and about ½ the length of the spathe; individual flowers indistinct; *spathe pure white*. Berries yellow (often not developing). Planted. L, F, C, T, G, P.

Dracunculus DRAGON ARUM

Tall, tuberous perennials with deeply palmately lobed leaves. Inflorescences with a long, prominent spadix (to 40 cm). Fruit a several-seeded berry.

1. *Arum italicum* subsp. *canariense* (with immature fruits)
2. *Arisarum simorrhinum*
3. *Arisarum simorrhinum* leaf (western island form)
4. *Arisarum simorrhinum* (cross section)
5. *Arisarum simorrhinum* (fruits)
6. *Zantedeschia aethiopica*
7. *Dracunculus canariensis*
8. *Dracunculus canariensis* (flowers)

Dracunculus canariensis TARAGONTÍA, TARAGUNTÍA An upright perennial to 1.5 m with 5–7-lobed leaves with long stalks 20–40 cm. Spathe cream to greenish-white; spadix pale yellow. Fruits borne as reddish-orange berries in dense spikes. Tracksides, verges, ravines and forests. F(?), C, T, G, H, P.

Colocasia

Robust, tuberous perennials with large, shield-shaped leaves with parallel veins. Tropical.

Colocasia esculenta ÑAMERA, ELEPHANT EAR An exotic-looking, robust perennial with many large, dark green, tilted, shield-shaped leaves to 60 cm. Spathes small (relative to leaves), pale greenish-yellow. Fruit a red berry (seldom produced). Native to Asia; cultivated and naturalised in ditches and other wet habitats. C, T, G, P.

ALISMATACEAE

Annual or perennial, hairless, aquatic herbs, rooted in mud, with erect or floating leaves; leaves stalked, with linear to rounded blades. Flowers cosexual and regular, arranged in panicles or racemes; sepals and petals 3; stamens 6–numerous; style absent, ovary superior. Fruit a head of achenes or few-seeded follicles.

Alisma

Aquatic perennials with leaves all basal, linear or narrowly-elliptic to oval. Flowers many, each with 3 petals, white or pale mauve with a yellow basal blotch; stamens 6; carpels numerous in 1 whorl, 1-seeded and curved inwards.

Alisma lanceolatum LLANTÉN DE AGUA, WATER PLANTAIN An erect, hairless, aquatic perennial with leaves and stems carried above the water, 20 cm–1.2 m high. Leaves all basal, long-stalked and oval–elliptic, attenuated at the base, the blade 50 mm–30 cm. Numerous loose sprays of pale flowers form a branched, pyramidal inflorescence; flowers 7–12 mm across, long-stalked. Fruits widest in the middle, with straight, erect styles. Muddy and damp habitats or shallow water. C.

DIOSCOREACEAE

A primarily tropical family including the yam. Leaves simple and alternate. Plants dioecious; flowers small and greenish borne in spikes or racemes in the leaf-axils; stamens 6 (vestigial in female flowers); style 1, stigmas 3. Fruit a several-seeded berry.

Dioscorea

A twining herb unusual in having dioecious, inconspicuous flowers and red berries (*Bryonia* shares these characteristics but with hairy, palmately lobed leaves and tendrils).

Dioscorea edulis (syn. *Tamus edulis*) NORSA, BLACK BRYONY A twining perennial climber to 5 m, dying back annually to a tuber; superficially similar to *Smilax* but spineless. Leaves glossy dark green, heart-shaped and long-stalked, 50 mm–15 cm x 40 mm–11 cm. Flowers greenish-yellow, 3–6 mm across, borne in loose racemes; male flowers with 6 stamens, female flowers with 6 minute lobes and a conspicuous ovary. Fruit a rounded red berry 10–13 mm across. Forests and thickets. C, T, G, H, P.

1. *Dracunculus canariensis* (fruits)
2. *Colocasia esculenta*
3. *Colocasia esculenta*
4. *Dioscorea edulis*
5. *Colchicum psammophilum*
6. *Colchicum psammophilum* (fruiting specimen)
7. *Colchicum hierrense* (LEOPOLDO MORO)

1. *Smilax aspera* subsp. *mauritanica*
2. *Smilax aspera* subsp. *mauritanica*
3. *Smilax canariensis*
4. *Smilax canariensis* (habit)
5. *Smilax canariensis* (tendril)

COLCHICACEAE

Herbaceous perennials with rhizomes or corms, previously included in the Liliaceae. Leaves few, alternate and sheathing below. Flowers 1–few, conspicuous; tepals and stamens 6; styles 3. Fruit a capsule.

Colchicum AUTUMN CROCUS

Cormous perennials with basal leaves appearing with, or after, the flowers. Flowers crocus-like with 6 spreading tepals; stamens 6; styles 3. Fruit a capsule, borne centrally in the leaves. Now includes *Androcymbium*.

Colchicum psammophilum (syn. *Androcymbium psammophilum*) CEBOLLÍN DE GATO, AJILLO A low, later winter-flowering cormous perennial with rosettes of broadly linear, pleated, tapering leaves to 15 cm. Flowers 25–30 mm across, borne 1–6; virtually stalkless; white with faint purple stripes, with yellowish glands at the base. Fruit a 3-valved capsule. Bare sand flats (jables) and coastal dunes. *L, F. *C. hierrense* (syn. *Androcymbium hierrense*) is larger with leaves to 35 cm and flowers 50–80 mm across. Rocky habitats. *G, H, P.

SMILACACEAE

Climbing lianas with heart-shaped leaves, hooked spines, and paired tendrils at the leaf bases. Flowers unisexual; tepals and stamens 6; style absent, stigmas 3. Fruit a berry with 1–several seeds.

Smilax

Woody vines with spines and tendrils. Flowers greenish with 6 tepals. Berries black when ripe, with 1–3 seeds.

Smilax aspera ZARZAPARRILLA, COMMON SMILAX A variable, creeping, scrambling or climbing shrub 6–10 m with angled, smooth or prickly stems. Leaves 42–68 mm (10.3 cm), dark shiny green and leathery, triangular to heart-shaped, often with prickles on the margins, and with a pair of tendrils at the base of the leaf stalk, and *7–9 main veins*. Flowers unisexual, green-white or yellowish or pinkish, scented, to 5 mm, borne in branched clusters. Berry red, ageing *blackish-red*. **Subsp.** *mauritanica* is the form on the islands. Woods and thickets. F, C, T, G, H, P. *S. canariensis* ZARZAPARRILLA CANARIA has few or no thorns, leaves with 3–5 main veins, and fruits ripening *black*. Mainly in the western laurel forests. F, C, T, G, P.

ORCHIDACEAE | ORCHID FAMILY

One of the largest families of flowering plants, with about 26,000 species worldwide. Perennials with flowers borne in spikes or racemes; flowers each with a bract, usually conspicuous, zygomorphic; perianth with 6 tepals in 2 whorls of 3: the outer 3 tepals ('sepals') all similar, the inner 3 ('petals') with the central the largest and distinct, known as the lip or labellum; ovary inferior; anthers and stigma together form a central column. Fruit a capsule. Some groups were taxonomically exaggerated before the availability of DNA data, so nomenclature is inconsistent in floras.

Orchis

Tuberous perennials with 2 ovoid tubers and several leaves in basal rosettes, those on the stem often sheath-like. Flowers often borne in short, dense spikes; upper 5 tepals incurved, the 2 lateral sepals incurved or erect to spreading; lip with 2 lateral lobes and 1 terminal lobe, the latter often larger and 2–3-lobed. Spur absent, short or long.

Orchis mascula ORQUÍDEA MACHO, EARLY PURPLE ORCHID A slender perennial with lanceolate, shiny-green leaves with *darker spots*. Bracts lanceolate and shorter than the ovary. Flowers mauve, the middle sepal curving with the petals to form a hood, the laterals deflexed; lip 8–16 mm long, 3-lobed and convex from the centre, the centre pale and spotted, the middle lobe notched, sometimes with white spots towards the base; lateral lobes scarcely deflexed; spur at least as long as the ovary (as long or 2 x length of lip), pointing upwards and thickened at the apex. Woods. P.

Orchis canariensis (syn. *Orchis patens* subsp. *canariensis*) ORQUÍDEA CANARIA A slender perennial, similar in form to the species above, with oval, dark green, slightly shiny basal leaves to 30 cm, unspotted. Flowers lilac or pink, sometimes greenish, with purple markings on the lower lip borne on reddish-purple stems. Malpaís, and rocky, mossy shelves in forests. *C, T, G, H, P.

Neotinea

Perennials with 2 ovoid tubers and leaves 2–4 at the base; those on the stem reduced. Flowers with upper 5 tepals incurved; lip with 2 large lateral lobes and a larger terminal lobe; spur short and rounded at the tip.

Neotinea maculata (syn. *Orchis intacta*) ORQUÍDEA MANCHADA, DENSE-FLOWERED ORCHID A small, pale perennial 10–25(40) cm, normally with densely spotted leaves (sometimes unspotted) forming a rosette. Inflorescence *small and dense; flowers very small* (appearing dwarfed by bracts) and scented; dull pinkish-white with purplish markings; *sepals forming* a *pale hood*; lip 3-lobed, 3–5 mm long only, the middle lobe rectangular and the laterals pointed. Mossy habitats; woods. C, T, G, H, P.

Himantoglossum

Tall, robust perennials with 2 ovoid tubers. Flowers with a very long, narrow lip with 2 lateral lobes; spur short.

Himantoglossum metlesicsianum ORQUÍDEA DE TENERIFE A robust perennial with stout stems to 50 cm. Foliage leaves matt green, to 30 cm long, sometimes with faint markings. Inflorescence dense with pink flowers, the lower bracts prominent; lateral sepals form a hood with the petals; lip distinctly 3-lobed (the middle lobe notched), the laterals with wavy edges. Rare, on malpaís at 700–1,300 m under pines. *C, T, H, P.

Ophrys BEE ORCHID

Perennials with 2 ovoid tubers. Flowers dupe male insects into attempting to mate with them to bring about cross-pollination (pseudocopulation). Sepals large, often greenish or pink; petals 2, smaller and hairy; lip large, hairy and variously patterned.

Ophrys bombyliflora ORQUÍDEA ABEJONA, BUMBLEBEE ORCHID A short, loosely clump-forming perennial 50 mm–35 cm. Basal leaves oval–lanceolate, forming a flat rosette, stem leaves erect and clasping the stem. Flowers few, borne in short spikes of 1–6; sepals green; petals green with a purplish base, triangular, and ⅓ the length of the sepals; *lip small*, 6–10 mm long, 3-lobed with the lateral lobes deflexed, brown with a central bluish, shield-shaped mirror. C, T.

1. *Orchis canariensis*
2. *Himantoglossum metlesicsianum* (habit)
3. *Himantoglossum metlesicsianum*
4. *Ophrys bombyliflora*

Serapias TONGUE ORCHID

Tuberous perennials with 2(–5) ovoid tubers, similar to *Ophrys* but with an elongated lip that is downward-pointing and tongue-like with 2 short, upturned lateral lobes.

Serapias parviflora ORQUÍDEA GALLO MENUDA, TONGUE ORCHID A short, tuberous, clump-forming perennial. Leaves narrowly lanceolate and grey-green. Flowers reddish-maroon; petals drop-shaped, lip narrow with 2 brownish-red ridges at the base, equalling the hood (15–22 mm long). Open, sunny habitats. L, C, T, H, P.

Gennaria

Perennials with rhizomes, stem with 2 alternate leaves. Flowers greenish.

Gennaria diphylla ORQUÍDEA DE DOS HOJAS, TWO-LEAVED GENNARIA A short, sometimes clumped perennial 16–36 cm high with 2 oval to heart-shaped leaves at the base, 45 mm–12 cm long. Inflorescence dense, 1-sided with 10–47(85) flowers; tepals all of similar length, 3 mm long, with backwardly curved tips; greenish yellow; lip 3-lobed, the middle lobe the largest; spur grooved. Rocky habitats and woods. L, C, T, G, H, P.

Habenaria BOG ORCHID

Tuberous perennials distinguished by the presence of two club-shaped projections on the stigma.

Habenaria tridactylites ORQUÍDEA DE TRES DEDOS A short tuberous perennial to 30(60) cm with oblong–lanceolate leaves to 18 cm long. Flowers *greenish-yellow*, borne in lax spikes, the upper segments forming a helmet and the lower lip deeply divided into three long lobes; spur long. Old terraced fields, rocky crevices. *L, C, T, G, H, P.

IRIDACEAE | IRIS FAMILY

Bulbous, tuberous or rhizomatous perennials, usually with linear leaves, all basal or alternate. Flowers with 6 tepals, the outer 3 often different from the inner 3, enclosed in 1–2, often papery, spathes when in bud; stamens 3; style often with 3 branches, ovary inferior. Fruit a 3-parted capsule.

Romulea

Low, cormous perennials with basal, linear leaves, often 4-grooved. Flowers normally enclosed in bud by 2 spathes; flowers *Crocus*-like, *borne on slender green stems*; tepals 6, contracting into a very short perianth tube; stamens 3, style solitary and 3-branched.

Romulea grandiscapa (incl. *R. hartungii*) AJO GATO A low, cormous perennial, with 2–5 erect to spreading, recurved leaves 0.6–1 mm wide. Flowers solitary or in groups of 2–3; corolla tube to 6.5 mm; tepals 20–30 mm, narrowly lanceolate (9–10 mm wide), long-pointed; flowers violet-purple to white, usually with a whitish-yellow to yellow throat, and striped purple; anthers to just 4 mm; stigmas usually overtopping the anthers. Malpaís and hillsides. *L, F, C, T, G, H, P.

Gladiolus

Cormous perennials with fans of flat, sword-shaped, ribbed leaves. Flowers borne on long, rigidly erect stems, each flower with a green bract; tepals 6, unequal, fused into a short tube at the base; stamens 3; style slender with 3 short branches.

IRIDACEAE | 65

Gladiolus italicus GLADIOLO A tall, cormous perennial to 80 cm, with *broad, sword-shaped leaves* 40–80 cm long, 17 mm wide, arising from a spherical corm with netted fibres. Flowers 6–15, borne in a 2-sided inflorescence, each flower with an equally long leafy bract; flowers 40–50 mm long, pinkish-red, the segments very unequal: the upper longer and broader than the laterals; anthers *longer* than the filaments. Seeds winged. Roadsides and cultivated land. L, F, C, T, G, H, P.

Chasmanthe

Herbaceous perennials with strap-shaped leaves.

Chasmanthe floribunda CRESTAGALLO A bulbous perennial to 60 cm. Leaves strap-shaped (linear–lanceolate), to 20 mm wide. Flowers orange-red, borne on leaning, 1-sided racemes; corolla with a slender lower portion to 15 mm and a cylindrical upper portion to 25 mm; tepals unequal; stamens arched. Native to South Africa, naturalised along roadsides, abandoned crops, slopes and ravines. C, T, G, P.

1. *Gennaria diphylla*
2. *Habenaria tridactylites* (habit)
3. *Habenaria tridactylites*
4. *Romulea grandiscapa*
5. *Chasmanthe floribunda*

ASPHODELACEAE (including Xanthorrhoeaceae)

Many genera formerly included in the Liliaceae (in its broader, traditional description), characterised by dense tufts of long, narrow leaves and stout, woody spikes of flowers; perianth of 6 tepals, often conspicuous; stamens 6; style 1 with minute or 3-lobed stigma. Fruit a capsule.

Aloe

Succulent shrubs with robust, spiny-margined leaves forming a rosette. Flowers in racemes; corolla tubular, often orange or red. Fruit a capsule. Native to Africa and Madagascar, widely cultivated.

Aloe vera SÁBILA, ALOE A stoloniferous perennial succulent with numerous basal leaf rosettes with grey-bluish or reddish-tinged leaves to 40–50 cm. Inflorescences to 1 m, with *yellow flowers* 25–30 mm with protruding stamens. Native to the Middle East, widely naturalised and farmed. L, F, C, T, G, P.

Aloe arborescens SÁBILA ARBOREA, TREE ALOE A *large, much-branched shrub* to 2(3) m. Leaves succulent and narrowly triangular–lanceolate, 50–60 mm long, crowded in large rosettes and deflexed-spreading; greyish without pale markings, and toothed along the margin. Inflorescence unbranched, *erect*; flowers *scarlet*, 35–45 mm long. Native to southern Africa, naturalised. L, C, T, P. *A. marlothii* MOUNTAIN ALOE is *single-stemmed*, has leaves with prickles and spreading racemes of orange flowers. Planted. L, F, C, T.

Asphodelus ASPHODEL

Robust, hairless, herbaceous and tuberous perennials (rarely annuals) with leafless stems and linear leaves in basal tufts. Tepals free, all similar; flowers borne on simple or branched stems; filaments hairless. Many species very similar; fruit shape an important diagnostic.

Asphodelus ramosus GAMONA A tall, robust herbaceous perennial 50 cm–1.6(2) m with numerous fleshy roots. Leaves flat, strap-shaped and grey-green. Flowers white and star-like, with tepals 11–20 mm, with brownish mid-veins, borne in tall, erect, much-branched inflorescences in which the *lateral branches are almost as long as the terminal*. Capsules ovoid, large, 6–12 x 4–9 mm. Hillsides, tracksides; relatively common. Regional forms include **subsp.** *ramosus* (L, F), and **subsp.** *distalis* (all islands). L, F, C, T, G, H, P.

Asphodelus fistulosus GAMONA, HOLLOW-LEAVED ASPHODEL A small, tufted *annual* or short-lived perennial 30–60 cm (1.5 m) with numerous fleshy roots. *Leaves linear and slender, 1–35 mm wide, hollow and cylindrical*. Flowers white, borne in a lax raceme of 10–15 on a long, smooth stalk with membranous, whitish bracts; tepals 8.5–12.5 mm, with brownish mid-veins. Capsule 5–6 x 4–6 mm. Disturbed habitats. L, F, C, T, G, H, P. *A. tenuifolius* GAMONILLA is similar but with fibrous roots, narrower leaves to 2.5 mm and smaller flowers with tepals 5.5–7.5(8) mm. Capsule 3–4 mm long. Common on malpaís. L, F, C, T, G, H, P. *A. ayardii* is a very similar perennial with the style 3–4 mm longer than the stamens (rather than equalling them as in the above two species). Distribution unclear due to confusion with the species above; possibly throughout.

1. *Aloe vera* (habit)
2. *Aloe vera*
3. *Aloe marlothii*
4. *Asphodelus ramosus*
5. *Asphodelus tenuifolius*
6. *Asphodelus tenuifolius* (fruits)

AMARYLLIDACEAE

Bulbous or rhizomatous (or cormous) perennials with leafless stems, similar to Liliaceae. Flowers solitary or in umbels, enclosed in papery bracts in bud; tepals 6, usually all petal-like; stamens 6; style 1, ovary 3-parted. Fruit a 3-parted capsule, often succulent.

Narcissus DAFFODIL

Bulbous, usually early-blooming perennials with basal leaves and hollow, leafless scapes. Flowers with 6 tepals; stamens 6, surrounded by a cup- or trumpet-like corona. Capsule 3-parted. A difficult genus with high variability and hybridisation.

Narcissus tazetta NARCISO, BUNCH-FLOWERED DAFFODIL A hairless, bulbous perennial 23–60 cm, with blue-green, flat, broad, strap-like leaves 6.2–12 mm wide, equalling the stems, present during flowering; flowers appear in winter. Flowers borne in umbels of 3–many (15); fragrant, white (sometimes cream to yellow), with a prominent yellow corona 3–6 mm long that is 2 x as wide as high, on a 2-edged, flattened scape. Spathe at the base of the inflorescence membranous. Grassy habitats. C, T.

Pancratium SEA DAFFODIL

Bulbous perennials with flowers borne on long-stemmed, terminal umbels; corolla funnel-shaped with 6 narrow tepals and a corona with 12 teeth.

Pancratium canariense LÁGRIMAS DE LA VIRGEN An autumn-flowering, bulbous perennial. Leaves linear, glaucous with scapes to 80 cm with umbels of 4–11 flowers. Flowers long-pedicellate, to 7 cm long. Tepals all similar; stamens 6, borne on the rim of the cone-like corona. Fruit a more-or-less obtriangular capsule. North-facing slopes. *L, F, C, T, G, H, P. *P. maritimum* AZUCENA DE MAR, SEA DAFFODIL Summer-flowering, clump-forming, with few subsessile or short pedicellate flowers (4–9), *large* and white, to 15 cm long, borne on scapes to about 25 cm. Fruit a 3-parted ellipsoid capsule. Rare and local on coastal dunes. F.

Allium

Distinctive bulbous perennials smelling of onion or garlic when crushed. Flowers borne in terminal, often spherical umbels; tepals 6, all similar; stamens 6. Fruit a 3-parted capsule. Some species produce *bulbils* in the flowerheads (and/or *bulblets* – offsets of the bulb).

(a) Innermost *stamens with 3-parted filaments* (filaments with teeth).

Allium ampeloprasum AJO PORRO, WILD LEEK A *robust, leek-like* perennial with a membranous bulb and stout flowering stems to 1(2) m with *numerous yellow-brown bulblets*. Leaves 4–10, pale blue-green, V-shaped in cross-section, rough along the margins and with a central keel beneath; withered by flowering. Flowerheads spherical with numerous lilac flowers with slightly protruding stamens, borne on long stalks; innermost stamens with 3-parted filaments. Hillsides. L, C, T, P.

Allium sphaerocephalum ROUND-HEADED LEEK A bulbous perennial 20–70(90) cm with scapes circular in cross-section. Leaves 2–4(5), semi-cylindrical with a groove, sheathing below, 1–2(3) mm wide; spathes 2-valved, shorter than the umbels; bulblets present. Flowers purple-red, each tepal with a greenish keel, borne in *dense, spherical heads* to 40 mm across; lateral teeth of 3-parted filaments only just *equalling* the central (fertile) part; stamens clearly *protruding*; flower stalks 5–20(30) mm long. Forests. T.

(b) Innermost stamens entire (without teeth) and leaves *cylindrical, subcylindrical or filiform.*

Allium vineale PUERRO DE VIÑA, CEBOLLÍN, AJO SALVAJE, CROW GARLIC A bulbous perennial 22 cm–1 m with rounded (not angular) scapes. Leaves 2–4(6), cylindrical, channelled and sheathing the base of the scape. Spathe 1-valved and beaked, papery, *not longer than the flowers* and soon-falling; flowers reddish or greenish and bell-shaped, 3–4 mm long, often few: *most or all of the flowers replaced by bulbils, often sprouting green shoots before falling*; stamens of flowers *protruding*. Dry grassy habitats. L, F, C, T, P.

Allium pallens A bulbous perennial 20–80 cm with 3–4 leaves that sheath the stem to almost halfway, hairless. Flowers borne in rounded, usually compact heads 20–40 mm across; whitish or yellowish-green with unequal spathes, *the longest much exceeding the flowerheads*; tepals 3.5–5 mm, dull pinkish-white; anthers just exserted. Disturbed, dry habitats. Dry, open habitats. L.

AMARYLLIDACEAE | 69

(c) Innermost stamens entire (without teeth) and leaves keeled to flat; *flowers white to pink*.

Allium canariense TARABASTE A small bulbous perennial with leaves *hairless* along the margins. Flowers white or flushed pink, with tepals 7.5 mm long. Hillsides and malpaís; common. *L, F, C, T, G, H, P. ***A. subhirsutum*** is similar (probably confused) but has leaves slightly *hairy along the margins* and white flowers (*not* flushed pink), borne on stalks 3–5 x longer than the tepals. L, C, T, G, H, P. ***A. roseum*** has entirely pink flowers with segments 9–13 mm long. L, F, C, T, G, H, P.

Allium nigrum AJO NEGRO A bulbous perennial, 60–90 cm. Leaves 2–5, all basal, *broadly linear*, to 25–33(90) mm wide, tapered to a point, much exceeded by the scapes, which are rounded in cross-section. *Spathes 2–3(4)-valved* with pointed tips, free to the base. Flowers 30–90, white or very pale lilac with green mid-veins, and *blackish ovaries*, borne in dense, spherical umbels; anthers yellow. Various dry, grassy habitats. C, T.

1. *Narcissus tazetta*
2. *Pancratium canariense*
3. *Allium canariense*
4. *Allium pallens*
5. *Allium subhirsutum*

Crinum

Perennials with long-necked bulbs and long, strap-shaped leaves. Flowers with 2 spathes; stamens 6. Fruit spherical.

Crinum × amabile A cultivated hybrid (***C. asiaticum × C. zeylanicum***) with long, strap-like leaves and showy flowers with narrow, deflexed, whitish tepals; stamens with long, purple filaments and yellow anthers. Planted.

ASPARAGACEAE

A large and variable family comprising many genera of shrubs or large perennials, formerly included in an even larger, broadly defined family, Liliaceae. Flowers cosexual (rarely dioecious, e.g. *Asparagus* and *Ruscus*); perianth with 6 tepals; stamens 6; styles 1–3. Fruit a capsule or berry with 1–numerous seeds.

Dipcadi

Bulbous perennials with tubular (not star-shaped) flowers, and numerous black, flattened, disk-like seeds.

Dipcadi serotinum TARABASTE GATO A late winter-flowering, hairless, bulbous perennial 20–40 (50) cm high. Leaves all basal and linear–lanceolate, 15–35(60) cm long. Inflorescence an erect, lax raceme of 6–20(23) flowers facing more or less 1 way; *perianth brownish-orange*; bracts exceeding their flower stalks. Easily distinguishable by its flower colour. L, F, C, T, G.

Dracaena

Robust, rather succulent, tree-like perennials with stout trunks. Flowers with 6 tepals and stamens. Fruit a fleshy berry or capsule. Native mainly to tropical and sub-tropical Africa and Macaronesia.

Dracaena draco DRAGO A slow-growing, tree-like perennial 8–12 m with a very *robust, swollen trunk*, eventually densely branched above. Leaves sword-shaped, thick, grey-green and flat, 40–90 cm long, 30–45 mm wide. Flowers borne in numerous clusters of 4–7 in large, hairless inflorescences to 1.2 m long; flowers 7–9 mm across, whitish-yellow with 6 tepals and 6 thickened, flattened stamens. Fruit a yellowish to reddish-brown berry. Rare as a wild plant, in ravines and cliffs; commonly planted. C, T, G, H, P. ***D. tamaranae*** DRAGO DE GRAN CANARIA is similar but with leaves narrower, rigid, grooved and with transparent margins. Flowers greenish-white, borne in terminal panicles. A rare island endemic. *C.

Asparagus ESPARRAGUERA, ASPARAGUS

Shrubby, hairless, rhizomatous perennials with tough stems (becoming woody) and with *cladodes* (rather than true leaves). Flowers small and inconspicuous, bell-shaped with 6 tepals. Fruit a small berry. Several non-native species occur that are omitted here.

1. *Crinum × amabile*
2. *Dipcadi serotinum*
3. *Dipcadi serotinum* (fruits)
4. *Dracaena draco* (cultivated form)
5. *Dracaena draco* (wild form)
6. *Dracaena tamaranae*
7. *Asparagus pastorianus* (habit); (insets) leaves and fruits

(a) Plant prominently spiny.

Asparagus pastorianus ESPINA BLANCA A scrambling, greyish shrub with woody stems bearing *large spines, 5–15(20) mm,* and cladodes 12–40(70) mm in clusters of 8–30 in the axils of the spines. Locally common on dry hillsides. L(?), F, C, T, G(?).

Asparagus horridus ESPÁRRAGO BORRIQUERO, CHAPARRO A grey-green, wiry shrub, but with stems bearing rigid, *spine-like, solitary* cladodes to 40 mm long. Flowers dull purplish-yellow. Hot, dry areas. L, F.

(b) Plant not prominently spiny. Cladodes <30 mm, needle-like or channelled.

Asparagus scoparius ESPARRAGÓN RABO DE BURRO An erect shrub with small, needle-like cladodes <20 mm, densely crowded (4–20) on alternate, ascending branches. Inflorescences many-flowered. F, C, T, G, H, P. *A. fallax* ESPARRAGUERA DE MONTEVERDE is similar but *densely leafy* and with few-flowered inflorescences; tepals 3–4 mm. *T, G.

Asparagus umbellatus ESPARRAGUERA COMÚN A shrub with pliable, spineless stems and short, flat cladodes. Flowers borne in umbels; tepals 5–7 mm, yellowish. F, C, T, G, H, P.

Asparagus nesiotes A shrub with wiry stems to 3(4) m with needle-like (not conspicuously spiny) cladodes, arranged in regular, *dense fascicles* of 7–25. Flower white, solitary or few. Fruit a berry ripening red. **Subsp.** *purpuriensis* ESPARRAGUERA MAJORERA is the form on the islands. Malpaís, roadsides, hillsides. *L, F.

Asparagus plocamoides ESPARRAGÓN COLGANTE is similar to *A. arborescens* (see below), but with *arching* branches with very fine, short, needle-like cladodes. Pine woods. *C, T, G, P.

(c) Plant not spiny. Cladodes long, >3.5 mm, fleshy.

Asparagus arborescens ESPARRAGÓN A tall, *erect* shrub with spineless stems. Cladodes long, spreading, rather fleshy, not prominently spiny. Flowers greenish-white. Dry slopes up to 1,000 m. *L, F, C, T, G, H, P.

Semele

Climbers with leaf-like, alternate cladodes. Inflorescences *borne on the margins or towards the centre* of the lower surface of the cladodes. Anthers in a ring. Fruit a 1-seeded berry.

Semele androgyna GIBALBERA, ALICACÁN A leathery climber with hairless lanceolate to ovate cladodes 10–12 cm long, 40–50 mm wide, often with wavy margins. Flowers small, borne in fascicles of 2–6 on the cladode margins. Fruit a berry, greenish to black, 10 mm. Laurel forests, shaded cliffs. C, T, G, H, P. *S. gayae* is similar, but with flowers and fruits borne on the central part of the cladodes – *not* the margins. *C.

1. *Asparagus horridus*
2. *Asparagus umbellatus* (immature specimen)
3. *Asparagus umbellatus* (fruits)
4. *Asparagus nesiotes* subsp. *purpuriensis*
5. *Asparagus nesiotes* subsp. *purpuriensis* (fruits)
6. *Asparagus plocamoides*
7. *Asparagus arborescens*
8. *Semele androgyna*
9. *Semele androgyna* (flowers)

Ornithogalum

Bulbous, spring-flowering perennials with basal leaves in rosettes. Flowers normally white and star-shaped, borne in racemes; each flower with 1 bract; tepals free and petal-like; stamens 6. Fruit a 3-valved capsule with many seeds.

Ornithogalum narbonense CEBOLLETA   A spring-flowering perennial 25–57 cm. Leaves 4–6, linear with sheathing bases, to 16 mm wide, persistent until after flowering. Flowers 20–30 mm across, white, scentless, borne erect on stalks the same length in a *tall, many-flowered, slender raceme* of up to 50, with tepals wide and spreading, with prominent external green stripes; anthers yellow; ovary to 5 mm with a flattened top, *distinctly exceeded by the narrow style*. A Mediterranean plant; occasionally naturalised. L, F, C, T, G.

Drimia   SEA SQUILL

Similar to *Ornithogalum* but with very large bulbs and numerous flowers borne in a long, terminal spike, in late summer to autumn.

Drimia maritima (syn. *Urginea maritima*) CEBOLLA ALMORRANA, SEA SQUILL   A stout, late summer-flowering perennial 60 cm–1.2(1.5) m with a very large bulb to 6–15(18) cm across, sitting close to the soil surface; bulb greenish with a red tunic. Leaves borne in winter–spring, all basal, large, broadly lanceolate, at least 20 mm wide, wavy-margined and shiny green. Flowers white and star-like, to 16 mm across, borne in a long, stout, spike-like inflorescence arising erect from the ground; filaments 4.5–5.5 mm long. L, F. *D. hesperia* is similar but considered distinct by some authors based on its narrower leaves and laxer inflorescences that have more or less horizontal flower stalks and recurved tepals. *T.

Scilla   SCILLA

Bulbous spring-flowering perennials with basal leaves and leafless stalks. Flowers star-shaped and usually blue, borne in racemes or solitary, with or without *1 bract at the base of each flower stalk*; tepals 6, separate and spreading. Fruit a 3-parted capsule.

Scilla latifolia VARA DE SAN JOSÉ   A robust, glaucous plant with a large (often exposed) bulb and leaves 3–6 in a rosette, broadly lanceolate, to 40 cm. Flowers 4 mm, lilac-blue, borne in congested racemes on stems to 50 cm; anthers bluish. Fruits berry-like. Widespread but rather rare and local. L, F, T(?), G, H, P. *S. haemorrhoidalis* CEBOLLA ALMORRANA MENOR   is similar but much smaller in all parts; bulbs just 10 mm, stems to 15 cm; flowers few and lax. *L(?), F(?), C, T, G, H, P.

Leopoldia   TASSEL HYACINTH

Bulbous perennials with dense to lax racemes. Fertile flowers constricted at the throat, often brownish or greenish, and apical sterile flowers blue, violet or pink, often highly reduced. Capsule 3-lobed, splitting and often compressed.

Leopoldia comosa (syn. *Muscari comosum*) JACINTO SILVESTRE, TASSEL HYACINTH   A bulbous perennial 20–60(90) cm. Leaves 2–3(5), linear, recurved, grooved, tapered gradually to the tip, to 40 cm long and 6–25 mm wide. Flowers borne in a lax raceme with small, blue, sterile flowers held *erect, on long, slender blue stalks* to 15 mm long, *forming a tuft at the apex*; fertile flowers brownish-green, *shortly tubular-bell-shaped* with 6 outwardly curved, short, cream-coloured or whitish teeth. Grassy habitats. L, F, C, T, G, H, P.

1. *Ornithogalum narbonense*
2. *Drimia maritima*
3. *Drimia maritima* (leaves)
4. *Drimia hesperia* (habit)
5. *Drimia hesperia*
6. *Scilla latifolia*
7. *Scilla latifolia* (fruits)
8. *Scilla haemorrhoidalis*
9. *Leopoldia comosa*

Agave

Large, imposing succulents with fibrous, very robust leaves in rosettes. Flowers cosexual and regular; stamens 6; ovary superior or inferior and 3-parted. Fruit a splitting capsule; seeds numerous. Native to the Americas.

Agave americana PITERA, CENTURY PLANT A very large and imposing succulent 6–8 m during flowering; plants take at least 10 years to flower, after which they die, but perennate by offsets. Leaves very large and succulent, to 2 m long and 15–22(30) cm wide, with a stout *black spine at the tip 30–50 mm long*; blue-green or variegated with yellow stripes. Flowers borne in summer on a *tree-like inflorescence* that persists for years, greenish-yellow, to 90 mm long, borne on at least ½ the length of the inflorescence scape. Fruit an oblong capsule. Common, particularly near developed areas; cliffs and roadsides. L, F, C, T, G, H, P. *A. sisalana* SISAL has straight, rigid leaves that are much narrower, 5–11 cm wide, which when mature are spineless along the margins; terminal spine 15–25 mm long. Flowers blue-green, and *mixed with leafy bulbils* which detach and form new plants. L, F, C, T, G, H, P. *A. fourcroydes* HENEQUÉN is similar to both the above species but *with a distinct basal trunk 50 cm–1(2) m* (not stemless). Planted and naturalised. L, F, C, T, G, H, P.

Agave attenuata CUELLO DE CISNE is similar to the above species but smaller, with less succulent leaves with a *pale blue-white waxy bloom*, and *flowers borne in an extremely dense, fox-tail-like inflorescence, drooping to the ground when mature*. Planted and naturalised near towns. L, F, C, T, G, H, P.

ARECACEAE | PALM FAMILY

Trees with large, pinnately or palmately divided leaves. Flowers unisexual or cosexual, small and greenish, usually borne in spikes or panicles with papery sheaths; petals 3; stamens 6. Fruit a berry or drupe.

Washingtonia

Trunk long and slender; leaves with long bare stalks terminating in a rounded fan of segments. Native to North America.

Washingtonia robusta PALMERA DE ABANICO A robust palm with a very stout, smooth trunk with diagonal furrows to 25 m tall and 40–70 cm wide, the upper part *clothed with the long-persistent withered leaves*. Leaves large and fan-like. Native to Mexico. Planted throughout and naturalised. *W. filifera* is similar but shorter, to 15 m, with a trunk to 1 m across and leaves with white threads pendent from the leaf intersections. Native to California. Planted throughout.

Hyophorbe

Trunk often swollen, ringed with leaf scars, grey. Leaves pinnate. Inflorescence solitary, branched in 3–4 orders.

Hyophorbe verschaffeltii An exotic-looking palm with a swollen, bottle-like smooth grey trunk to 6 m. Leaves erect, arching, with large sheathing bases. Widely planted.

---

1. *Agave americana*
2. *Agave americana* (inflorescence)
3. *Agave sisalana*
4. *Agave fourcroydes*
5. *Agave attenuata*
6. *Washingtonia robusta*
7. *Hyophorbe verschaffeltii*
8. *Bismarckia nobilis*
9. *Phoenix canariensis* (naturalised specimen)
10. *Phoenix canariensis* (cultivated specimen)

Bismarckia

Trunk swollen, to 45 cm across. Leaves *circular* with up to 20 segments. Inflorescence pendent.

Bismarckia nobilis An exotic-looking palm to 10 m with a swollen, trunk with the remains of sheathing leaf bases. Leaves *fan-like*: circular, *silvery*. Widely planted.

Phoenix

Trunk long and stout; leaves pinnately cut into numerous leaflets. DNA analysis shows species are freely hybridising, so accurate identification of true species in the field may not be possible.

Phoenix canariensis PALMERA CANARIA, CANARY PALM A tall palm to 20 m with a solitary trunk 50 cm–1.2 m across and crown with green fronds 1.8–6 m long, with pinnately divided leaflets 21–89 cm. Fruits 15–23 mm, orange ripening to orange-brown or purple-brown. Uncommon as a native, in rocky ravines. Very commonly planted. *L, F, C, T, G, H, P.

Phoenix dactylifera PALMERA DATILERA is similar (especially when immature) but taller, to 30 m with a *narrower trunk 20–50 cm across* and smaller crown. Leaves glaucous. Fruit (the edible date) larger, 25–75 mm long, orange-brown and sticky when ripe. Planted (possibly over-recorded in error through confusion with the species above). L, F, C, T, G, P.

STRELITZIACEAE

Herbaceous (though often tree-like) perennials similar in form to plants in the banana family. Leaves sheathing, sometimes forming a 'trunk'. Leaves simple. Flowers borne in inflorescences, zygomorphic; stamens 5 or 6; style 1, stigmas 1 or 3. Fruit a capsule.

Strelitzia BIRD OF PARADISE

Evergreen perennials with long-stalked leaves arranged in 2 ranks (fan-like) and flowers borne in horizontal inflorescences emerging from robust, folded, boat-like spathes.

Strelitzia reginae AVE DEL PARAÍSO, BIRD OF PARADISE An erect, evergreen, rhizomatous perennial to 2 m. Leaves borne in 2 ranks, appearing fan-like, broadly-oval, 25–70 cm long, 10–30 cm wide, borne on *long stalks* to 1 m. Flowers distinctive in shape, borne above the leaves; spathe held horizontally, greyish, ending in a long point; sepals *bright orange*; petals *purple-blue to white*. Commonly planted in gardens and roadsides throughout (seldom naturalised).

MUSACEAE | BANANA FAMILY

Herbaceous (though tree-like) perennials with leaves that have persistent, overlapping basal sheaths forming a 'trunk'. Monoecious with flowers borne in clusters (effectively a raceme of cymes); stamens 5–6; carpels 3. Fruit a berry.

Musa BANANA

Herbaceous, robust perennials with trunk-like stems and 6–20(-numerous) leaves in a canopy. Flowers borne 12–20 per cluster, unisexual. Fruit the familiar banana.

Musa acuminata PLATANERA, PLANTAIN, CULTIVAR BANANA Herbaceous (but tree-like), tropical-looking perennials 2–9 m. Leaves very large and broad, to 2 m long and 50 cm across with a prominent midrib, sometimes splitting at the margins; pinnately veined. Flowers unisexual, borne in large pendulous clusters. Fruit an elongated berry (a banana), 20–30 cm long. Widely farmed.

CANNACEAE

A family with one genus of perennials native to the American tropics. Petals and sepals 3; staminodes (modified stamens) conspicuous.

Canna indica A rhizomatous, tropical-looking perennial 50 cm–2 m with broad, alternate, simple leaves to 60 cm long, with parallel veins. Flowers borne in spikes; corolla tubular, orange or scarlet, 45–75 mm. Fruit a warted capsule; seeds black. A garden plant, naturalised in thickets and ditches; local. C, T, G.

1. *Strelitzia reginae*
2. *Canna indica* (habit)
3. *Canna indica*
4. *Canna indica* (habit)

JUNCACEAE

Erect, reed-like annuals or perennials with white, pith-filled stems. Differing from grasses and sedges in having regular flowers with 6 similar tepals and (3)6 stamens; style 0 or 1; stigmas 3. Fruit a capsule with 3–numerous seeds.

Juncus

Annuals or perennials with 1–2-faced leaves; hairless. Flowers with 6 tepals (unlike in other grass-like families); stamens 6. Capsule with numerous seeds. Damp habitats.

Juncus acutus JUNCO COMÚN, SHARP RUSH A perennial 70 cm–1.8 m with *densely tufted*, sharply pointed leaves. Flowers reddish-brown, borne in dense, rounded inflorescences exceeded by their bracts; anthers up to 5 x longer than the filaments; tepals more or less equal. *Capsule 3.2–6 mm long, much longer than the tepals; inner tepals with membranous margins extended into lobes.* Coastal flats, saltmarshes. **Subsp.** *leopoldii* is the form on the Islands. L, F, C, T, G, P. *J. maritimus* JUNCO MARINO, SEA RUSH is similar but shorter and laxer, with *capsules triangular–ovoid and pointed, 2.5–3.5 mm long, equal to, or slightly exceeding the tepals*. L, C, G, P. *J. capitatus* JUNCO CABEZÓN, DWARF RUSH is a dwarf *annual* to just 50 mm. C, T, G, H, P. *J. effusus* JUNCO FINO, SOFT RUSH has soft-tipped leaves and flowers with a *long, narrowly sheathed bract*. C, T, G, P. *J. inflexus* JUNCO AZUL is similar but *blue-grey* with *strongly ribbed stems* (12–18 clear ribs) and an interrupted pith. C.

Luzula

Tufted perennials with flat or channelled, white-hairy leaves. Perianth segments papery; stamens 6. Fruit a capsule with 2–3 seeds.

Luzula canariensis LÚZULA CANARIA A robust perennial with flat, somewhat reddish leaves with white hairs at the base. Flowering stems leafy; inflorescence dense, flat-topped to elongated; *perianth segments silvery-white*. *T, G. *L. forsteri* LÚZULA DE FORSTER is small with more or less 1-sided inflorescences with a *greenish-brown, star-like* perianth. Seeds with an erect appendage. C, T, G.

CYPERACEAE

A large family of hairless, herbaceous sedges with rhizomatous root systems; leaves grass-like but stems solid, often triangular. Flowers wind-pollinated, borne on spikelets, highly reduced, the perianth often in the form of bristles; monoecious or sometimes dioecious; stamens 1–3; style 0 or short; stigmas 2–3. Fruit a small, 2–3-angled nut. Only a small subset is described here.

Carex SEDGE

Rhizomatous, spreading or tufted perennials with stems triangular in section. Inflorescence of 1-flowered spikelets grouped in spikes; lowest bract leaf-like or glume-like; flowers *unisexual*, either mixed in 1 spike or dioecious (often the upper spikes male and the lower female); perianth bristles 0; stamens and stigmas 2–3.

Carex canariensis CUCHILLERA CANARIA A tufted, grass-like perennial to 1 m with flat, conspicuously rough, arching leaves, 3–5 mm wide. Spikelets to 20 mm long, with male and female

1. *Juncus acutus*
2. *Luzula canariensis* (habit)
3. *Luzula canariensis*
4. *Carex canariensis*
5. *Carex divulsa*
6. *Carex perraudieriana*
7. *Carex perraudieriana*
8. *Carex paniculata* subsp. *calderae*

flowers. Forests. C, T, G, H, P. *C. divulsa* CUCHILLERA has leaves 2–3 mm wide and *lax, interrupted* inflorescences with very short spikelets. Forest margins. L, F, C, T, G, H, P. *C. otrubae* CUCHILLERA DENSA has very *dense*, oblong to sub-cylindrical inflorescences to 50 mm. Wet habitats. C, T, G. *C. perraudieriana* CUCHILLERA ANCHA has shiny, arching leaves to 12 mm wide that are conspicuously veined, and *pendulous*, tuft-like inflorescences. Rare; forests. *F, T, G, P.

Carex paniculata A large and densely tufted sedge, forming clumps to 1 m across, with 3-angled, greenish to blackish-brown stems to 1.5 m high. Spikes all similar in appearance, often close together, with both male and female flowers, borne stalkless and overlapping. **Subsp. *calderae* CUCHILLERA DE LAS CAÑADAS** is the form in the region. Springs in high mountain; local. *T.

Cyperus

Tufted annuals or perennials with stems triangular in section and grass-like leaves. Inflorescence an umbel or umbel-like raceme with many-flowered, grass-like spikelets clustered in dense heads; lowest 2–10 bracts leaf-like and exceeding the inflorescence; flowers cosexual without a perianth; stamens 1–3; stigmas (2)3. Often in wet habitats.

(a) Inflorescence in a *single, dense capitulum* exceeded by long, tough bracts.

Cyperus capitatus JUNQUILLO A small, tough, hairless, blue-grey rhizomatous perennial with few, wiry leaves; stems solitary, 30(40) cm. Leaves blue-green, becoming yellow with age, with inrolled margins. Bracts leaf-like, erect and exceeding the inflorescence; later brown and withered. *Flowers borne in dense, brown, terminal spikelets*; stamens 3. Coastal dunes. L, F, C, T, P.

(b) Inflorescence compound. Stamens (2)3.

Cyperus laevigatus JUNCIA CLARA A short, tufted perennial to 50 cm with mat-forming rhizomes; stems erect and bunched or solitary, rounded or triangular in section. Leaves few, reduced, 30–60 mm. Inflorescence a fascicle of 2–4(9) stemless spikelets with greenish-brown flowers; stamens 3; *bracts 2, erect, forming an apparent elongation of the stem*, and exceeding the inflorescence. Stream margins and other wet habitats. L, F, C, T, G, P.

Cyperus teneriffae JUNCIA NEGRA An annual with fibrous roots and few, weak leaves, 1(–2) mm wide. Inflorescence an *umbel* of 3–10 *flattened*, reddish-brown, green-margined spikelets. Stamens 3. C, T.

Cyperus alternifolius subsp. *flabelliformis* (syn. *C. involucratus*) PARAGÜITAS A robust, clumped perennial to 1.3 m with tough, woody rhizomes to 13 mm wide and numerous erect stems. *Leaves reduced to brownish sheaths* (sometimes extended into a spine tip). Inflorescence terminal with numerous radiating leafy rays to 20 cm long and greenish to brownish spikelets; stamens and stigmas 3. Widely cultivated and naturalised. C, T, G, P.

Cyperus longus JUNCIA LARGA A rather robust perennial 37–78 cm (1 m) with thick, far-spreading rhizomes 3–10 mm wide. Leaves to 2–5 mm wide, shorter than or equalling the stems. Inflorescence diffuse: a simple or compound umbel with 6–10(12) rays *of brownish or reddish spikelets* 12–30 cm; stamens 3; bracts 2–6, the outer exceeding the inflorescence. Damp places and pool margins. F, C, T, G, P.

Cyperus esculentus CHUFA, EDIBLE CYPERUS A perennial to 60 cm (1 m) with *underground tubers* and leaves to 2–10 mm wide. Inflorescence an umbel of 3–8 *straw-coloured* spikelets 10–12 cm long forming lax clusters on the branches; glumes densely overlapping; stamens and stigmas 3; bracts 2–3 x longer than the inflorescence. G, P.

1. *Cyperus capitatus*
2. *Cyperus laevigatus*
3. *Cyperus alternifolius* subsp. *flabelliformis*
4. *Cyperus teneriffae*
5. *Cyperus eragrostis*

(c) Inflorescence compound. Stamens 1.

Cyperus eragrostis JUNCIA AMOROSA A shortly rhizomatous perennial with erect, often solitary stems to 60(80) cm. Flowers borne in greenish-yellow to brownish spikelets 8–13 mm long, rather compact; *stamen 1*. Native to tropical America. C, T, G, P.

Eleocharis

Perennials with rounded to ridged stems and no leaf blades. Inflorescence a terminal spikelet, the lowest bract glume-like; flowers cosexual; perianth of 0–6 bristles; stamens 3; stigmas 2 or 3.

Eleocharis palustris PALILLO, COMMON SPIKE-RUSH A hairless, aquatic perennial with creeping rhizomes with solitary (1st year) and later numerous, *tufted*, leafless stems to 60(75) cm, bearing cylindrical spikelets 5–30 mm long; *stigmas 2*. Stems reddish at the base with *leafless sheaths*; sheaths pale brown. Stems with approximately equal air canals in cross-section. Marshes and seasonally flooded habitats. C, T, G. *E. multicaulis* is very similar, differing mainly in having *3 stigmas*. Nuts 3-angled. Similar habitats. T.

POACEAE | GRASS FAMILY

Annual or perennial, often rhizomatous or creeping plants. Leaves alternate, linear and sheathing the stem, generally with a membranous ligule at the base of limb. Inflorescences very variable, often a spike or panicle; flowers not brightly coloured, wind-pollinated and with (1)3(6) stamens; pistil with normally 2 styles, enclosed within 2 bracts; the whole called a floret; florets arranged into spikelets with 2 empty bracts (glumes) at the base. A large family divided into several subfamilies; it is well beyond the scope of this book to describe all species on the islands; only a small subsection is included.

SUBFAMILY POOIDEAE

The largest subfamily of grasses with numerous genera. Stems usually hollow. Spikelets normally cosexual with 1–many female florets; lemmas with or without awns; stamens 1–3; styles and stigmas 2.

Arrhenatherum

Tufted perennials with well-branched, diffuse panicles; spikelets with 2(–5) florets, the lowest male, those above cosexual; glumes unequal; lemmas 7-veined with long, bent, dorsal awns.

Arrhenatherum calderae MAZORRILLA DEL TEIDE An erect, tufted grass with long flowering stalks to 1 m, overtopping the leaves; leaves to 20 cm, linear, flat, glaucous, hairless above, rough below and along the margins; ligule blunt and slightly toothed. Panicles diffuse, narrowly ovoid, 80–90 mm long; anthers yellow to whitish. High-elevation rocky slopes; local. *T, P.

Agrostis

Annual to perennial grasses with slender, compound inflorescences; spikelets small, 1-flowered; glumes papery, 1-veined; lemma ovate, 3–5-veined with a short awn (or absent).

Agrostis castellana GREÑA A rhizomatous perennial to 80 cm; ligules pointed or rounded, 2–3 mm. Inflorescence a diffuse, later contracted panicle; lemma with lateral veins extended at the ends into minute awns. Grown as a lawn grass in amenity and sports areas, escaping. L, F, C, T, G, H, P.

Festuca

Tufted perennials. Leaf-blades flat or folded. Inflorescences an open or contracted panicle; glumes subequal; lemmas 3–5-nerved.

Festuca agustinii CERRILLO DE RISCO A tufted perennial with *wiry, narrow*, blue-grey leaves. Inflorescence a diffuse panicle, 60 mm–16 cm. Mainly on wet cliffs. *C, T, G, H, P.

Melica

Clumped, perennial grasses with short rhizomes that have spike-like or sparsely-branched panicles; spikelets with 1–3 cosexual florets; glumes 3–5-veined; lemmas 7–9-veined.

Melica canariensis TRIGUERILLA A rather sparse, tufted perennial 80 cm–1 m with flat leaf blades to 17 cm, rough. Inflorescence a contracted, scarcely interrupted panicle 90 mm–21 cm; plumose, arching and turning straw-brown with age. Cliffs and rocky slopes. L, F, C, T, G.

1. *Arrhenatherum calderae*
2. *Agrostis castellana*
3. *Festuca agustinii* (habit)
4. *Festuca agustinii*
5. *Melica canariensis* (habit)
6. *Melica canariensis*
7. *Nassella neesiana*

Nassella

Perennials with terminal panicles; spikelets with 1 fertile floret; glumes shorter than spikelet, lemma various.

Nassella neesiana FLECHILLA, CHILEAN NEEDLE GRASS A tussocky grass to 1 m with flat, coarse leaves 1–5 mm wide. Panicles diffuse; glumes 16–26 mm, purplish; awns 6–9 mm, twisted. Native to South America, naturalised in disturbed habitats, especially road margins. C, T, G, P.

Bromus ACEITILLA, ESPIGUILLA, BROME

Annuals. Inflorescence with flattened, long-stalked spikelets with many overlapping florets; glumes unequal, often awned, 3–7(9)-veined; lemma 7–9(11)-veined, minutely split at the apex. Dwarf specimens in dry areas are not reliable for identification.

Bromus madritensis ACEITILLA, COMPACT BROME A short, tufted annual with erect stems. Leaves tapered to 5 mm wide, softly white-hairy. Inflorescence a rather lax, erect, wedge-shaped panicle with short, bunched branches; spikelets to 60 mm; lemma with a long bristle-tip to 16 mm, hairy or hairless; glumes 1–3 veined; lemma 5–7-veined, 10–19 mm. Common on fallow and cultivated ground. L, F, C, T, G, H, P. *B. diandrus* has a long, very lax, nodding inflorescence (not wedge-shaped) with spreading spikelets; lemma 22–45 mm, 7-veined. L, F, C, T, G, H, P. *B. hordeaceus* has an erect, rather short and dense inflorescence with spikelets on short stalks (exceeded by their spikelets) with a slightly inflated appearance; lemma 6.5–11 mm long, 2.5–5 mm wide, 7–9-veined. Regional forms include **subsp. *hordeaceus*** (T, P) and **subsp. *molliformis*** (throughout). *B. squarrosus* has a *deflexed* awn, at right angles to the lemma; glumes 3–9-veined; lemma 8–11 mm long, 5–9 mm wide, 7–11-veined. Dry ground and wasteland. F.

Phalaris

Sterile lemmas shorter than fertile floret; glumes winged.

Phalaris canariensis ALPISTE A tufted annual grass with erect stems 20 cm–1.2 m, 5–6-noded. Ligule membranous. Leaves to 31 cm, 4–16 mm wide. Inflorescence a solid and *congested, spike-like, ovoid panicle* 15–63 mm long; spikelets strongly compressed, greenish-white. L, F, C, T, G, H, P.

Avena OAT

Annuals. Inflorescence a compound, diffuse panicle, branched with large, long-stalked and drooping spikelets with 2–3 florets; glumes papery and exceeding the florets, 7–11-veined; lemma leathery and 7–9-veined with a stout, bent, long awn. A variable and difficult genus.

Avena barbata BALANGO, BEARDED OAT An erect grass with solitary or clumped stems to 1 m. Leaves linear, to 15 mm wide, and hairless or slightly hairy on the margin; ligule membranous, to 5 mm. Inflorescence a 1-sided, very lax panicle with spikelets drooping on slender stalks; lowest lemma 12–18 mm with *2 bristles* at the tip, 3–5 mm long. Fallow and cultivated ground and roadsides. L, F, C, T, G, H, P. *A. canariensis* BALANGO CANARIO is a similar Canary Island endemic, with lax, unilateral inflorescences with spikelets 15–20 mm with a *hairy* lemma with two *short teeth* at the tip. Dry, interior habitats. *L, F, T.

1. *Bromus diandrus*
2. *Bromus hordeaceus* subsp. *molliformis*
3. *Phalaris canariensis*
4. *Avena barbata*
5. *Avena canariensis*
6. *Briza maxima*
7. *Briza minor*

Gaudinia FRENCH OAT-GRASS

Annuals to short-lived perennials with long, spike-like inflorescences; spikelets with 4–11 florets, all fertile; lower glume half as long as the upper, 3–5- and 5–11-veined respectively; lemmas 5–9-veined.

Gaudinia fragilis AVENA FRANCESA, FRENCH OAT-GRASS A softly-hairy annual to 40 cm (1 m) with flat leaves with hairy margins; lower leaves with spreading hairs on the sheaths. Inflorescences long, slender and green, about 10 cm; spikelets 9–20 mm, borne in 2 alternate rows, arranged edgeways to the axis; lower glume 3 mm, 3-veined, the upper 7 mm, 7–9-veined; lemma 7–11 mm, toothed, with a twisted, bent awn 5–13 mm long. Grassy habitats. C, T.

Lagurus HARE'S TAIL

Annuals. Inflorescence very compact, compound and spike-like, densely silkily hairy; spikelets with single florets, falling as a unit when ripe; glumes bi-lobed, 1-veined and awned, longer than the obscurely 5-veined lemma.

Lagurus ovatus RABILLO DE CONEJO, HARE'S TAIL A softly hairy, grey-green annual to 60 cm. Leaves linear–lanceolate and flat; ligule hairy and membranous, to 3 mm. Inflorescence distinctive; egg-shaped, dense, 'fluffy' soft and white, 5–20 mm long; lemma semi-transparent and with awns 8–20 mm. Common in sandy coastal environments and rocky slopes inland. Doubtfully native to the Canary Islands. C, T, G, H.

Rostraria

Small annuals with spike-like panicles that have very short branches; florets cosexual; glumes unequal (the lowermost 1-veined, the uppermost 3-veined); lemmas 5-veined, toothed.

Rostraria cristata CAÑOTILLA, MEDITERRANEAN HAIR GRASS A short, erect grass to 20(60) cm with hairless to hairy stems. Flowers borne in spikelets 3–8 mm long with awns 1–3, in green, cylindrical, spike-like panicles 10 mm–10 cm long and 4–10 mm wide. Disturbed habitats. L, F, C, T, G, H, P.

Briza

Annuals or perennials. Inflorescence distinctive: spikelets flattened and inflated, ovoid- to heart-shaped and awnless, pendulous with 4–30 overlapping, all cosexual florets.

Briza maxima TEMBLADERA, LARGE QUAKING GRASS A hairless, low annual grass with often solitary, erect stems. Leaves flat and linear, to 4 mm wide. Inflorescence a lax panicle, with up to 15 large, drooping papery spikelets appearing inflated, on slender stalks, green then purplish, ripening pale brown, 8–25 mm long. Common on fallow ground and roadsides. L, F, C, T, G, H, P. *B. minor* is similar but with numerous (>20), smaller spikelets 2.5–5 mm long. Open habitats. C, T, G, H, P.

Gastridium NIT-GRASS

Annuals with compact, spike-like inflorescences with ascending to adpressed branches; spikelets laterally compressed, solitary with a single cosexual floret; glumes unequal and exceeding the florets, 1-veined and papery at the tips; lemma membranous, 5-veined and awned or not (awns shorter than the lemmas).

Gastridium ventricosum CAÑOTA, NIT-GRASS A small annual to 50(90) cm with flat, hairless leaves, often withered during flowering; ligules to 3 mm long, and pointed. Flowers borne in green, strongly erect (at least at first), more or less bilaterally symmetrical and laterally compressed panicles 5 mm–10(16) cm long; spikelets short, 3–5 mm long with a single floret. L, F, C, T, G, P. *G. phleoides* is native to Asia but naturalised as a casual weed, distinguished by its very dense panicles with longer spikelets 5–8 mm long; lemma densely pubescent, with an awn 4–7(8) mm, often exceeding the glumes. F, H, P.

Lolium ACEBÉN, BALLICO, RYE GRASS

Annuals or perennials. Inflorescence simple, unbranched and spike-like with stalkless spikelets alternately arranged edgeways onto a jointed axis, flattened; glumes solitary in lateral spikelets (2 in terminal); lemma 5–9-veined and awned or not.

Lolium perenne CÉSPED INGLÉS, RYEGRASS A hairless, tufted, wiry perennial to 50(90) cm. Stems smooth and slender, bent below. Leaves narrow, up to 3 mm wide, and folded until mature. Ligule to 1 mm, abruptly pointed. Inflorescence simple, to 15 cm long with compressed, oval spikelets to 15 mm; lower lemmas 3.5–9 mm and almost always awnless. L, F, C, T, G, P. *L. multiflorum* is similar but an annual with awned lemmas (awns to 15 mm); glume *shorter* than the spikelet. L, F, C, T, G, H. *L. canariense* has lemma with awns to 20 mm and glumes *equalling* the spikelets. Local but widespread. L, F, C, T, G, H, P.

Dactylis COCK'S FOOT

Perennials. Inflorescence a more or less 1-sided panicle or compound with spikelets crowded into dense clusters at the ends of the side branches; spikelets flattened, short-stalked with 2–5 florets; glumes keeled and 3-veined; lemma keeled and 5-veined, very shortly awned or awnless.

Dactylis glomerata JOPILLO, COCK'S-FOOT A perennial, bluish, clump-forming grass with erect or spreading stems to 1.4 m. Leaves rough, with ligules 2–10 mm. Inflorescence an erect, rather unequal and 1-sided tufted panicle of laterally compressed spikelets borne in dense clusters on lateral branches, often with prominent yellow stamens. Grassy habitats. C, T, H, P. *D. smithii* has a *lax, shrubby habit* with branching culms that have *numerous nodes*. L, F, C, T, G, H, P. *D. metlesicsii* is a rare endemic of the *Spartocytisus supranubius* vegetation of Tenerife. Plant 70 cm–1 m. Leaves glaucous; ligules 3–7 mm. Inflorescence dull reddish-purple. High elevation. *T.

Lamarckia GOLDEN DOG'S TAIL

Annuals. Inflorescence with spikelets of 2 kinds: the upper with 1 fertile floret and 1 rudimentary floret, the lower with several pairs of overlapping, blunt, sterile lemmas in 2 ranks.

Lamarckia aurea CEPILLITO DORADO, GOLDEN DOG'S TAIL A more or less hairless, low annual grass with tufted, erect stems to 20(30) cm. Leaves linear, 2–6 mm wide with hairy margins; ligule membranous, 5–10 mm long, pointed or blunt. Inflorescence 30–90 mm long, dense, rather 'fluffy' and 1-sided with the outer spikelets sterile, greenish and later golden. Common on fallow and cultivated ground and roadsides. L, F, C, T, G, H, P.

Cynosurus

Annuals or perennials. Inflorescence a compact spike-like panicle of fertile spikelets with (1)2–5 cosexual florets and sterile spikelets with sharp-pointed lemmas in a herringbone arrangement.

Cynosurus echinatus COLAPERRO, ROUGH DOG'S-TAIL A short to tall hairless annual with erect or spreading stems to 75 cm (1 m) and flat leaves 2–10 mm wide. Inflorescence a dense, plume-like, 1-sided, oblong panicle 10–40(80) mm long of shiny green or purplish spikelets; the outer spikelet of each pair comb-like with several pairs of spreading, long-awned, sterile lemmae; inner spikelet fertile and wedge-shaped. Dry rocky and grassy scrub. L, C, T, G, H, P.

Catapodium

Annual grasses with a simple or little-branched, stiff, 1-sided, spike-like inflorescence; spikelets rather compressed with many (3–14) florets in 2 ranks; glumes 2, nearly equal and papery; lemma blunt and leathery, 5-veined and awnless.

Catapodium rigidum (syn. *Desmazeria rigida*) GRAMILLA, FERN-GRASS A small, stiff, hairless, tufted, bluish annual with erect stems to 15(60) cm with several to numerous erect or spreading stems. Leaves often purplish, fine-pointed, to 2 mm wide and flat or with inwardly rolled margins. Inflorescence a more or less 1-sided panicle to 80 mm long, often branched below, with sparse, tiny spikelets to 7 mm, each with 5–10 minute florets; glumes 1.3–2 mm and pointed; lemmas longer, 2–2.6(3) mm and blunt. Coastal sands, dry, bare habitats and walls and rocks. L, F, C, T, G, H, P.

Parapholis HARD GRASS

Annuals with very slender, whip-like inflorescences with alternate spikelets arranged broadside and set into hollows of the axis, each with a single floret; glumes equal and 3–5-veined; lemma finely 3-veined.

Parapholis incurva BALLICO PETUDO, CURVED SEA HARD-GRASS A distinctive, short, tufted annual with spreading, curved stems 10–20 cm long. Leaves flat or inrolled, linear and pointed, to 2 mm wide, rough above and along the margins with reddish sheaths. Inflorescence 10–80 mm (15 cm), often not exserted from its sheath, slender, rigid, cylindrical, strongly curved and jointed with spikelets adpressed to the stem; spikelets to 7 mm long, a little longer than the joints of the axis; glumes equal, closing the cavities of the axis. Saline coastal habitats. L, F.

Piptatherum (including *Oryzopsis*)

Shortly rhizomatous perennial grasses with transparent ligules. Spikelets short and dorsally compressed; lemma with a long, straight, often falling terminal awn. A much confused genus; *Oryzopsis miliacea* now established to be genetically distinct from *Piptatherum*.

Oryzopsis miliacea (syn. *Piptatherum miliaceum*) CERRILLÓN FINO A tall, erect, perennial to 1.5 m with hairless stems, and leaves rough above. Flowers borne in light green to straw-coloured panicles to 40 cm long; 1-sided, drooping and lax, with several (to 20) branches at each node along the stalk; glumes 3–4 mm, lemma 2–2.5 mm, hairless, stiff, membranous (not leathery) with awns 3–5 mm. Dry open habitats among shrubs and other vegetation. L, F, C, T, G, H, P.

Piptatherum coerulescens (syn. *Oryzopsis coerulescens*) A tall, erect, perennial to 70 cm (1 m) with hairless stems. Leaves 15–31 cm long and 1–12 mm wide, rough on both surfaces, with long ligules to 11 mm long, transparent. Flowers borne in panicles 30 mm–15 cm long with spikelets 5–14 mm; glumes sub-equal, exceeding the florets, 3–9-veined; lemma 2.6–6.5 mm long, leathery, with awns 1–15 mm long, falling. L, C, T, G, H.

1. *Gastridium phleoides*
2. *Lolium canariense*
3. *Dactylis smithii*
4. *Dactylis metlesicsii*
5. *Lamarckia aurea*
6. *Cynosurus echinatus*
7. *Catapodium rigidum*
8. *Parapholis incurva*
9. *Piptatherum coerulescens*

Stipa

Annuals or perennials with ligule a membrane fringed with hairs; lemma with forward-pointing bristles and a long, terminal, persistent awn.

Stipa capensis CHIRATE An annual or biennial to 20 cm, with leaves to 15 cm long with revolute margins; blue-grey and hairy or hairless. Inflorescence a dense, slender panicle; enclosed at the base by a subtending leaf, to 15 cm long and 10 mm wide; spikelets solitary, the fertile spikelets stalked; glumes persistent and all more or less similar, 15–20 mm long and exceeding the florets, and 3-veined; awns 70 mm–10 cm. Arid habitats. L, F, C, T, G, H, P.

Aegilops

Inflorescence with stalkless spikelets arranged broadside along the axis in a distinctive compact ovoid or cylindrical cluster; glumes large, tough and strongly veined with 2–4 awns at the apex; lemmas also awned. Many similar species co-occur.

Aegilops geniculata ROMPESACOS, TRIGO GUANCHE A low, tufted annual with erect stems to 40 cm. Leaves with a flat blade 2–3 mm wide, finely hairy on the upper surface; ligule very short. Inflorescence congested and an inverted cone-shape, *not more than 2 x as long as broad*, with 1–2 vestigial spikelets at the base of the fertile ones; fertile spikelets often just 2–4, *broadest at, or just below the middle*; lemma with a long bristle 15–25(30) mm, *as long as those of the glumes*. Common on bare and fallow ground and in olive groves. L, T, G, P. **A. neglecta** is similar but with glumes of lateral spikelets with 2–3 (not 3–5) awns, the spike (excluding awns) *at least 5 x as long as broad*, and the awns of the lemmas about *half as long* as those of the glumes. L, T, P.

Hordeum

Annuals, sometimes perennials. Inflorescence spike-like, dense and long-awned with spikelets in clusters of 3 arising from each joint in the axis; glumes narrow, long-awned and 1–3-veined; lemmas 5-veined with long awns. Other, very similar species occur.

Hordeum murinum CEBADILLA, WALL BARLEY An annual 10–60 cm with tufted, erect, smooth stems. Leaves linear, to 4 mm wide and hairy on both surfaces with shiny sheaths; ligule membranous and small, to 1 mm long. Inflorescence a more or less bilaterally symmetrical, bristly, dense, spike-like panicle, lemmas with awns 10–45 mm long; glumes with awns 10–30 mm long. L, F, C, T, G, H, P.

SUBFAMILY **ARISTIDOIDEAE**

Inflorescence with spikelets with one floret and three lemma awns joined basally in a column.

Aristida

Tufted annuals or perennials with long-awned panicles.

Aristida adscensionis RABO DE BURRO An annual forming erect or sprawling tufts to 1 m. Leaves linear, to 20 cm long, 3 mm wide, flattened or folded. Inflorescence a very slender panicle to 30 cm, variously diffuse to contracted, often arching slightly; spikelets green or purple; glumes unequal; *awns long, upward-swept*. Bare habitats; common. L, F, C, T, G, H, P.

---

1. *Stipa capensis*	3. *Aristida adscensionis*	5. *Phragmites australis*	7. *Cynodon dactylon*
2. *Hordeum murinum*	4. *Arundo donax*	6. *Eleusine indica*	8. *Tragus racemosus*

SUBFAMILY ARUNDINOIDEAE

Stems hollow, plants often reed-like. Spikelets cosexual with 1–many female florets; lemmas with 1–3 awns; stamens 1–3; styles and stigmas 2.

Arundo

Robust perennials. Inflorescence large, compound, feathery with numerous spikelets, each 1–7-flowered; glumes lanceolate, papery and keeled, 3-veined; lemma papery with dense silvery hairs.

Arundo donax CAÑA, GIANT REED An extremely robust, rhizomatous perennial with bamboo-like stems to >6 m. Leaves grey-green, to 60 mm wide. Inflorescence a large panicle to 60 cm long; silky and silvery, with spikelets 12–20 mm long, each with usually 3(4) florets; lemma notched at the apex with a short bristle-tip; glumes papery, keeled and 3-veined, usually (not always) longer than florets. Common in a range of habitats. L, F, C, T, G, H, P.

Phragmites

Spreading perennials. Inflorescence a large, feathery and compound panicle with numerous slender spikelets with numerous 2–6(10) florets with hairy stalks; glumes unequal, 3–5-veined; lemma hairless and 1–3-veined, awnless.

Phragmites australis (syn. *P. communis*) CARRIZO A bed-forming, large, reed-like grass, rather similar to *Arundo donax* with tall, rather slender stems to 3.5 m that do not overwinter. Leaves to 50 cm long and 50 mm broad, grey-green and tapered to the tip; sheaths smooth and hairless, surrounding the leaf nodes. Flowers borne in drooping, 1-sided, more or less cylindrical, bunched greenish or purplish panicles with spikelets 8–16 mm long, each with up to 10 florets. Aquatic habitats such as lake margins. L, F, C, T, G.

SUBFAMILY CHLORIDOIDEAE

Stems hollow or solid; spikelets cosexual with 1–many female florets; lemmas with 1 or more awns; stamens 1–3; styles and stigmas 2.

Eleusine

Tufted annuals or tussocked perennials. Leaves linear. Inflorescence digitate to shortly racemose with clustered spikes. Glumes 1–several-nerved; lemmas 3-nerved.

Eleusine indica PATA DE GALLINA A tufted annual to 85 cm. Leaves folded, 50 mm–35 cm long. Inflorescence star-like (digitate) with 1–10(17) slender, ascending spikes to 15 cm long, sometimes with a few below the main cluster. Pastures. F, C, T, G, P.

Cynodon BERMUDA GRASS

Perennials. Inflorescence a compound umbel of 3–6 slender branches with stalkless spikelets arranged in 2 rows; glumes 1-veined; lemma 3-veined.

Cynodon dactylon PATA DE GALLINA, GRAMA, BERMUDA GRASS A spreading, short perennial with creeping stems to 30 cm. Leaves linear and flat, hairless or hairy along the margins. Inflorescence distinctive and star-like: 3–6 spikes 20–50 mm long, outwardly spreading from a single central axis; spikelets 2–3 mm long, stalkless. Native to tropical Africa. L, F, C, T, G, H, P.

Tragus

Annuals. Inflorescence a spike, or spike-like, with 2–5 spikelets on very short branches at each node, each with 1 cosexual floret; glumes unequal, the upper longer and 5–7-veined, each vein with hooked bristles; lemma 3-veined.

Tragus racemosus ACEITILLA DE CALCETINES, STALKED BUR GRASS A creeping, branched, spreading annual, rooting at the nodes with erect stems to 40 cm and short, flat leaves to 3 mm wide with spines along the margins. Inflorescences spike-like, long, cylindrical, purple; spikelets 3–5 per node, the upper glumes with 7 rows of fine-crooked bristles on the backs. Dry sandy areas, dunes, waste places and olive groves. L, F, C, T, G, H, P.

Eragrostis LOVE GRASS

Annuals or tufted perennials with ligules a ring of hairs. Inflorescence a lax, diffuse panicle; spikelets with 3–many florets, narrow; glumes usually 1-veined; lemmas 3-veined, keeled and blunt.

Eragrostis barrelieri PAJUCO A tufted annual 10–60 cm with flat or inrolled leaves 20 mm–10 cm long, 20–55 mm wide, bluish; ligule a fringe of hairs. Inflorescence an open, diffuse panicle 30 mm–20 cm, the stiff branches with spikelets almost to the base; spikelets 5–20 mm long with 5–30 fertile florets; all glumes 1-keeled, 1-veined; lemma 3-veined, keeled, membranous with distinct lateral veins. Bare, dry habitats. L, F, C, T, G, H, P.

SUBFAMILY **PANICOIDEAE**

The second largest subfamily of grasses with numerous species. Stems often solid. Spikelets unisexual or cosexual; lemmas sometimes awned; stamens 3; styles and stigmas 2.

Sorghum

Annuals or perennials. Inflorescence large and much-branched with paired, shiny spikelets with 1 stalkless fertile floret and 1 or more stalked, sterile or male florets; glumes all compressed and 3-pointed at the tip.

Sorghum halepense SORGO, JOHNSON GRASS A large, erect, deeply rhizomatous perennial to 1.5 m with stems silkily-hairy at the nodes. Leaves hairless with rough margins, <20 mm wide. Inflorescence a large, terminal, rather lax, pyramidal panicle to 30 cm long; spikelets shiny, to 5.5 mm long; lemma notched at the apex and with an awn to 12 mm long; sterile floret stalked, violet, hairy and lanceolate. L, F, C, T, G, H, P. *S. bicolor* is larger (to 2 m), not rhizomatous, with broader leaves (>20 mm) and lacks creeping stems; inflorescence dense, ovoid, to 50 cm long. C, G, P.

Melinis

Spikelets with awns arising from between apical lobes of upper glume and lower lemma.

Melinis repens YERBARUBÍ An annual or short-lived perennial, tufted. Stems erect or ascending, 20 cm–1.5 m, rooting from lower nodes. Leaves 2–10 mm wide, to 30 cm long, flat; ligule a fringe of hairs. Inflorescence a *reddish, soft-bristly* compound, open panicle 50 mm–20 cm, with branches to 60 mm. Glumes dissimilar, thinner than fertile lemma. Native to Africa; introduced. T.

Pennisetum

Perennials. Inflorescence a dense, narrow panicle with fascicles of spikelets interspersed with bristles.

Pennisetum villosum (syn. *Cenchrus longisetus*) RABO DE GATO BLANCO, LONG-STYLED FEATHER GRASS A large, rather exotic-looking, sparsely clump-forming perennial grass to 70 cm tall. Leaves to 30 cm long and rigid. Inflorescence terminal and rather like those of *Lagurus ovatus* but larger; 'fluffy' and white and broadly cylindrical, borne in larger numbers terminally, drooping on slender stems; spikelets to 14 mm, interspersed with bristles. Native to eastern Africa; a casual weed of humid habitats C, T, P. *P. setaceum* (syn. *C. setaceus*) RABO DE GATO, FOUNTAIN GRASS is similar but has a densely tufted tussock with inflorescence cylindrical to 30 cm, normally tinged reddish. Native to tropical and subtropical areas of North Africa to India. Dry ravines, slopes and rocky habitats; invasive. L, F, C, T, G, H, P.

Hyparrhenia

Inflorescence branched with paired, slender, spike-like clusters arising from leaf-like bracts.

Hyparrhenia sinaica CERRILLO A tufted perennial to 1.2 m with smooth stems. Leaves linear, to 3 mm wide, more or less hairless. Inflorescence a panicle of paired racemes, each enclosed in leaf-like bracts; lemma to 4.5 mm with a stout bristle-tip to 20 mm, twisted and hairy below. Dry, rocky habitats. L, F, C, T, G, H, P.

Andropogon

Annuals or perennials. Inflorescence usually with paired racemes; spikelets compressed, paired, with 1 fertile floret; lower glumes with 1–11 veins, upper glumes with 1–3 veins; lemmas less firm, 1–3-veined.

Andropogon distachyos A clumped, rhizomatous perennial to 20 cm. Leaves 70 mm–20 cm long, 1–5 mm wide. Flowers borne in erect to divergent, slender paired racemes 40 mm–14 cm; spikelets in pairs, adpressed; glumes dissimilar, exceeding the florets, the lower with 7–11 veins; the upper 1-keeled and awned; lemma 2-veined and awned. G, T.

Cenchrus SANDBUR

Annuals or tufted perennials. Ligules a fringe of hairs. Inflorescence spike-like with groups of 1–few spikelets on short stalks enclosed by a spiny bur of fused spines and bristles; spikelets with 2 florets, the lower sterile; glumes unequal; lemma 5-veined, awnless.

Cenchrus ciliaris GRAMA A tufted perennial with short rhizomes and keeled leaves 50–85 mm. Inflorescence a dense, cylindrical, bristly panicle; spikelets with 1 basal sterile floret and 1 fertile floret, 2–5.5 mm long; lower glume 1-veined, the upper glume 1–3-veined; lemma 5-veined; anthers 3, stigmas 2. Disturbed and semi-arid habitats. L, F, C, T, G, H, P.

---

1. *Eragrostis barrelieri*
2. *Sorghum halepense*
3. *Melinis repens*
4. *Pennisetum setaceum*
5. *Hyparrhenia sinaica*
6. *Cenchrus ciliaris*
7. *Setaria adhaerens*
8. *Setaria parviflora*

Setaria BRISTLE GRASS

Annuals or perennials. Inflorescence a cylindrical and spike-like panicle with numerous, densely clustered, stalkless spikelets with 2 florets; glumes unequal; lemma 5-veined, awnless. Naturalised.

Setaria viridis PEGA-PEGA, GREEN BRISTLE GRASS A loosely tufted annual to 50 cm (1 m) with wide, flat, hairless leaves; ligule a ring of hairs. Inflorescence dense, very densely hairy, cylindrical and erect; bright green, up to 12(17) cm long; spikelets 2–2.5(2.7) mm long; upper glume as long or almost as long as the spikelet. Waste ground. C, T. *S. adhaerens* is similar but with the main axis of the inflorescence sparsely bristly with hairs <0.2 mm (not densely hairy with hairs >0.5 mm). L, F, C, T, G, P. *S. parviflora* is tussocky, shortly rhizomatous; leaves with hairless margins. Inflorescences very narrow, to 6 mm wide, with spikelets to 2.7 mm. Rocky habitats. C, T, G, P.

Panicum

Annuals. Inflorescence a compound, diffuse and much-branched panicle with slender branches; spikelets flattened, with 2 florets, the upper fertile and the lower male or sterile; glumes unequal; lower lemma 5–11-veined, awnless.

Panicum repens PANIZO, TORPEDO GRASS A perennial with creeping underground stems and stiff, erect stems to 80 cm. Leaves in 2 ranks, blue-grey and stiff, to 6 mm wide, the uppermost more or less equalling the inflorescence. Inflorescence erect and narrow with slender ascending branches with numerous whitish branches of spikelets, each to 2 mm, without awns; glumes unequal, the upper exceeding the lower. Damp, sandy habitats, particularly on the coast. L, C, T, G. *P. miliaceum* is an annual superficially similar to *Sorghum* before flowering: leaves to 20 mm wide and sheaths with long hairs. Inflorescence rather dense and flopping, to 20 cm long, with numerous slender branches of plump (bead-like), often purplish spikelets each 4.5–5.5 mm long; glumes unequal, the upper exceeding the lower. F, P.

SUBFAMILY **DANTHONOIDEAE**

Very robust, often woody-based perennials; stems often solid. Spikelets unisexual or cosexual with 1–many female florets; lemmas with 1 awn; stamens 3; styles and stigmas 2.

Cortaderia PAMPAS GRASS

Densely tufted perennials. Inflorescence a large, plume-like, spreading panicle; spikelets laterally compressed, with 2–7 florets; glumes slightly unequal and 1-veined; lemma silky-hairy, 3–5-veined, acuminate and long-awned.

Cortaderia selloana PLUMERO, HIERBA DE LA PAMPA, PAMPAS GRASS A very tall, dense perennial to 3 m tall and >1 m across. Leaves long and slender, to 2 m long and 1 cm wide, with sharp serrated edges. Inflorescence plume-like, overtopping the leaves; panicle open; ovoid and dense, 25 cm–1 m; spikelets solitary, lanceolate and stalked, laterally compressed, 12–16 mm, each comprising 3–7 fertile florets with diminished florets at the apex; glumes all similar, exceeded by their spikelets. Native to South America. L, F, C, T, G, H, P.

CERATOPHYLLACEAE | HORN-WORT FAMILY

A family with a single genus with few species, characterised by a submerged aquatic habit and minute unisexual flowers with a superior ovary; stamens 10–25; style 1. Fruit an achene.

Ceratophyllum

Submerged, aquatic, rootless perennial herbs with *whorled, forked leaves*.

Ceratophyllum demersum LIMPIATUBOS, HORN-WORT A submerged aquatic perennial with slender, flexible stems to 1 m with whorls of brittle, dark green leaves *forked x 1 or x 2*. Flowers minute, arising from the axils though rarely formed; unisexual, green and stalkless. Fruit a 1-seeded nut 4–5 mm with a pair of spreading basal spines and a solitary terminal spine. Slow-moving or still water. T.

EUDICOTS

A major group of plants, typically with 2 cotyledons (seed leaves), netted veins radiating from a central main vein (rarely parallel veins), flower parts in multiples of 4 or 5 (rarely 7), and unlike most monocots, often with secondary growth, forming trees and shrubs.

PAPAVERACEAE | POPPY FAMILY

Annuals or perennials with milky or watery sap. Leaves shallowly 1–2-pinnately lobed. Flowers solitary or in racemes. Sepals 2(3), petals 4(6), often crumpled when newly opened; stamens numerous; style 1. Fruit a splitting capsule.

Papaver AMAPOLA, POPPY

Annuals or perennials often with a white latex. Flowers solitary with red, mauve or white petals; stigma a stalkless, 4–20-rayed, flat disk.

Papaver somniferum ADORMIDERA, OPIUM POPPY A vigorous, erect, *whitish* to blue-grey annual to 50 cm (1 m) with pinnately divided, oval leaves. Lower leaves with a short stalk, the upper leaves clasping the stem. Flowers large, petals 25–50 mm long, pale purple with a dark centre, anthers yellow. Capsule hairless. L, F, C, T, G, H, P.

Papaver rhoeas AMAPOLA, COMMON POPPY An erect, bristly annual 60(80) cm. Leaves pinnately lobed, to 15 cm with pointed segments, often 2-pinnately divided. Flowers solitary on long stalks with long, bristly, *spreading* hairs; petals bright red–crimson with or without a dark centre, 30–45 mm long; anthers bluish. Capsule more or less *ovoid* (narrower at the base), *and hairless,* ‹20 mm (not ›2 x as long as wide: see below). Common on disturbed ground. L, F, C, T, G, H, P. *P. dubium* LONG-HEADED POPPY is similar but with *addressed* hairs on the upper parts of the stem, the leaf segments blunt, not pointed, flowers paler or more orange, usually without a dark centre; petals 15–35 mm long; capsule *oblong*, somewhat widened towards the apex (2–4.5 x as long as wide); anthers *violet,* exceeded by the stigmatic disk. Disturbed ground. L, F, C, T, G, H, P.

Papaver argemone AMAPOLA ESPINOSA, PRICKLY POPPY A bristly annual to 45 cm with adpressed hairs on the stem, and pinnately divided leaves. Flowers scarlet, petals 15–25 mm long, often with a dark centre, and the *petals not overlapping*; anthers bluish. Capsule <25 mm, shortly *cylindrical*, ribbed, and *sparsely bristly*. Dry, disturbed ground. L, F. ***P. hybridum*** is similar but smaller with darker crimson-red flowers with petals 10–25 mm long, and an ovoid to spherical capsule *densely covered in pale, stiff bristles*. Similar habitats. L, F, C, T, G, H.

Glaucium

Annuals or perennials with a watery latex. Flowers solitary with 4 petals; stamens numerous. Fruits *very long and narrow,* splitting into 2 parts.

Glaucium corniculatum CORNETA, RED-HORNED POPPY An erect, hairy, blue-grey annual 30–40 cm. Leaves 10–25 cm long with oblong, toothed lobes, those below stalked, the uppermost stalkless. Flowers borne on rather *short* stalks 25–40 mm long, *orange to red flowers* (sometimes yellow); petals 30–40(50) mm long; sepals hairy. Capsule cylindrical, long and narrow, 10–22 cm, *with adpressed hairs*. Dry, rocky habitats. L, F, C, T, G, P.

Glaucium flavum AMAPOLA CORNUDA, YELLOW HORNED-POPPY A blue-grey, branched biennial to perennial 30–90 cm. Leaves 15–30 cm with oblong, wavy and pinnately lobed leaves, the upper leaves clasping the stem; stems with a yellowish latex when cut. Flowers *bright yellow*; petals 30–40 mm long. Fruit narrowly cylindrical, long and narrow, 15–30 cm; curved, and *hairless* but with small whitish tubercles. Coastal sands and shingle, or disturbed habitats inland. C, T, P.

Eschscholzia

Annuals or perennials with deeply-cut leaves and clear, colourless to orange sap. Flowers with 4 petals; stamens numerous.

Eschscholzia californica AMAPOLA DE CALIFORNIA, CALIFORNIA POPPY A blue-grey, spreading annual 25–65 cm with feathery, finely cut leaves. Flowers yellow to orange with 4 petals, to 70 mm across; sepals fused, hooded. Fruit >10 x longer than wide. Of North American origin; naturalised on roadsides and slopes; invasive. L, C, T, H, P.

Fumaria MELLORINA, PAMPLINA, FUMITORY

Trailing or scrambling annual herbs with 2–4-pinnately divided leaves and distinctive leaf-opposed racemes of tubular 2-lipped flowers with 2 small sepals, 2 outer petals and 2 narrower inner petals; stamens 2. Fruit a more or less spherical 1-seeded achene. A hand lens is essential to identify in the field. Flower length important.

(a) Flowers whitish to pinkish-red, ≥9 mm long.

Fumaria capreolata WHITE RAMPING-FUMITORY A hairless, blue-green, scrambling annual with stems 30 cm–1(2) m. Leaves wedge-shaped with blunt, narrowly oblong terminal lobes. Flowers 10–25(30), *held sub-erect*, on racemes equalling or shorter than their stalks; corolla 2-lipped,

1. *Papaver rhoeas*
2. *Papaver rhoeas* (fruit capsule)
3. *Papaver hybridum* (fruit capsule)
4. *Glaucium corniculatum*
5. *Glaucium flavum*
6. *Eschscholzia californica*
7. *Fumaria bastardii*
8. *Fumaria muralis*
9. *Fumaria montana*
10. *Fumaria parviflora*
11. *Fumaria coccinea*

10–13(14) mm, the *upper petal compressed with upturned margins not concealing the keel, creamy white*, often tinged with pink, and *tipped with reddish black*. Capsule 2–2.3 mm, its stalk *strongly curved*. Disturbed ground. L, F, C, T, P.

Fumaria bastardii TALL RAMPING-FUMITORY A trailing annual with flowers (8)9–11(12) mm, very pale to pink, with 10–22 borne in clusters *longer* than their stalks; sepals 2–3(3) mm, with *sharp, forward-pointing teeth all around the margin*. Fruit wrinkled when dry. L, F, C, T, G, H, P. *F. muralis* is similar, with pinkish flowers 9–11(12) mm borne in clusters *as long* as their stalks, and longer sepals (3–5 mm) with teeth in the basal part only (or absent); fruits scarcely wrinkled to smooth when dry. L, F, C, T, G, H, P. *F. montana* has 8–19 pink, dark-tipped flowers, each c. 10 mm long, in racemes equalling their stalks. Sepals toothed, wider than the corolla. L, F, C, T, G, H, P.

(b) Flowers whitish to pinkish-red, *small*, <9 mm long.

Fumaria officinalis COMMON FUMITORY A delicate, hairless, blue-green, scrambling annual with broad, flat, oval-lanceolate leaf segments. Flowers 7–8(9) mm, *numerous*, 10–45, *mauve*, tipped with blackish purple on the wings of the upper petal and apex of the inner petals, borne in a raceme longer than the *short stalk*. Fruits 1.8–2.5 mm, much *wider than long*, warted (*not* shiny). Disturbed ground. L, F, C, T, P. *F. parviflora* has *channelled leaf segments*. Flowers small, 5–6 mm long and *pallid* (white, flushed very pale pink) in almost stalkless racemes; sepals minute, 0.5–1.5 mm long; *bracts at least equalling fruiting stalks*. Fruit scarcely keeled. Dry to arid habitats; common. L, F, C, T, G, H, P. *F. vaillantii* is similar to *F. parviflora* but with *pink flowers* and bracts *shorter* than the fruiting stalks. L, F, C, T.

(c) Flowers *red*.

Fumaria coccinea is distinguished by its narrower linear *red* flowers. *L(?), F(?), C(?), T, G, H, P(?).

RANUNCULACEAE | BUTTERCUP FAMILY

Herbaceous annuals and perennials or woody climbers with alternate, simple or compound leaves. Flowers typically regular (sometimes zygomorphic); sepals and petals 5; stamens numerous. Fruit an aggregate of achenes or follicles (or a berry or capsule).

Ranunculus BUTTERCUPS

Terrestrial or aquatic herbs with entire or lobed leaves. Flowers borne solitary or in cymes; sepals and petals normally 5, the petals often shiny and white or yellow; stamens numerous (or 5–10); carpels numerous. Fruit a head of 1-seeded achenes. Many species occur; not all are described.

(a) Flowers yellow. Carpels *spiny*.

Ranunculus muricatus RANÚNCULO CENTELLA, ROUGH-FRUITED BUTTERCUP A short, usually hairless annual with a stout, much-branched stem 50 mm–30(40) cm and *kidney-shaped* lower leaves with 3–7-shallow-toothed lobes; upper leaves wedge-shaped with up to 5 lobes or occasionally entire. Flowers pale yellow, 6–16 mm across; sepals deflexed. Achenes *strongly keeled, with spines to 1 mm on the faces* (not the grooves) and tapered into an abruptly curved beak 2–2.5 mm. C, T, G, H, P.

Ranunculus arvensis RANÚNCULO CAMPESTRE, CORN CROWFOOT A (typically) hairless annual 10–45(60) cm with spathulate, simple or more commonly toothed to dissected leaves with narrowly lanceolate to linear lobes. Flowers *pale* greenish to lemon-yellow, 4–12 mm across, borne in branched clusters on slender stalks. Achenes 5–6(8) mm with *prominent, long, rigid spines >1 mm long;* beak 2–3 mm. Disturbed land. C.

(b) Flowers yellow. Carpels smooth or wrinkled.

Ranunculus cortusifolius MORGALLANA A robust, tuberous, hairy perennial to 60 cm. Basal leaves large, to 30 cm across, circular to heart-shaped, with shallow lobes. Flowers yellow, to 50 mm across, borne in sub-corymbose inflorescences. Achenes smooth, black when ripe, with a persistent style. Damp, rocky habitats and under walls or boulders. L, F, C, T, G, H, P.

Ranunculus sardous RANÚNCULO SARDO A hairy *annual* with a scarcely swollen underground stem base. Leaves 3-lobed and shiny. Flowers 12–25 mm across and *pale yellow*; sepals with dark markings along the margins (best seen in bud). Achenes smooth except for a row of small tubercles surrounded by a green border. Damp, grassy habitats. C, T.

(c) *Flowers pink or white* **and plant** *aquatic.*

Ranunculus peltatus RANÚNCULO DE AGUA, POND WATER-CROWFOOT An annual or perennial aquatic herb with floating and submerged leaves with divergent leaf segments with *rounded tips*. Flowers to 30 mm across. Fruit stalk to 15 cm long (*longer* than stalk of opposite leaf). Aquatic habitats. T.

Ficaria

Tuberous perennial herbs with heart-shaped leaves. Flowers yellow, with 7–12(13) petals; stamens and carpels numerous. Fruit a head of achenes.

Ficaria verna (syn. *Ranunculus ficaria*) RANÚNCULO DE ALMORRANA, LESSER CELANDINE A variable (with several subspecies described), hairless perennial 50 mm–25 cm with long-stalked, triangular *heart-shaped*, wavy-margined to shallowly lobed, fleshy, dark green leaves. Flowers 10–30 mm across, shiny and yellow, turning white on ageing with 7–12(13) narrow petals and *3 sepals*. T.

Nigella LOVE-IN-A-MIST

Annual herbs with solitary flowers and feathery, pinnately divided leaves with narrow segments. Flowers solitary (or few); sepals petal-like, petals in the form of clawed nectaries; stamens numerous; carpels usually 5, variably fused along their inner margins and many-seeded. Fruit a capsule.

Nigella damascena ARAÑUELA, LOVE-IN-A-MIST An erect, hairless annual to 15–40(50) cm with alternate, finely 2–3-pinnately divided leaves with narrow segments, the *uppermost forming a feathery whorl just below the flowers*. Flowers solitary, sky-blue, 15–30(35) mm across with 5 petals and a central cluster of stamens and carpels; carpels joined over most of their length. Fruit capsule spherical, *strongly inflated*, 10-celled when dissected. Disturbed ground. T, G, P.

Delphinium

Annuals or perennials with broadly palmately divided leaves and (typically) blue flowers borne in erect racemes, each with 5 petal-like outer segments, the uppermost spurred at the back, and 4 inner petal-like segments, the 2 uppermost with spurred nectaries. Fruit made up of 3–5 follicles.

Delphinium staphisagria ESPUELA MATAPIOJOS A *hairy biennial* 30 cm–1 m with alternate leaves along the stem, 1-pinnately divided into 5–7 lobes (sometimes 2-pinnate); leaves 10–15 mm across. Flowers borne in lax racemes, deep blue, large to 25 mm across; upper sepal with *short, blunt, down-turned spur* 30–50 mm long (shorter than the petals). Fruit with inflated follicles to 22 mm, bearing few (3–6) large seeds to 6 mm. Woods. T, G.

1. *Ranunculus cortusifolius*
2. *Ranunculus cortusifolius*
3. *Delphinium staphisagria*
4. *Adonis microcarpa* (fruiting head)
5. *Adonis microcarpa*
6. *Crassula lycopodioides*
7. *Crassula lycopodioides*
8. *Crassula ovata*

Consolida LARKSPUR

Annuals similar to *Delphinium* with palmately divided leaves that have *numerous thread-like segments*. Flowers in terminal racemes; sepals 5, the upper long-spurred, petals 4. Fruit a *solitary follicle*.

Consolida ajacis ESPUELA, LARKSPUR A downy annual to 1 m with a simple or branched stem and deeply dissected lower leaves; leaves persistent (not withered) in flower. Flowers rather large and few, borne in lax inflorescences, typically *bright blue* (pink, white and pale blue forms are cultivated), the upper petal with a *long* backwardly projecting spur 13–18 mm long; flower stalks *long,* 15–40 mm. Follicles tapered at the apex. Disturbed habitats. T, H.

Consolida regalis ESPUELA, FORKING LARKSPUR A slender, rather downy, widely branching annual to 50 cm. Leaves divided into linear lobes, *withered* during flowering. Flowers violet-blue to dark blue, to 28 mm, borne in lax *panicles*; spur 12–18(25) mm. Follicles 7–11 mm, hairless; seeds black. Disturbed habitats. L, C, T, H.

Aquilegia COLUMBINE

Herbaceous perennials with leaves spirally arranged. Flowers regular with 5 sepals and petals, each with a backward-pointing spur; stamens numerous; carpels 5(10). Fruit a follicle.

Aquilegia vulgaris AGUILEÑA COMÚN, COLUMBINE A variable, erect, often branched, hairy perennial to 60 cm (1 m). Leaves stalked and toothed or lobed. Flowers large, to 50 mm long, *blue-purple* (sometimes white or pink) with 5 similar tepals, the *petal-like segments elongated into erect, curving spurs* 15–22 mm long. Follicles 15–20 mm. Woods. T.

Adonis

Annuals with 1–3-pinnately divided leaves, often with linear segments. Flowers with 5(–8) petal-like sepals; petals 3–20, glossy; stamens numerous. Fruit a head of numerous achenes, elongating when mature.

Adonis microcarpa OJO DE PERDIZ An erect annual 10–30 cm, similar to *A. annua* in form, but branched from the base. Leaves 3-pinnately lobed, the terminal lobes linear. Flowers solitary, flat, and *small* with petals 4–10 mm long, *yellow or bright red,* often with a *blackish spot* at the base. Fruiting head 8–25 mm across, cylindrical to slightly conical; achenes *small,* 2.3–3 mm long with a prominent convex hump. L, F, C, T, P.

CRASSULACEAE

Typically succulent annuals and perennials with alternate, opposite or whorled leaves. Flowers regular, star- or bell-shaped with 3–18(20) sepals and petals and an equal number of stamens, or 2 x as many. Fruit a cluster of follicles.

Crassula

Aquatic or terrestrial annuals and perennials, usually hairless with succulent leaves, often in fused pairs. Flowers with 3–5–numerous sepals, petals and stamens.

Crassula tillaea MOSSY STONECROP A tiny, often *dark red* moss-like annual just 10–50 mm. Leaves 1–2 mm, oval and crowded. Flowers borne in small groups in the leaf axis, *virtually stalkless*, 1–2 mm across with petals shorter than the sepals; white or pale pink; *sepals and petals 3* (rarely 4). Capsules with 1–2 seeds. Rocky or stony habitats. L, F, C, T, G, H, P.

Crassula lycopodioides PINITO A small, spreading succulent to 25 cm with squarish stems with minute leaves packed densely and regularly. Flowers yellow. Native to South Africa; naturalised around gardens, on roofs, and in ravines. L, C, T, G, H, P.

Crassula ovata PLANTA DE JADE, JADE PLANT A succulent shrub to 2 m with thick stems and opposite, dark green, glossy leaves 30–90 mm. Flowers white or pinkish, starry, with petals to 7 mm. Commonly grown; seldom naturalised. L, C, T, G.

Sedum STONECROP

Succulent annuals or perennials with flat or cylindrical leaves. Flowers with 4–9–numerous petals and 2 x as many stamens. Numerous similar species; many rare or local species not included here.

Sedum rubens HIERBA JABONERA, RED STONECROP A *very small, erect annual*, usually to just 30 mm–9(15) cm, often sticky and glandular above, with linear alternate leaves 7–16(20) mm long, greyish-green, tinged with red. Flowers *whitish or pink* with darker mid-veins, borne in small, compact clusters with erect-spreading to recurved branches; *petals 5, 4–5(6) mm long, 2 x the length of the sepals*; stamens usually 5. Fruiting heads spreading. Rocky habitats and walls. L, F, C, T, G, H, P. *S. nudum* subsp. *lancerottense* PELOTILLA DE LANZAROTE is a perennial plant with *trailing to hanging* stems and pale, fleshy leaves and *yellow* flowers. *L.

Umbilicus NAVELWORT

Perennial herbs with *round leaves joining their stalks at the centre of the blade* (peltate). Flowers borne in spike-like racemes; petals 5, stamens 2 x as many, fused to the corolla.

Umbilicus gaditanus SOMBRERILLO A fleshy, hairless perennial 15–30(60) cm, with fleshy, *circular basal leaves borne on long stalks*; upper leaves smaller, crowded, linear. Flowers with 5 petals and 2 x as many stamens, whitish-green, tubular, borne in long, tapered racemes on stalks 13–70 cm. Damp, rocky places. L, F, C, T, G, H, P.

Aeonium BEJEQUE, VEROL, VERODE, TREE HOUSELEEK

Succulent perennial herbs and shrubs with leaves in dense rosettes. Now contains the formerly separate genus *Greenovia*. The genus is an example of an adaptive radiation across the islands. Habit as well as minute leaf characteristics, including marginal cilia, are important.

(a) Vegetative stems *simple* (or scarcely branched); plant hairless, or virtually so.

Aeonium appendiculatum Stems *smooth* and simple; rosettes 30–35 cm across, somewhat compressed; leaves glaucous, oblanceolate to subobovate, hairless. Similar to *A. urbicum,* with hairless inflorescences of pink-tinged, *white* flowers with styles divergent from the base. *G.

1. *Sedum rubens*
2. *Sedum nudum* subsp. *lancerottense*
3. *Umbilicus gaditanus*
4. *Aeonium urbicum*
5. *Aeonium urbicum* var. *meridionale* (habitat)
6. *Aeonium urbicum* var. *meridionale*
7. *Aeonium urbicum* var. *meridionale* (rosette)
8. *Aeonium urbicum* var. *meridionale* (flowers)

Aeonium urbicum BEJEQUE PUNTERO DE TENERIFE is a similar, unbranched shrub; young stems *scaly*. Inflorescences hairless, large, pyramidal, with pinkish to greenish-white flowers. **Var. urbicum** has green leaves and styles divergent near the tips. *T (north). **Var. meridionale** has glaucous leaves and pinkish-white flowers with styles divergent from the base. *T (south, west). *A. pseudourbicum* is similar but with narrow, oblanceolate leaves with scarce, short cilia along the margins and somewhat *globose, hairy* inflorescences. *T (west).

Aeonium undulatum An unbranched shrub with oblanceolate–spathulate or oblong–spathulate, hairless leaves with ciliate margins. Flowers yellow. *C.

Aeonium hierrense SANJORA An unbranched shrub with thick, obovate to oblanceolate, glaucous leaves with ciliate margins. Flowers with 6–9 whitish, often pink-tinged petals borne in *pyramidal, hairy* inflorescences. *H, P.

Aeonium nobile BEJEQUE ROJO An unbranched shrub with very succulent, broadly obovate to circular leaves. Flowers *deep red*. Dry slopes; local. *P.

(b) Vegetative stems *branched*; plant hairless, or virtually so.

Aeonium castello-paivae A small, branched shrub; rosettes small, 30–70 mm across; leaves obovate–spathulate, hairless, pale yellowish-green, often variegated red, the innermost sub-erect. Flowers greenish-white. Damp cliffs. *G.

Aeonium decorum A branched shrub; rosettes small, to 10 cm across; leaves obovate to oblanceolate and pointed, hairless with slightly ciliate margins, reddish. Flowers few, whitish-pink, borne in lax, hairy inflorescences. *T, G.

Aeonium percarneum A branched shrub with large rosettes (>50 mm), hairless; leaves with curved cilia along the margins. Flowers white or pink with darker streaks; calyx pubescent. *C.

Aeonium arboreum A robust, branched shrub with large, *dense, conical inflorescences* of yellow or yellowish-pink flowers. Variable with geographically defined forms described: **subsp. arboreum** (*C) and subsp. *holochrysum* (*T, G, H, P). *A. rubrolineatum* (*G), which has markedly red-striped, narrowly lanceolate-spathulate leaves, is widely treated as distinct but also considered a mere form of *A. arboreum*; *A. holochrysum*, *A. manriqueorum* and *A. vestitum* are also widely treated as distinct but synonymised under this name. *C, T, G, H, P.

Aeonium ciliatum A branched shrub with obovate-spathulate, hairless, glaucous, rather dark to yellowish green leaves with straight or slightly curved cilia along the margins. Flowers with whitish petals, greenish beneath. A rare endemic. *T.

Aeonium lancerottense A branched shrub with rosettes >50 mm across, hairless; leaves with shortly and weakly ciliate margins. Flowers pink, borne in large, broadly conical inflorescences. Malpaís and rocky slopes and cliffs; a locally common island endemic. *L.

---

1. Aeonium undulatum
2. Aeonium castello-paivae (habit)
3. Aeonium castello-paivae
4. Aeonium decorum
5. Aeonium percarneum
6. Aeonium percarneum (left) and A. arboreum subsp. arboreum (right)
7. Aeonium arboreum subsp. arboreum (rosettes)
8. Aeonium arboreum subsp. arboreum
9. Aeonium rubrolineatum (habit)
10. Aeonium rubrolineatum
11. Aeonium holochrysum (habit)
12. Aeonium holochrysum

Aeonium balsamiferum FARROBA A branched shrub with *aromatic* leaves with ciliate margins, *smelling of balsam* when crushed. Flowers yellow, borne in small inflorescences. *L, F.

Aeonium gomerense A branched shrub to 2 m with compressed rosettes 20–28 cm across; leaves reddish, obovate to oblanceolate, wedge-shaped at the base, hairless, glaucous, with ciliate margins. Flowers whitish or pinkish, borne in ovoid inflorescences to 40 cm. *G.

Aeonium smithii A small, branched shrub with hairy stems. Leaves *densely hairy beneath*, with *reddish glands above*. Flowers yellow. *T.

Aeonium davidbramwellii A branched shrub with hairless, glaucous, red-brown-tinted leaves with finely ciliate margins. Flowers white. *P.

Aeonium haworthii A branched shrub with small rosettes, 60 mm–11 cm across; leaves obovate, hairless, glaucous, with red, cilliate margins; inner leaves sub-erect. Flowers cream to pink. T. The widely recognised *A. volkeri* has also been considered a subspecies of the former. The form shown is subsp. *paucifolium*. *T.

Aeonium sedifolium A small, branched, sticky shrub with *many small, well-spaced rosettes*; leaves short, very succulent with dull red stripes. Flowers yellow. *T, G, P.

(c) Vegetative stems *branched*; plant *hairy*, at least in part.

Aeonium lindleyi GOMERETA A branched, sticky shrub with thick, hairy, rhomboid-shaped leaves. Flowers yellow. Subsp. *lindleyi* occurs on Tenerife, and subsp. *viscatum* occurs on La Gomera. *T, G.

Aeonium goochiae MELERA A branched shrub with small rosettes; leaves green, rhomboid, glandular-hairy without ciliate margins. Flowers pink. *P.

Aeonium saundersii A branched shrub with almost circular leaves, densely glandular-hairy above and below. Flowers few, pale yellow, borne in simple inflorescences. *G.

Aeonium spathulatum A small, branched shrub with smooth stems and *sticky* (glandular) short, obovate–spathulate leaves with glands beneath, and ciliate margins. Flowers golden yellow. Var. *spathulatum* and var. *cruentum* are recognised; the latter on EL Hierro and La Palma. *C, T, G, H, P.

Aeonium valverdense A branched shrub with obovate, thick, minutely hairy leaves with ciliate margins. Flowers with 7–9 whitish, pink- or red-tinged petals. *H.

(d) Vegetative stems short to absent (rosettes appearing to grow directly out of the rock).

Aeonium canariense GÓNGARO A sparingly branched shrub with short stems and yellowish-green, hairy leaves. Flowers greenish-white to yellow. The widely recognised *A. subplanum* (G) is considered a form of this species, as is *A. palmense*. Four geographically defined subspecies are recognised: subsp. *canariense* (*T), subsp. *christii* (*H, P, T), subsp. *latifolium* (*G) and subsp. *virgineum* (*C). *C, T, G, H, P.

Aeonium tabulaeforme PASTEL DE RISCO A herbaceous perennial with *flat rosettes*; leaves green with long-ciliate margins, *densely overlapping*. Flowers pale yellow. Sea cliffs. *T.

1. *Aeonium ciliatum* (habit)
2. *Aeonium ciliatum*
3. *Aeonium lancerottense* (rosettes)
4. *Aeonium lancerottense* (habit)
5. *Aeonium lancarottense* (flowers)
6. *Aeonium balsamiferum*
7. *Aeonium balsamiferum* (leaf margins)
8. *Aeonium volkeri* subsp. *paucifolium* (habit)
9. *Aeonium volkeri* subsp. *paucifolium*
10. *Aeonium sedifolium*
11. *Aeonium lindleyi* (rosettes)
12. *Aeonium lindleyi*

Aeonium cuneatum GÓNGARO DE ANAGA A short, herbaceous perennial with hairless, glaucous and long spatulated leaves without glands, borne in *cupped* rosettes. Flowers golden yellow. *T.

Aeonium simsii FLOR DE PIEDRA A woody-based herbaceous perennial with dense clusters of rosettes; leaves with glands on the underside and *long, translucent cilia* along the margins. Flowers yellow. *C.

Aeonium aizoon (syn. *Greenovia aizoon*) BEA DE GÜÍMAR A perennial with multiple rosettes 30–50 mm across; leaves green, oblong–spathulate, 35 mm long, densely glandular-hairy. Flowers 17–21 parted, deep yellow, 10–40, borne in inflorescences 10 cm wide on leafy stems 80 mm–12 cm. Rocky shelves. *T.

Aeonium aureum (syn. *Greenovia aurea*) BEA A perennial with 3–5 rosettes 80 mm–25 cm across; leaves to 11 cm, obovate–spathulate, glaucous and *hairless* with translucent (non-ciliate) margins. Flowers (25)30–32(35)-parted, deep yellow, borne in inflorescences to 45 cm across on stems 15–35 cm. Rock faces. *C, T. The widely recognised *A. diplocyclum* (syn. *Greenovia diplocycla*) which is smaller, virtually hairless, with solitary rosettes and (18)19–20(24)-parted flowers, is now included within the former species. *G, H, P.

Aeonium dodrantale (syn. *Greenovia dodrentalis*) BEA DE TENERIFE A perennial with *numerous, small* rosettes 30–60 mm across; leaves obovate–spathulate, 20–30 mm long, glaucous with pale margins with glandular hairs, later hairless. Flowers (18)30–34-parted, deep yellow, borne in inflorescences to 10 cm across. *T.

Aichryson GONGARILLO

Small annual to perennial herbs with alternate, entire, rather fleshy leaves. Calyx 6–12-parted, green, fleshy; flowers yellow; petals as many as calyx segments, stamens twice as many.

(a) Perennial shrublets.

Aichryson tortuosum PELOTILLA A small, compact shrublet to 20 cm with woody-based stems. Leaves crowded at the stem tips, *stalkless*, spathulate or obovate, fleshy, *sticky*-hairy with blunt tips. Inflorescences short, with yellow flowers. Cliffs. *L. *A. bethencourtianum* GONGARILLO MAJORERO is similar but with *short-stalked* leaves which are hairy but not sticky. *F.

(b) Annual or short-lived herbs; hairy.

Aichryson laxum An erect, soft-hairy annual or biennial to 30 cm with suberect branches. Leaves *densely hairy*. Inflorescences rather flat and diffuse; flowers yellow, 9–12-parted. Forests and shady cliffs, walls and banks. Locally common. *F, C, T, G, H, P. *A. porphyrogennetos* is similar but with wide-spreading branches, and purplish leaves, broadest above the middle. Ravines. *T, C. *A. bollei* has black glands along the leaf margins. *P.

Aichryson brevipetalum is *small*, to just 12 cm, densely hairy, and has very small 6–7-parted flowers with petals shorter than their sepals. P. *A. parlatorei* is similar but with petals *longer* than their sepals. *C, T, G, H, P.

1. *Aeonium lindleyi* subsp. *viscatum* (habit)
2. *Aeonium lindleyi* subsp. *viscatum*
3. *Aeonium canariense* (habit)
4. *Aeonium canariense* (rosette)
5. *Aeonium tabulaeforme* (habit)
6. *Aeonium tabulaeforme*
7. *Aeonium cuneatum*
8. *Aeonium simsii*
9. *Aeonium aizoon*
10. *Aeonium aureum*
11. *Aeonium aureum* (rosette)

1. *Aeonium diplocyclum* (summer habit)
2. *Aeonium diplocyclum* (leaf margins)
3. *Aeonium dodrantale*
4. *Aichryson tortuosum* (habit)
5. *Aichryson tortuosum*
6. *Aichryson bethencourtianum*
7. *Aichryson laxum*
8. *Aichryson laxum* (growing as a chasmophyte)
9. *Aichryson laxum* (left) and *A. pachycaulon* subsp. *immaculatum* (right)

1. *Aichryson punctatum*
2. *Aichryson pachycaulon* subsp. *gonzalezhernandezii*
3. *Kalanchoe daigremontiana*
4. *Kalanchoe beharensis*
5. *Monanthes wildpretii*
6. *Monanthes brachycaulos* (habit)
7. *Monanthes brachycaulos*
8. *Monanthes brachycaulos* (flower)

(c) Annual or short-lived herbs; hairless or virtually so.

Aichryson punctatum A *hairless or virtually hairless* (except inflorescences) annual or biennial herb. Leaves with *dark purple crenulations* along the margins. Flowers yellow, 6–8-parted. Damp cliffs. *F, C, T, G, H, P. **A. *pachycaulon*** is similar but more robust with leaves *without* dark purple crenulations. Subspecies include: subsp. *pachycaulon* (*F), subsp. *gonzalezhernandezii* (*G), subsp. *immaculatum* (*T), subsp. *parviflorum* (*H?, P), subsp. *praetermissum* (*C). Sea cliffs and forests. *F, C, T, G, H, P.

Kalanchoe

Succulents native to Madagascar, commonly grown in gardens. Now includes *Bryophyllum* (notable for having plantlets on the fringes of the leaves).

Kalanchoe daigremontiana (syn. *Bryophyllum daigremontianum*) A grey-waxy succulent perennial to 1 m with leaves to 20 cm with *plantlets along the margins* that fall and root. Flowers reddish, pendulous, bell-shaped, borne in terminal clusters. Planted and naturalised. F (Lajares), C, T, P.

Kalanchoe beharensis A *grey-felted* succulent shrub to 1 m. Leaves triangular–lanceolate, crimped, to 10 cm. Flowers borne in branched corymbs; calyx 7 mm; corolla tube 7 mm. Planted. L.

Monanthes PELOTILLA

Small succulent perennials with leaves often borne in compact, globular rosettes.

(a) Small perennials. Leaves borne in congested rosettes.

Monanthes wildpretii A small, spreading perennial with rosettes 10–30 mm across; leaves hairy, spathulate, slightly attenuated at the tips. Inflorescences borne laterally, with reddish, glandular-hairy stems and pinkish flowers; calyx 6–7-parted; petals with reddish longitudinal streaks. Mossy rocks. *T. *Monanthes minima* has long stalked leaves with *ciliate edges*. Flowers small, to 6 mm; calyx purplish, cobweb-hairy. *T.

Monanthes brachycaulos A small perennial with rather flat, loose, often purplish rosettes of oblanceolate–spathulate, *papillose leaves*. Stems short <10 mm. Rocky habitats. *C, T. **M. *pallens*** is similar, with dense, flat to concave rosettes >15 mm across and unbranched stems. *T, G. **M. *polyphylla*** has small, dense rosettes with globose or elongated, pale green to pinkish leaves and terminal inflorescences with few flowers. *C, T, G, P.

(b) Small annuals to perennials. Leaves loose to crowded but *not* in distinct rosettes.

Monanthes laxiflora A variable species with fleshy, hairless or papillose, *more or less opposite leaves,* not borne in rosettes. Flowers dull pale yellow to pink. Sometimes recognised forms include: var. *laxiflora* (*C, T, G, P), var. *chlorotica* (*T), and var. *microbotrys* (*L, F, C). *L, F, C, T, G, H, P. **M. *anagensis*** is similar but with *alternate, long, narrowly elliptic leaves*, 4–6 x as long as wide, usually reddish. *T.

Monanthes icterica A tiny, short-lived annual herb 20–60 mm. Leaves rather loose, alternate, clustered near the branch tips, 5–9 mm, obovate, hairless and slightly papillose. Flowers sparse, 3–4 mm across. *T, G.

(c) A small subshrub.

Monanthes muralis A small, *shrubby perennial with rosettes borne at the end of woody, branched stems*; purplish. *H, P.

1. *Monanthes pallens*
2. *Monanthes polyphylla*
3. *Monanthes laxiflora*
4. *Monanthes laxiflora* (red form) habit
5. *Monanthes laxiflora* (red form)
6. *Monanthes laxiflora* var. *laxiflora*
7. *Monanthes laxiflora* var. *microbotrys* (habit)
8. *Monanthes laxiflora* var. *microbotrys* (green form)
9. *Monanthes anagensis*
10. *Monanthes icterica*

CYNOMORIACEAE

A family (with 1 genus) of root parasites of members of the Amaranthaceae. Flowers unisexual or cosexual, borne in a dense, brush-like inflorescence; tepals (1)3–6(8); stamen 1. Fruit a small, 1-seeded nut.

Cynomorium coccineum MALTESE FUNGUS A highly distinctive, blackish-red plant sprouting as a club-shaped structure to 25(30) cm from an extensive underground rhizome system, with lanceolate scale leaves. Inflorescence cylindrical with very dense, tiny flowers; tepals 3–6(8); stamens solitary, exserted. Parasite of shrubby Amaranthaceae on coastal dunes. Formerly on eastern islands, now apparently reduced to a single population. L (La Graciosa).

VITACEAE | GRAPE FAMILY

A large, primarily tropical family of climbers with leaf-opposed tendrils. Flowers small, borne in clusters; petals and stamens 5; style 1. Fruit a 1–4-seeded berry.

Vitis

Leaves simple, palmately lobed. Flowers with fused petals, falling as the flowers open.

Vitis vinifera PARRA, VIÑA, GRAPE A climbing shrub to >10 m with alternate leaves and tendrils opposite; tendrils branched. Leaves long-stalked and palmately 5–7-lobed, coarsely toothed. Flowers small, greenish, cosexual, borne in clusters, pendent when mature. Fruit a berry (grape). Widely grown. L, F, C, T, G, H, P.

ZYGOPHYLLACEAE

A mostly tropical family of plants with pinnately divided leaves with stipules. Flowers regular, with a 5-parted perianth and 8 or 10 stamens. Fruit fleshy or a capsule (sometimes splitting into 5 mericarps).

Tetraena

Shrubs and herbs (succulents in the Canaries). Leaves opposite with 1–2 pairs of leaflets. Flowers solitary or in pairs, regular; stamens 10; ovary 3–5-parted. Fruit dry, splitting when ripe (a *schizocarp*).

Tetraena fontanesii UVA DE MAR A distinctive shrublet with succulent, almost grape-like leaves, glaucous turning orange. Flowers with 5 whitish-pink petals. Fruits 5–7 mm, initially similar to the leaves, later corky, breaking into 5 valves. Coastal rocks and dunes; common. L, F, C, T, G, H. *T. gaetula* has rather flattened, greyish leaves, small pinkish flowers and *elongated, cylindrical-bell-shaped* reddish fruits 7–15 mm long. Apparently native on F (**subsp.** *gaetula*) and introduced on F, and C (**subsp.** *waterlotii*).

---

1. *Cynomorium coccineum*
2. *Cynomorium coccineum* (subterranean stem)
3. *Cynomorium coccineum*
   MATIAS HERNANDEZ GONZALEZ
4. *Vitis vinifera*
5. *Tetraena fontanesii* (habit)
6. *Tetraena fontanesii*
7. *Tetraena fontanesii* (fruit)
8. *Tetraena gaetula* (in fruit)
9. *Tetraena gaetula* subsp. *gaetula*

Tribulus

Prostrate, annual herbs with leaves that have 5–8 pairs of leaflets. Flowers yellow. Fruits generally spiny.

Tribulus terrestris ABROJO, SMALL CALTROPS A prostrate, hairy annual with long, trailing stems to 50(80) cm with pinnately divided leaves that have 5–8 pairs of elliptic leaflets 6–10 mm, somewhat silver-hairy. Flowers yellow, with 5 petals 8–14 mm long, borne in the leaf axils borne on short stalks (4–5 mm). Fruits 8–10 mm, *with prominent, robust spines*. Dry, bare areas. F, C, T, P.

Fagonia

Perennial herbs, often woody-based with opposite, often trifoliate leaves. Flowers borne solitary in the leaf axils with 5 free sepals and petals and 10 stamens. Fruit a 5-parted capsule.

Fagonia cretica RAPASAYA A short, hairless to hairy spreading, prostrate perennial 60–70 cm, woody at the base. Leaves 6–25 mm, trifoliate, leathery, with spine tips. Flowers bright reddish-purple with 5 free petals 8–9.5 mm, borne solitary between pairs of spine-tipped stipules. Fruit a 5-angled, egg-shaped capsule 7–9 mm. Dry habitats. L, F, C, T, G, H, P.

LEGUMINOSAE (FABACEAE) | PEA FAMILY

The third largest family of flowering plants, with 6 subfamilies. Herbs or trees, usually with trifoliate or pinnately compound leaves. Flowers zygomorphic, with an upper petal (standard), 2 lateral wings that lie on the side of the 2 lower, typically fused petals (keel), concealing the 10 stamens and single style; stamens 9 (sometimes all 10), fused into a basal tube. Fruit (legume), highly variable, often splitting and pod-like, or a nut.

SUBFAMILY CAESALPINIOIDEAE

Trees and shrubs, often with tendrils and 2-pinnately divided leaves. Flowers borne in racemes; calyx 2–5-lobed or sepals free; petals 5 (rarely 0, 2 or 6); stamens 10. Fruit a splitting pod.

Paraserianthes

Trees or shrubs with pinnate leaves with 5–22 opposite segments. Flowers borne in spikes or racemes, 5-parted. Fruit a straight, flat pod.

Paraserianthes lophantha A shrub or small tree to 6 m with velvety stems. Leaves with 14–24 leaflets, each with 40–60 ultimate linear segments to 10 mm, silky beneath. Flowers borne in paired spikes to 50 mm, whitish or yellowish. Native to Australia, planted and naturalised. F, C, T, G, H, P.

Acacia ACACIA

Trees or shrubs with either 2-pinnately divided true leaves, or leaves reduced to a single blade (phyllodes) and flowers borne in dense clusters; stamens numerous, free. Fruit a splitting or non-splitting pod-like legume. Native to South Africa and Australia; numerous forms and cultivars, often difficult to distinguish.

1. *Tribulus terrestris*
2. *Fagonia cretica*
3. *Paraserianthes lophantha*
4. *Acacia cyclops* (habit)
5. *Acacia cyclops* (splitting fruit)
6. *Acacia saligna*
7. *Acacia saligna*
8. *Leucaena leucocephala*

(a) Leaves reduced to phyllodes (leaf-like blades); *pinnately divided leaves absent.*

Acacia cyclops ACACIA MAJORERA A shrub 2–4 m. Phyllodes elliptic to lanceolate, blue-green and rather short, *broad and straight*, typically 40–90 mm long and >12 mm wide. Flowers yellow, borne in clusters of 30–40, 4–6 mm wide, solitary or in groups of 2–3 in the leaf axils. Fruit a pod 40 mm–10 cm, splitting to reveal large brown-black seeds to 70 mm across with a bright *orange-red* casing. Native to Australia, widely planted and naturalised. L, F, C, T.

Acacia saligna (syn. *A. cyanophylla*) ACACIA AZUL, BLUE-LEAVED WATTLE A short tree to 8 m with phyllodes 5–50 mm wide, and flowers dark yellow, borne in *long, drooping branches*; heads 6–8 mm. Fruit distinctly constricted between the seeds, 50 mm–14 cm. L, F, C, T, G, P.

(b) Leaves (at least some) *2-pinnately divided.*

Acacia dealbata ACACIA PLATEADA, SILVER WATTLE A bushy tree 12–15(30) m with smooth grey bark and silvery-hairy twigs and young leaves. Leaves *2-pinnately divided* with 10–26 pairs of primary divisions; leaflets 2–5 mm long, stipules rudimentary. Flowers *pale, bright yellow* and fragrant, borne in heads 5–6 mm across that form large terminal panicles that exceed the leaves. Fruit to 10 cm, linear-oblong and laterally flattened, bluish-brown. T, H.

Acacia farnesiana AROMO A *spiny, deciduous* tree 1.5–4 m with *short spines* 10–30 mm. Leaves 2-pinnately divided into 10–21 pairs of leaflets 2–7 mm long and just 0.75–1.75 mm wide. Flowers bright yellow, 2.5–3 mm long, borne in *dense, spherical clusters* 10–15 mm across, solitary or in groups of 2–5 on stalks 35 mm long. Fruit *broadly* sub-cylindrical, rather straight and rigid, 40–70 mm. L, F, C, T, G.

Acacia melanoxylon ACACIA NEGRA, BLACKWOOD ACACIA A dense, pyramidal tree to 40 m *with both true compound leaves and phyllodes present*; phyllodes elliptic–lanceolate (not long and linear), greyish. Flowers borne in small clusters in the leaf axils; *creamy-white*. L, T.

Leucaena

Acacia-like trees and shrubs with 2-pinnately divided leaves. Flowers borne in spherical heads. Fruiting pods flattened.

Leucaena leucocephala AROMO BLANCO A shrub or small tree to 18 m with greyish bark, superficially similar to *Acacia*. Leaves rather large, to 35 cm long, oval in outline, 2-pinnately divided with 4–9 pairs of 11–22 leaflets 8–16 x 1–2 mm. Flowers *pale whitish-yellow,* numerous, borne in spherical heads 20–50 mm across. Fruits 14–26 cm long, borne in pendent clusters; *seeds clearly visible,* maroon-brown when mature. Seeds 18–22 per fruit, 6–10 mm long. Widely planted and invasive. L, F, C, T, G, H, P.

Delonix

Trees with 2-pinnately divided leaves. *Flowers large and showy, borne in terminal corymbs.* Fruiting pod woody and flattened.

Delonix regia FLAMBOYÁN A tree 10–15(18) m with a robust trunk. Leaves 20–60 cm long with stout stalks, 2-pinnately divided into 10–25 pairs, in turn divided into 12–40 pairs of leaflets 5–20 mm x

---

1. *Delonix regia* (flowers)
2. *Delonix regia* (fruits)
3. *Caesalpinia spinosa*
4. *Ceratonia siliqua* (fruiting plant)
5. *Ceratonia siliqua* (male flowers)
6. *Senna didymobotrya*
7. *Senna bicapsularis*

3 mm. Flowers numerous and conspicuous (the whole canopy appearing red when in bloom), borne in numerous corymbs 15–30 cm long; flowers 30 mm–13 cm across, *bright scarlet;* petals 50–65 mm long, the uppermost paler with dark red markings. Fruit flexuous, turning dark brown and *woody and persistent; large*, 30–75 cm long, 50–76 mm wide with 30–45 seeds. Planted throughout.

Caesalpinia

Trees and shrubs with 2-pinnately divided leaves. Flowers yellow or red, showy, borne in elongated clusters. Fruits flattened and sickle-shaped.

Caesalpinia gilliesii (syn. *Erythrostemon gilliesii*) BARBA DE CHIVO A spineless shrub 1–4 m with 2-pinnately divided leaves 10–15 cm long, divided into 3–10 pairs, in turn with (6)7–10 pairs of leaflets 5–6 x 2–4 mm. Flowers borne in extended racemes to 20 cm long; flowers showy, with 5 yellow petals 20–35 mm and 10 *long-protruding, red stamens* 70 mm–12.5 cm, at first upward swept, later drooping. Fruit a flattened, sickle-shaped, splitting pod, covered in short, red glandular hairs, with few seeds. Scaped from gardens. L, F, C, T, G, H.

Caesalpinia spinosa (syn. *Tara spinosa*) TARA A shrub to small tree with prickles on the leaves and stems; leaves divided into 2–5, terminating in a pair of leaflets, each 8–20 mm. Flowers borne in congested racemes; sepals unequal; petals yellow, 6–7 mm; stamens very unequal. Fruit an oblong pod 60 mm–10 cm. Naturalised in ravine beds. F, C, T.

Ceratonia CAROB

Shrubs and trees. Normally dioecious. Flowers inconspicuous (greenish), *regular;* stamens 2–8. Fruit a non-splitting, pod-like legume.

Ceratonia siliqua ALGARROBO, CAROB TREE An evergreen shrub or tree to 10 m with leaves to 24 cm, pinnately divided into 1–5 pairs of dark green, rounded, leathery, untoothed leaflets; terminal leaflet absent. Flowers green or reddish and small with 5 sepals but without petals, borne in lateral racemes *directly from the trunk* in early autumn. Fruit large, 45 mm–20 cm, linear–oblong and laterally flattened, bluish-brown when ripe and pendent. Roadsides, field boundaries and cultivated. L, F, C, T, G, H.

Senna

A mainly tropical genus of herbs, shrubs and trees with pinnately divided leaves. Flowers typically yellow with 5 sepals and petals, petals not fused in a tube; stamens 10 (7 fertile, all free). Fruit pod-like, several-seeded.

Senna didymobotrya (syn. *Cassia didymobotrya*) FLOR DE GOFIO, CASIA An evergreen, vigorous and robust shrub to 5(9) m with leaves about 50 cm long with *14–18 pairs* of broadly-elliptic leaflets. Flowers bright yellow, borne in large, dense, terminal *racemes with numerous blackish-brown buds* at the apex. Cultivated; possibly naturalised. F, C, T.

Senna corymbosa (syn. *Cassia corymbosa*) RAMA NEGRA An evergreen or semi-evergreen, bushy and leafy shrub to 1(2) m with pinnately divided leaves with *2–3 pairs* of broadly elliptic leaflets, rounded at the base. Flowers numerous, bright yellow, borne in loose terminal clusters, each to 30 mm across with 5 unequal petals; stamens curved and exceeded by their petals (lower central anther plump). C. *S. bicapsularis* (a commonly misapplied name) has leaves with *3–4 pairs* of leaflets and flowers with a *shrunken lower central anther*. F, C, T, G, H.

SUBFAMILY PAPILIONOIDEAE

A diverse subfamily of trees, shrubs and herbs, often with tendrils. Leaves variously pinnately or palmately divided, or with 1 or 3 leaflets. Flowers solitary or in racemes or panicles; sepals (3)5, fused; petals (0)5(6); stamens typically 10. Fruit variable, often pod-like.

Anagyris

Large shrubs with trifoliate leaves. Yellow flowers borne in short racemes; stamens 10. Fruit a pod-like legume.

Anagyris latifolia ORO DE RISCO A shrub to 3(5) m. Leaflets lanceolate to ovate, 50–60 mm. Flowers large, yellow, with reddish lines on the standard. Fruit large, non-splitting, *segmented*, with each seed separated by a partition. Very rare and local; low altitude scrub. *C, T, G, P.

Genista (including *Rivasgodaya*) RETAMÓN

Shrubs with (mostly) trifoliate leaves, often with stipules. Flowers yellow, borne in lateral and terminal clusters; calyx tubular-bell-shaped, 2-lipped, the upper with 2 teeth, the lower with 3. Fruits narrowly oblong, compressed. Species here formerly all classified under *Teline*.

Genista canariensis (syn. *Teline canariensis*) An erect or spreading, much-branched shrub to 3 m; *young branches silky*. Leaflets small, rather rounded and almost hairy above, densely hairy beneath, with flat margins. Fruit short-hairy. Forests. *C, T. *G. microphylla* (syn. *T. microphylla*)* is similar but more compact, with elliptical leaflets with inrolled margins; young branches *woolly*. Locally common in mountains around Tejeda. *C.

Genista stenopetala (syn. *Teline stenopetala*) GACIA A tall shrub to 6 m with *hairless* branches. All leaves trifoliate; leaflets elliptical to lanceolate, 3–18 mm, more or less hairless above, densely silky-hairy beneath. Flowers borne 10–26 in elongated terminal racemes. Fruit oblong, densely hairy. A polymorphic species with five subspecies described. Beneath the laurel forest zone or pine forest. *T, G, H, P (naturalised in C). *G. osyrioides* (syn. *T. osyrioides*) has *unifoliate* upper leaves; leaflets more or less *linear* (to 2 mm wide). Shaded cliffs in the west and south of Tenerife. *T.

Genista pallida (syn. *Teline pallida*) A variable, small shrub to 2 m with virtually *stalkless leaves*; leaflets narrowly elliptical, silky, with *downturned margins*. Flowers borne 4–20 in short, dense, terminal heads; standard densely silky. The following forms are recognised: **subsp.** *pallida* (T), **subsp.** *gomerae* (G), and **subsp.** *silensis* (T). Rocks and cliffs in laurel forests and sclerophyllous woodlands. *T, G.

Genista rosmarinifolia (syn. *Teline rosmarinifolia*) A small shrub with *linear* leaflets with *downturned margins (superficially rosemary-like)*. Flowers rather few but borne in compact clusters; standard uniformly silky. Rocky ravine slopes. *C.

Genista nervosa (syn. *Teline nervosa*) An ascending shrub to 4 m with silky young branches. Leaves trifoliate with oblanceolate to elliptic leaflets. Flowers borne up to 56 in dense terminal racemes; calyx tubular-bell-shaped, 9–10 mm, silky-hairy. Rocky cliffs and scrub; local. *C (northeast). *G. benehoavensis* RETAMÓN much-branched shrub, densely tomentose, with profuse yellow-orange flowers. High altitudes. *P.

Ulex

Spiny, densely branched shrubs with adult leaves reduced to weak spines. Flowers yellow; calyx yellowish and divided to the base into 2 lips; stamens 10, fused in a tube. Fruit (1)2–6(8)-seeded.

Ulex europaeus TOJO A stout, erect, very spiny shrub to 2(2.5) m (unless in exposed habitats), densely branched above and bare below. Young twigs and spines glaucous; twigs somewhat hairy. Leaves highly reduced, to 8 mm; *spines stout, straight, deeply furrowed and hairless*; calyx yellowish, 2-lipped and persistent, 10–16(20) mm, flowers large, the standard (12)14–18 mm, *pale yellow* with straight wings longer than the keel. Introduced and naturalized. T.

Cytisus BROOM

Spineless shrubs with leaves that have 1 or 3 leaflets. Flowers yellow (sometimes white), borne in racemes; all 10 stamens in a tube. Fruit a legume, several–many-seeded.

Cytisus scoparius RETAMA NEGRA, BROOM A spineless, much-branched shrub 1–2.5 m, with long, slender and flexible stems, normally *5-angled and ridged*, hairy or not. Leaves small, trifoliate (1-foliate and stalkless on young branches); oval-elliptic, 11–14 mm *with short stalks* (13 mm). Flowers large, 15–20 mm, golden yellow, *solitary or paired*. Fruit oblong, compressed, 25–50 mm and *hairy along the margins only*. Damp woods. T, G.

Chamaecytisus ESCOBÓN

Tall shrubs with trifoliate leaves. Flowers white, borne in lateral clusters. Calyx tubular, 2-lipped, teeth small. Fruit black.

Chamaecytisus proliferus A tall shrub. Leaves with rather large, lanceolate, acute leaflets. Flowers white, borne in clusters of 1–4 arising from the axils; calyx deeply bilobed, variably silky-hairy. Fruit compressed, black. Locally abundant in hill forests. The following forms are recognised: **subsp.** *proliferus* occurs throughout the range with various varieties described (e.g. **var.** *canariae* (C), **var.** *hierrensis* (H), **var.** *palmensis* TAGASASTE (P, cultivated elsewhere)); **subsp.** *angustifolius* (T, G) and **subsp.** *meridionalis* (C). *C, T, G, H, P.

Spartium

Spineless, broom-like shrubs with stiff, rush-like branches; leaves simple, often absent or inconspicuous. Flowers with stamens 10, fused in a tube. Fruit many-seeded.

Spartium junceum RETAMA AMARILLA, SPANISH BROOM A large, spineless, broom-like shrub to 3 m with cylindrical, blue-green, rush-like stems. Leaves sparse, linear–oval and soon-falling, 15–30 mm. Flowers large, 20–28 mm, and bright yellow, solitary but in large numbers; sweetly scented; *calyx spathe-like*. Fruit 40 mm–12 cm, flattened. Cultivated and naturalised in open stands. C, T, G, H, P.

---

1. *Anagyris latifolia*
2. *Genista canariensis*
3. *Genista microphylla*
4. *Genista pallida* subsp. *gomerae*
5. *Genista pallida* subsp. *pallida*
6. *Genista rosmarinifolia*
7. *Genista nervosa*
8. *Genista benehoavensis*
9. *Ulex europaeus*
10. *Chamaecytisus proliferus* subsp. *angustifolius*

Spartocytisus

Erect, much-branched shrubs with small, soon-falling trifoliate leaves and white flowers; calyx with short and obscure teeth. Legume black, 4–6-seeded.

Spartocytisus filipes RETAMA FINA A small shrub with dense, slender, flexuous branches and long, interrupted clusters of white, fragrant flowers; flower stalks much longer than the calyx. Lower altitude forests; locally frequent. *T, G, H, P. *S. supranubius* RETAMA DEL TEIDE is similar but stout with *thick, glaucous stems* with flowers borne on the upper parts; flower stalks equalling or much shorter than the calyx. Subdominant in higher mountain zones. *T, P.

Retama

Rush-like, branched, spineless shrubs with slender, alternate branches and simple leaves. Flowers 10 stamens, fused in a tube. Fruit ovoid, 1–2-seeded, (normally) non-splitting.

Retama rhodorhizoides RETAMA A tree-like, fragrant, white-flowered shrub with flexuous greyish branches and small, linear, soon-falling leaves and ovoid fruits; similar to *Spartocytisus* but with a small, *deeply toothed* upper calyx lip and small, unsplitting 1(2)-seeded fruit. Abundant on south-facing slopes. *F, C, T, G, H, P.

Adenocarpus CODESO

Spineless shrubs with trifoliate leaves. Flowers lemon-yellow; stamens 10, fused in a tube. Fruit many-seeded, splitting.

Adenocarpus viscosus A densely leafy, sticky shrub. Leaves densely clustered; leaflets with inrolled margins. Flowers yellow; calyx glandular; standard petal virtually hairless. Fruit with sparse hairs. High mountain zones. *T, P. *A. foliolosus* has leaflets *without* inrolled margins, calyx *without* glandular papillae and *silkily hairy* standard petal. Fruit very *densely glandular*. Locally common on dry, forested slopes. Two forms recognised: **var. *foliolosus*** (throughout the range); **var. *villosus*** (C, P). *C, T, G, H, P. *A. ombriosus* is similar but laxer, with narrower leaflets, long (10 mm) petioles (1–3 mm in the former) and fruits *sparsely* glandular. Very rare, in pine scrub. *H.

Lupinus CHOCO, LUPIN

Herbs with palmate leaves. Flowers borne in long, conspicuous terminal racemes; stamens 10, fused in a tube. Fruit with 2–many-seeds, splitting.

Lupinus luteus YELLOW LUPIN A robust, hairy annual to 1 m. Leaves palmately lobed with 5–9 oblong leaflets, flowers bright *yellow*, 15–18 mm, borne in long whorls along the raceme, scented. Fruit 40–60 mm, densely hairy, black when ripe. Sandy places, and cultivated for fodder. T.

Lupinus pilosus A soft-hairy annual with brownish hairs, especially below, and leaflets *broadly oblong to lanceolate*. Flowers borne in whorled racemes; corolla bright *blue* to bi-coloured, 15–20 mm; calyx with a 2-lobed upper lip. Fruit short and broad, to 50 mm long. Seeds few and large, 9–12 mm. T, P.

---

1. *Chamaecytisus proliferus* subsp. *proliferus* var. *palmensis*
2. *Spartium junceum* (habit)
3. *Spartium junceum*
4. *Spartocytisus filipes*
5. *Spartocytisus supranubius* (OLIVER WHITE)
6. *Spartocytisus supranubius*
7. *Retama rhodorhizoides*
8. *Retama rhodorhizoides*

Lupinus angustifolius An annual to 1 m with 5–9 narrow leaflets (2–8 mm wide), linear to spathulate, slightly hairy beneath. Flowers dark blue, 12–16 mm long, borne alternately along the raceme. Fruit 40–70 mm with short hairs, yellow to black when ripe. Cultivated ground. C, T, H.

Lupinus albus WHITE LUPIN An erect, sparsely hairy annual to 50 cm (2) m with palmate leaves with 5–7(9) leaflets, hairless above. Flowers *alternate*, white to pale blue, *with a dark blue-tipped keel*, 18–20 mm long, borne in lax, terminal racemes; corolla 12–16 mm; upper lip of calyx just *shallowly* toothed. Fruits long-hairy, 60 mm–13 cm. Cultivated. C, T, G, H, P.

Astragalus

Annual or perennial herbs with pinnately divided leaves terminating in a single leaflet. Flowers borne in lateral clusters; keel blunt (not toothed or pointed); stamens 10, 1 free. Fruit variable (sometimes inflated). A very large genus of which only the most widespread or conspicuous are described here.

(a) Flowers whitish to yellow.

Astragalus solandri CHABUSQUILLO A prostrate, spreading annual, branched from the base with white-hairy stems 50 mm–45 cm. Leaves pinnately divided. Flowers pale greenish-yellow, borne in dense heads on long stalks. *Pods strongly arched*, compressed and tinged red-brown, rather shiny with minute white hairs. Dunes and bare ground near the sea. L, F, C, T, G.

Astragalus hamosus A spreading annual, branched from the base with stems to 40 cm. Leaves divided into 6–12 pairs of widely spaced oblong to broadly lanceolate leaflets. Flowers borne in racemes of 2–15(12); calyx 5–6(7) mm with white and dark appressed hairs; corolla inconspicuous, pale yellow, 7–11 mm. Fruit 30–40 mm, *cylindrical, strongly sickle-shaped* and long-persistent, densely appressed-hairy. Dry, bare habitats. L, F, C, T, G, H, P.

Astragalus boeticus CHABUSQUERA A hairy, erect annual with pinnately divided leaves with 6–13 pairs of oval leaflets, notched at the tip, hairless above, and slightly hairy below. Flowers 2–5, *pale sulphur yellow-white*, 8–11(14) mm, borne on dense racemes on stalks as long as the leaves or slightly shorter; wings longer than keel. Fruit 20–35(60) mm, oblong, triangular in section, shortly hairy, borne in dense clusters erect at first, then drooping. Sandy places and cliff-tops. L, C, T, H.

(b) Flowers whitish to violet-mauve.

Astragalus mareoticus A *prostrate, spreading annual*, branched from the base. Leaves with 6–8 pairs of leaflets, with short white hairs above and long white hairs below. Flowers few; corolla dull mauve. Fruits spreading, thick, *strongly arched and grooved*, purple-tinged and with minute, appressed hairs. Sandy and bare places. L, F. **A. edulis** has *squat, inflated, angled*, red-tinged fruits, triangular in cross-section. L, F.

Astragalus sinaicus A *small*, sprawling annual branched from the base with stems to 12(20) cm, similar to *A. stella* but sparsely hairy and with fewer flowers. Leaves divided into 6–10 pairs of leaflets. Flowers 3–7 borne virtually *stalkless* (rarely stalked to 40 mm) in the leaf axils; calyx 4–6 mm with black and white hairs; corolla pale mauve or violet, 6.5–8.8 mm. Fruits 8–12(16) mm, spreading to form a star, *with long, spreading hairs*. Bare places. L, F, C.

1. *Adenocarpus viscosus*
2. *Adenocarpus foliolosus*
3. *Lupinus albus*
4. *Astragalus solandri*
5. *Astragalus solandri* (fruits)
6. *Astragalus boeticus*
7. *Astragalus mareoticus*
8. *Astragalus sinaicus*

LEGUMINOSAE (FABACEAE), SUBFAMILY PAPILIONOIDEAE

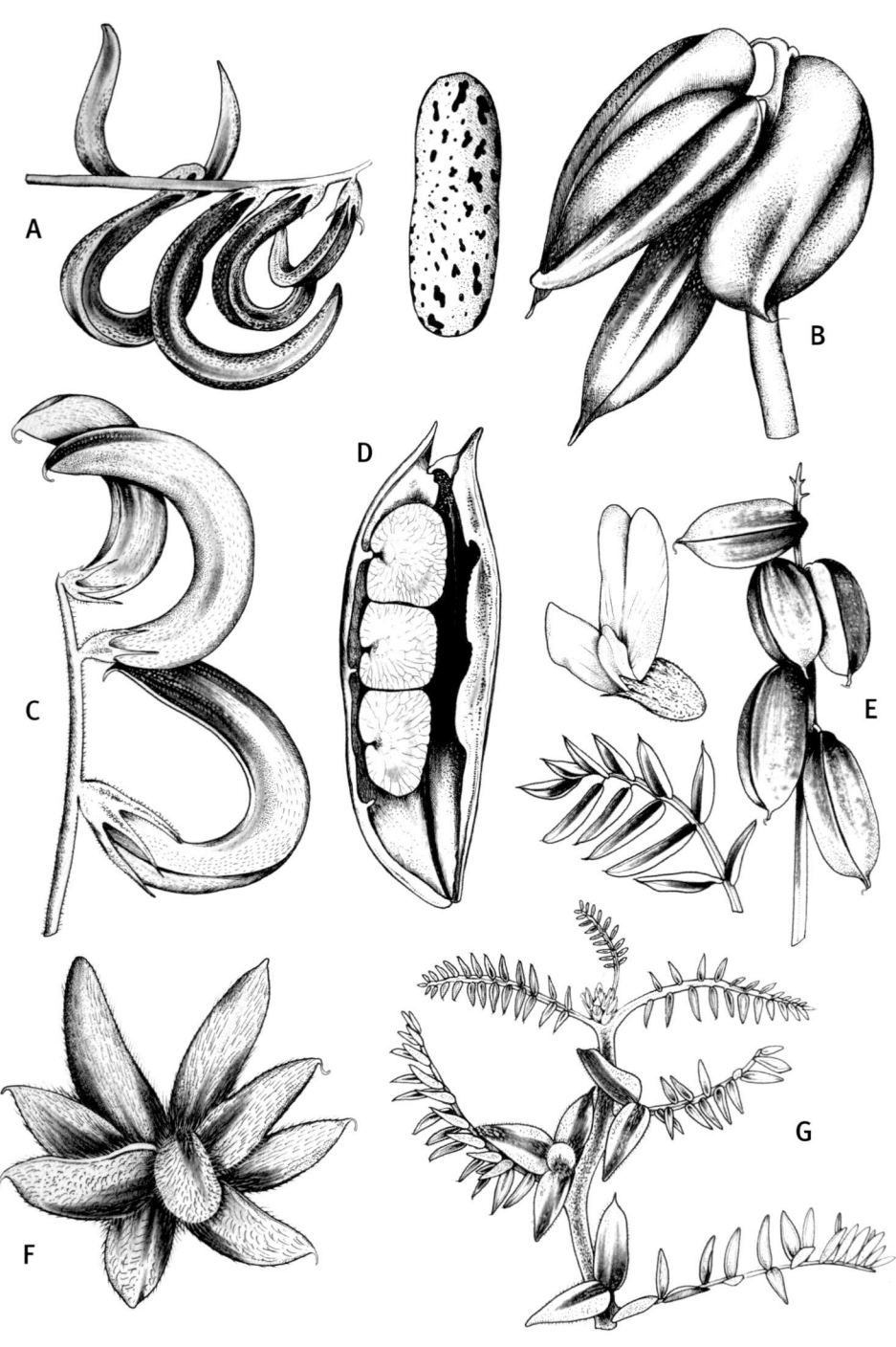

Astragalus **species: A.** *Astragalus solandri*; **B.** *Astragalus boeticus*; **C.** *Astragalus mareoticus*; **D–E.** *Astragalus edulis*; **F.** *Astragalus stella*; **G.** *Astragalus sinaicus*.

Astragalus stella A short, sprawling, densely hairy annual with long-spreading to ascending stems to 35 cm. Leaves with 5–11 pairs of oval leaflets, hairy on both surfaces. Flowers 6–14, yellowish or reddish-violet, (5.5)8–11 mm, borne on dense racemes on rather straight stalks *between half the length of the leaves* and the same length as the leaves; calyx 5–7 mm. Fruit 10–15 mm, more or less erect, almost straight, laterally compressed, *borne in star-shaped clusters*, each swollen at the base. Grassy habitats; local. L, F.

Biserrula

Sprawling annuals like the former genus (within which it is included in some floras), but with fruits long, strap-shaped, with conspicuous *saw-like* margins.

Biserrula pelecinus A branched, sprawling to ascending annual to 25(40) cm, sparsely white-hairy. Leaves with 7–15 elliptic leaflets. Flowers 2–5(11), borne on stalks rather shorter than their subtending leaves; calyx 3.5–4.5 mm; corolla *small,* just exceeding the calyx, blue-mauve. Fruit 15–40 mm, strongly flattened with conspicuous *saw-like margins*, hairless or sparsely hairy. L, F, C, T, G, H, P.

Bituminaria

Lax, spineless herbs *smelling of tar* when crushed, with trifoliate leaves. Flowers bluish, borne in dense, spherical heads; stamens 10, fused in a tube. Fruit 1-seeded, non-splitting.

Bituminaria bituminosa TEDERA An erect, branched perennial 20 cm–1 m. Leaves long-stalked, trifoliate with narrow to broad untoothed leaflets 12–90 mm long. Flowers 11–18 mm long, blue-violet, borne on long-stalked clover-like clusters. Fruit ovoid, 13–19 mm, flattened with a sickle-like beak. Regional forms include **var. *albomarginata*** with silvery leaf margins (*L), and **var. *crassiuscula*** (*T). Common in a range of habitats. L, F, C, T, G, H, P.

Cicer CHICKPEA

Annual or perennial herbs with glandular hairs. Calyx swollen basally.

Cicer canariense GARBANCERA CANARIA A *soft-hairy*, scrambling annual with leaves that have *numerous, long, narrow leaflets*. Flowers pale violet. Fruits conspicuously glandular white-hairy; seeds 6–8. Rare and local. *T, P.

Vicia VETCH

Climbing or scrambling annual or perennial herbs with unwinged stems, often with tendrils; leaves pinnately divided. Flowers with 10 stamens, 9 in a tube, 1 free. Fruit flattened and splitting.

(a) *Flowers small*, 2–10(15) mm, borne in clusters of 1–9(20) on stalks equalling or longer than the leaves.

Vicia hirsuta HAIRY TARE A slender, hairy to nearly hairless annual to 80 cm with leaves bearing 4–10 pairs of linear to oblong leaflets, each to 12 mm, notched at the tip and with a fine point; tendrils branched. Flowers white, tinged purple and small, 3–5 mm long in 1–9-flowered racemes; *calyx teeth equal*. Fruit hairy, 6–11 mm, black when ripe; seeds mostly 2. Grassy habitats. C, T, G, H, P. The following are similar: **V. *disperma*** is sparsely hairy with pale purple flowers 3–5 mm and broad, *hairless fruits* (except sometimes at the margins) 14–19 mm, with 1–2 seeds. C, T, G, H, P. **V. *parviflora*** has leaves with 2–4(5) pairs of leaflets, and 1–4(5)-flowered racemes that greatly exceed the leaves; flowers 6–9 mm with limb of standard longer than its claw. Fruit hairless,

12–17 mm; seeds 4–6(8). L, F, C, T, G, P. **V. ervilia** has 10–16 pairs of linear leaflets with a short terminal point in place of a tendril. Flowers 6–8 mm, whitish or pale pink with darker veins. Fruit hairless and constricted between the seeds; 14–22 mm; seeds 2–4. C, T.

Vicia tetrasperma A more or less hairless annual with 3–6(8) pairs of linear leaflets and entire stipules. Racemes with *1 or 2(6) flowers, about equalling (not exceeding) the leaves*; corolla pale purple; 12–15 mm, the calyx teeth unequal. Fruit hairless, 20–35 mm, *with 3–10 seeds*. Grassy habitats. L, F, C, T, G, H, P. **Vicia pubescens** is similar but *slightly hairy throughout*, with up to 6-flowered racemes and hairy fruits. F, C, T, G, H, P.

(b) Flowers 10–20(24) mm, borne in clusters of 6–30 on stalks longer than the leaves.

Vicia benghalensis CHINIPA A hairy annual to 80 cm with leaves that have 59 pairs of linear to elliptic leaflets; tendrils branched; stipules toothed or not. Flowers 2–20, *reddish to burgundy with blackish tips*, 15–18 mm long, with calyx convex at the base, borne in racemes longer than the leaves. Fruit hairy, 21–40 mm with 2–5 seeds. Hillsides, path-sides. L, F, C, T, G, H, P.

Vicia villosa FODDER VETCH A variable annual herb to 2 m, usually scrambling or climbing. Leaves pinnately divided with 4–12 pairs of linear–elliptic leaflets, and *unlobed*, hairy stipules and branched tendrils. Flowers 10–12(20) mm, violet to purple, often with creamy wings borne in conspicuous racemes of 10–30 flowers; stalk shorter than the subtending leaf; *calyx with swollen base; standard petal with a basal stalk-like part, 2 x the length of the blade*. Fruit brown and hairless, 20–40 mm with 2–8 seeds. L, F, C, T, G, H, P.

Vicia sativa ALVERJANA, COMMON VETCH A sprawling, hairy annual to 1.5 m. Leaves with 3–8 pairs of linear to heart-shaped leaflets, *tendrils branched*; *stipules toothed*. Flowers purplish-red, *solitary or paired* (up to 4), and often large, *20–30 mm long*; calyx teeth equal. Fruit yellowish or blackish, long, 35–68 mm, with 4–9 seeds. Grassy and disturbed habitats. L, F, C, T, G, H, P.

Vicia aphylla CHICHARILLA CANARIA A scrambling herb with *intricate, slender* stems. Leaflets to 30 mm, in 2–4 pairs, linear to linear–lanceolate. Tendrils coiled. Flowers rather few, 3–12; *calyx teeth minute* (the upper absent); corolla white with purplish veins, to 12 mm. Shaded slopes. *C, T, G, H, P. The following are similar: **V. filicaulis** has leaves in 1–2 pairs and flowers normally solitary with equal calyx teeth. *C. **V. scandens** CHICHARILLA MAYOR is more robust with leaflets to 38 mm, flowers borne 16–30; calyx teeth prominent, unequal, hairy; corolla to 18 mm. Laurel woods. *T. **V. vulcanorum** has 4–5(6) pairs of leaflets and 1–5(7) flowers. *L, F. **V. nataliae** CHICHARILLA GOMERA has *short leaflets, to just 18 mm*, up to 18 flowers per inflorescence, calyx teeth equalling their tube; corolla 15 mm. *G, T. **V. voggenreiteriana** has (3)4 pairs of leaflets 18–43 mm long and whitish-pink flowers with standard petal 12–14 mm. *G.

(c) Flowers *yellow or white*, <6 per cluster, on stalks *much shorter than the leaves*.

Vicia lutea CHINIPA AMARILLA, YELLOW VETCH A tufted, prostrate, hairless to hairy annual to 50 cm with leaves with 3–9 pairs of linear to oblong, bristle-tipped leaflets; tendrils simple or branched. Flowers 17–25 mm, hairless, borne in groups of 1–3, dull *yellowish-white* (sometimes flushed purple) with a hairless standard; calyx teeth unequal. Fruit yellow-brown and densely hairy, 25–43 mm with 2–4 seeds. Grassy habitats. L, F, C, T, G, H, P.

Vicia chaetocalyx A scrambling herb with leaflets in 6–7 pairs. Flowers *solitary*, almost stalkless; calyx teeth long; corolla *pale yellow with a purplish keel*. Grassy and disturbed habitats; rare. *T, C.

---

1. *Biserrula pelecinus*
2. *Bituminaria bituminosa*
3. *Bituminaria bituminosa*
4. *Bituminaria bituminosa* var. *albomarginata*
5. *Bituminaria bituminosa* var. *crassiuscula*
6. *Cicer canariense*

Lathyrus CHÍCHARO, PEAS

Similar to *Vicia* but typically with angled or *winged stems*, and often fewer, parallel-veined leaflets, sometimes reduced to a simple blade or tendril. Flowers with 10 stamens, 9 in a tube, 1 free. Fruit a pod-like legume.

(a) Flowers yellow.

Lathyrus annuus CHÍCHARO AMARILLO, ANNUAL YELLOW VETCHLING A climbing or scrambling annual to 1 m with leaves that have 1 pair of lanceolate leaflets, arrow-shaped stipules and branched tendrils. Flowers 12–18 mm, borne in erect, long-stalked racemes of 1–3, corolla *yellow to orange* and red-veined; calyx teeth equal. Fruit pale brown and glandular when young, 40–80 mm. Disturbed habitats. F, C, T, G, H, P.

Lathyrus aphaca CHÍCHARO CUCHILLERO, YELLOW VETCHLING A hairless, waxy grey-green scrambling annual to 40 cm (1 m) with unwinged (angled) stems. Mature plants with *leaves reduced to tendrils, but stipules large and leafy, broadly-triangular and paired*, 6–50 mm. Flowers borne solitary on erect stalks, yellow, 10–13 mm long. Fruit hairless, 17–35 mm long. Grassy and disturbed habitats. C, T, H, P.

Lathyrus ochrus CHÍCHARO ALBERJANA, WINGED VETCHLING A hairless, pale greyish annual to 1.5 m with *very broadly winged stems*, and leaves borne on *leaf-like stalks*; lower leaves oval–oblong, upper leaves with 2–3(4) pairs of leaflets. Flowers *pale yellow*, usually solitary, 13–21 mm long. *Fruit 39–65 mm with 2 wings* along the upper edge. Grassy habitats. C, T.

(b) Flowers reddish, purple, pink or white. Leaves simple or with just 2 leaflets; stems with wings <1 mm or wingless.

Lathyrus sphaericus CHÍCHARO ALTRAMUZ A more or less hairless, spreading annual to 30 cm. Leaves with 2 linear or lanceolate leaflets to 65 mm. Flowers solitary; calyx with lobes 3–4 mm; corolla *orange-red* (rarely white). Fruit linear–oblong, compressed, 40–60 mm, beaked, with parallel veins. C, T, G, H, P.

Lathyrus cicera RED VETCHLING A slender, hairless annual to 80 cm (1 m) with *narrowly winged stems*. Leaves with 1 (rarely 2) pair(s) of lanceolate to oval leaflets, with simple or branched tendrils on the upper leaves. Flowers 12–18 mm, solitary on stalks to 30 mm, *red, orange or brownish*; calyx teeth equal. Fruit hairless, 25–50 mm, and with 2 keels on the upper edge. Grassy habitats. L, C, H, P.
L. setifolius NARROW-LEAVED RED VETCHLING is similar but has scarcely winged stems and smaller orange-red flowers 9–13 mm with slightly *unequal calyx teeth*. Fruit 19–33 mm. T.

(c) Flowers reddish, purple, pink or white. Leaves simple or with just 2 leaflets; *stems broadly winged*.

Lathyrus tingitanus CHICHARACA A clambering, hairless annual herb with winged stems to 1.8 m. Leaves with *1 pair* of linear–lanceolate leaflets, tendrils branched; stipules arrow-shaped at the base; oval. *Flowers large and bright red-purple*, 25–35 mm long borne in racemes of 2–3(4) on stalks 28 mm–16 cm. Fruit hairless and shiny, 80 mm–11 cm. L, F, C, T, G, H, P.

Lathyrus odoratus CHÍCHARO DE OLOR, EVERLASTING PEA A clambering annual to 2.5 m, slightly hairy. Leaves with a single pair of ovate to elliptic, slightly undulate leaflets. Flowers 1–3(4),

---

1. *Vicia parviflora*
2. *Vicia benghalensis*
3. *Vicia scandens*
4. *Vicia nataliae*
5. *Vicia lutea*
6. *Lathyrus annuus*
7. *Lathyrus aphaca*
8. *Lathyrus ochrus*

bi-coloured, pink or purple, 20–35 mm and *sweetly scented*, borne in long-stalked racemes (stalks 14–20 cm). Fruit 55–65 mm. C, T, G, H, P.

Lathyrus angulatus A small, clambering, grey-green annual with winged stems and *long, paired, linear leaflets. Flowers small and pale blue or purple*, 9–14 mm, borne on flower stalks 25–85 mm, much longer than their leaves, and projecting in a point beyond the point of flower attachment. Fruit 25–45 mm, scarcely veined. C, T, G, H, P.

(d) Flowers red, purple, pink or white. *Leaves with 4–12 leaflets; stems broadly winged.*

Lathyrus clymenum CHÍCHARO MORADO A scrambling, hairless annual to 1 m with *broadly-winged stems and leaves with winged, leaf-like stalks;* leaves with 4–8 pairs of narrower, *linear–elliptic* leaves 0.5–14 mm wide. Flowers variably pink, dull yellow or crimson, 12–25 mm, borne in racemes of 1–2(3). Fruit *grooved* on the upper side. L, F, C, T, H, P. *L. articulatus* is often considered a form of the above: leaflets narrow; stems winged; lower leaves reduced and without leaflets; corolla with white or pink wings; fruit channelled on the lower margin. L, F, C, T, G, H, P.

Ononis RESTHARROW

Herbs or subshrubs with simple or trifoliate leaves (often both), the veins of the leaflets ending in marginal teeth; tendrils absent. Flowers with calyx deeply toothed with nearly equal teeth; all 10 stamens forming a tube. Fruit straight, 1–many-seeded, splitting.

(a) Shrubs with pink flowers.

Ononis christii TABOIRE DE JANDÍA A shrub to 90 cm. Leaves trifoliate. Racemes with 1–3 flowers; corolla rose-pink, hairless. Fruits borne on reflexed stalks; fruit linear to oblong. Seeds with tubercles. Dry scrub; rare. *F (Jandía).

(b) Shrubs with yellow flowers.

Ononis angustissima TABOIRE FINO A shrub with *long, coarsely toothed* leaflets. Flowers yellow, sometimes red-lined. Two forms are recognised: **subsp.** *angustissima* (*C, T) and **subsp.** *longifolia* on the other islands. *L, F, C, T.

Ononis hesperia CODESO, COYESO A shrub to 60 cm with hairless to glandular-hairy stems. Leaves rather rigid, toothed, folded, glandular-hairy. Flowers yellow with reddish-purple veins; calyx teeth 1.5–2.5 x their tube; flowers borne in erect to ascending racemes. Sand dunes; common in the east. L, F, C.

(c) Herbaceous annuals or perennials with yellow flowers.

Ononis hebecarpa TABOIRE AMARILLO A herbaceous annual with toothed leaflets 5–9 mm long. Flowers yellow, borne in pairs; corolla 10–12 mm, hairless; calyx conspicuously glandular-hairy. Fruits hairless with 8–20 *smooth* seeds. *L (Famara). *O. catalinae* is very similar but has fruits with *conical hairs* and *rough, triangular-kidney-shaped* seeds. Locally abundant after winter rain. *L, F.

Ononis sicula A short annual to 15(25) cm with narrowly oblong–linear, sharply toothed, rather erect leaflets, glandular-hairy flowerstalks without long spreading hairs; corolla 5–10 mm, exceeded by the calyx lobes. Fruit 9–14 mm, hairy, with 10–20 seeds. L, F.

---

1. *Lathyrus sphaericus*
2. *Lathyrus tingitanus*
3. *Lathyrus clymenum*
4. *Lathyrus articulatus*
5. *Lathyrus articulatus* (with fruit)
6. *Ononis christii*
7. *Ononis angustissima*
8. *Ononis hesperia*

(d) Herbaceous annuals or perennials with pink flowers.

Ononis pendula TABOIRE ROSADO A herb to 40 cm. Leaves beneath the flowers unifoliate. Flowers borne singly in the axils; corolla 1.5–2 x the length of the calyx with a hairless standard, pink or bluish with slightly darker veins and white or yellowish-white wings and keel; calyx teeth linear–lanceolate, entire. L, F, C.

Ononis tournefortii MELOSA DE ARENAS A herb with terminal, dense, multi-flowered racemes of flowers. Corolla 6–7.5 mm, slightly shorter than the calyx; standard glandular, pink with a whitish-yellow keel; calyx with simple and glandular hairs. Fruits included or scarcely exserted from the calyx, short glandular hairy with 6–8 seeds. L, F, T.

Ononis reclinata MELOSA RECLINADA A slender, *spreading* annual with *shaggy-hairy* stems ‹15 cm; middle leaves trifoliate, the lowermost and uppermost simple. Flowers *small* – just 5–10 mm – pink and solitary forming loose, leafy inflorescences, *abruptly recurved* at the apex; corolla *equalling* the calyx; calyx with *entire teeth*. Fruit 8–14 mm with numerous, *10–15(20) seeds*. Coastal habitats. L, F, C, T, G, H, P.

Ononis diffusa MELOSA PÁLIDA A *sticky, glandular-hairy* annual to 40(60) cm. Leaves trifoliate, with *toothed* margins. Flowers small, 8–14 mm, the corolla exceeding the calyx; pink with a whitish keel, borne singly at each node in a *dense, oblong spike*; calyx 7–8 mm with narrow teeth, longer than their tube. Fruit 4.8–5.5 mm with 3–4 seeds to 2 mm across, slightly curved. Coastal habitats. L, F, C, T, G, H, P.

Ononis serrata MELOSA ASERRADA A sticky annual to 30 cm. Flowers solitary, terminal; corolla pinkish-white, *small*, just 6–8 mm, not exceeding its calyx; calyx glandular-hairy. L, F, C, T, G, H, P.
O. mitissima MELOSA DE DAMAS is taller and more erect, scarcely hairy annual, to 60 cm, with trifoliate bracts with conspicuous white stipules. Flowers pink; corolla 10–12 mm, *exceeding* the calyx. Fruit 5–6 mm with 2–3 seeds. C, T, G, H, P.

Melilotus MELILOT

Annuals or biennials with trifoliate leaves often with toothed leaflets. Flowers small and sweet-smelling, borne in elongated racemes; stamens 10: 9 in a tube, 1 free. Fruit straight, 1–2-seeded, (normally) non-splitting.

(a) Flowers yellow.

Melilotus indicus SMALL MELILOT An erect, branched or simple annual to 40 cm with trifoliate leaves with toothed leaflets and stipules with *few, papery teeth* at the base. *Flowers tiny*, 2–3.5 mm, pale yellow to whitish and borne in *dense, many-flowered spikes* (10–40); wings and keel equal, and shorter than the standard. Fruit more or less spherical, pale brown when ripe, 1.5–3 mm and hairless. Disturbed ground. L, F, C, T, G, H, P.

Melilotus segetalis An annual with racemes *equalling or exceeding* their leaves, with *15–100 larger* flowers (4)5–7.5 mm; standard shorter than the keel. Fruit *yellow, egg-shaped* and with concentric grooves; 4–5 mm. Damp habitats. T. *Melilotus sulcatus* TRÉBOL DE OLOR, FURROWED MELILOT is very similar but with small flowers 2.8–4.5 mm, and a more or less *stalkless*; *inflorescence shorter than its leaf*. Fruit globose, 3–4.5 mm. L, F, C, T, G, H, P.

1. *Ononis hebecarpa*
2. *Ononis catalinae*
3. *Ononis pendula* (habit)
4. *Ononis pendula*
5. *Ononis tournefortii*
6. *Ononis serrata*
7. *Melilotus indicus*
8. *Melilotus sulcatus*

(b) Flowers white.

Melilotus albus WHITE MELILOT A vigorous annual to 1.5 m with slender branches and entire stipules. Flowers borne in rather lax racemes, corolla *white,* 4–5 mm. Fruit 3–5 mm, ridged, brown and hairless. Dry, disturbed habitats. C.

Trigonella FENUGREEK

Annuals with trifoliate, toothed leaves. Flowers borne solitary in leaf axils or in lateral racemes; stamens 10: 9 in a tube and 1 free. Fruit straight or curved with 1–many seeds in 2 rows; eventually splitting.

Trigonella stellata ALFOLGA OLOROSA A prostrate, sprawling annual. Leaves long stalked; leaflets narrowed at the base, toothed and notched at the tips. Flowers borne in dense clusters of 3–15, nearly *stalkless, congested at the base* of the plant and in the leaf *axils* along the stems. Fruit 5–8 mm long, cylindrical, divergently spreading in a star, with a netted surface. Bare, arid habitats. L, F, C, T.

Medicago MEDICK

Annual or perennial herbs typically with trifoliate leaves. Flowers yellow (sometimes purple) borne in small lateral clusters; stamens 10: 9 in a tube and 1 free. Fruit highly variable but important for identification; often curved, spiralled or spiny, 1–many-seeded.

(a) Fruit sickle- or bean-shaped.

Medicago lupulina BLACK MEDICK A spreading, hairy annual to 80 cm with leaves having 3 round to diamond-shaped leaflets, often notched at the tips. Stipules toothed or entire. Flowers small, 2–3 mm, yellow, and borne in ball-like, dense clusters of 20(50) flowers. Fruit 1.5–3 mm across, coiled, black when ripe and net-veined. Common in waste places. L, T, P.

(b) Fruit spineless, with a spiral of 1 or more turns.

Medicago sativa ALFALFA, LUCERNE A variable (with many forms described), hairy perennial to 90 cm. Leaflets oblong to linear, toothed at the apex. Stipules toothed at the base. Flowers 5–12 mm, violet or blue, borne in rounded, dense racemes of 10–30(50) flowers. Fruit 5–9 mm across, spiralled with 2–3(4) turns and a hole through the centre, not spiny. Common on waste places and on roadsides. L, F, C, T.

Medicago arborea LUCERNA, TREE MEDICK A silvery-grey leafy *shrub to 2 m* with silky-white younger branches. Leaflets narrowed at the base, slightly toothed at the apex. Stipules untoothed. Flowers yellow, 9–13 mm, borne in dense racemes of 8–20. Fruit 9–15 mm across, slender and coiled with 1 turn, hairless and net-veined. Disturbed habitats. F, H.

Medicago orbicularis LARGE DISK MEDICK A short, hairless or slightly hairy annual 30–60 cm. Leaflets oval–wedge-shaped, toothed at the apex. Stipules deeply toothed. Flowers 5 mm, yellow, in racemes of 2–4(5). Fruit *large and disk-like*, 10–20 mm across, smooth and spiralled anticlockwise with 4–7 turns, without a central hole. Waste places. L, F, C, T, G, H, P.

(c) Fruit often with spines or projections, with a spiral of 1 or more turns.

Medicago laciniata PELOTILLA A spreading annual with leaflets *deeply toothed or lobed*, and flower stalks rather longer than their subtending leaves' stalks. Flowers with wing petals shorter than the keel. Fruit 6–8 mm, spiny, hairless, with 4–6 turns. Desert and semi-deserts; locally common. L, F, C, T, G, H, P.

Medicago arabica SPOTTED MEDICK A low, prostrate and spreading, more or less hairless annual to 60 cm with trifoliate leaves. Leaflets notched and with a *conspicuous dark spot*. Flowers yellow, 4–6 mm borne in racemes of 1–5. Fruit cylindrical, 4.5–5 mm across, hairless, and with 3–5(6) anticlockwise turns with stout *spreading spines* 2–3.5 mm and a margin with 3 grooves. Fallow land and woods. C, T, P.

Medicago intertexta (syn. ***M. ciliaris***) A short annual 15–70 cm with spreading stems. Leaflets sometimes with a dark spot and finely toothed. Stipules toothed. Flowers dull yellow, 5–8 mm borne in clusters of 1–4(7). Fruit large, 10–17 mm across, *glandular-hairy to hairless pods, spiny*, with 5–8(10) anticlockwise spirals. Damp waste places. F, C, T.

Medicago littoralis A short, spreading, hairy annual with purplish stems to 80 cm. Leaflets oval to heart-shaped, toothed at the apex and *hairy on both surfaces*. Stipules toothed. Flowers yellow to orange, 5–7 mm long in clusters of 2–5(8). Fruit 3.5–5(6) mm long, cylindrical with 3–7 turns, shiny, spiny or not, and hairless, with a groove along the margin. Maritime sands. L, F, C, T, G, H.
M. soleirolii is similar with a discoid, *spineless* fruit with 2–4(5) *tightly appressed* clockwise coils with 5–7 *indistinct veins forming a fine network*. L, C.

1. *Trigonella stellata*
2. *Medicago sativa*
3. *Medicago arborea*
4. *Medicago laciniata*
5. *Medicago polymorpha*
6. *Medicago minima*

Medicago polymorpha HAIRY MEDICK A variable, prostrate to ascending, hairy or hairless annual with purplish stems to 50 cm and trifoliate leaves, *toothed at the apex*. Stipules *deeply toothed with slender teeth*. Flowers yellow, 6–6.5 mm long, borne in clusters of 2–9. Fruit 6–7 mm across, variable, typically flattened with 4–6 anticlockwise turns, with 2 rows of *grooved spines* 0.5–0.8 mm, the surface netted (*not shiny*) and the apical spiral larger than the rest. Grassy places and cultivated fallow land. L, F, C, T, G, H, P. The following are similar: **M. italica** (syn. **M. tornata**) has fruits 5–8 mm with 2 (1–6) spirals and *spines 2 mm, not grooved*; margins of the spirals sharply defined. **M. truncatula** has fruits 7–10 mm, with *much-thickened spines, swollen at the base*, 1.3–5 mm, and margins of the spirals thickened. L, C, T, G, H, P.

Medicago minima SMALL MEDICK A prostrate to ascending, densely hairy annual, similar to the *M. polymorpha* group, but with *stipules entire or slightly toothed, the teeth if present short*. Flower stalks as long as or a little longer than their leaves' stalks; flowers 2.5–4.5 mm. Fruits borne in clusters of 4–6, ball-like, spiny, 2.5–3.5(5) mm across (rather large relative to the plant) with 3–5 lax, sharp-edged spirals and spines to 1 mm, weakly hooked; not distinctly flattened, and often rather shiny (less netted) on the surface. L, F, C, T, G, H, P.

(d) Fruit with spines or projections, with a spiral of 1 or more, *greatly thickened* turns, appearing turban-like.

Medicago turbinata A robust, prostrate or ascending, *densely hairy* annual to 50 cm with reddish stems and toothed, trifoliate leaves. Flowers yellow-orange, 6–6.5 mm long, borne in clusters of 2–9. Fruit barrel-shaped, 6–7.5 mm with 4–6(7) clockwise or anticlockwise turns with *7–10 veins ending in a veinless zone*, a single groove along the margin, often with *short, broad spines* <1 mm long. Various, often disturbed habitats. L, F, C, T, P.

Lotus BIRD'S-FOOT TREFOIL

Herbs with leaves that have 5 leaflets (the lower 2 resembling stipules). Flowers often yellow, borne in flat-topped clusters; stamens 10, 1 free. Fruit several-seeded, splitting, smooth, or keeled and square in cross-section.

(a) Keel with a long beak; flowers yellow to red.

Lotus berthelotii PICO PALOMA A cliff-dwelling perennial with a woody stock. Leaflets linear, 10–18 mm, silky-hairy. Flowers *scarlet* with a long beaked keel and wing petals. Forest cliffs; very rare and local (also cultivated). *T. *L. maculatus* PICO DE EL SAUZAL is similar but with broader leaves and *yellow* petals with orange-red stripes and a darker, extended beak. Very rare and local. *T (north; also cultivated). *L. pyranthus* PICO DE FUEGO has dull orange-yellow flowers with shorter beaks. A rare endemic. *P. *L. eremiticus* PICO CERNÍCALO is a creeping shrub with similar, dull orange flowers; calyx with purple spots at the base. Basaltic rocks; very rare and local. *P.

(b) Spreading annuals to perennials; flowers yellow (many similar species).

Lotus pedunculatus (syn. *L. uliginosus*) CUERNECILLO GRANDE A vigorous, leafy perennial to 1 m, with *hollow and creeping, rooting stems*. Flowers 10–18 mm. Fruit 15–35 mm. Wet, marshy habitats. T.

Lotus edulis CUERNECILLO COMESTIBLE, EDIBLE LOTUS A short, spreading, slightly hairy annual to 50 cm with oval leaflets. Flowers yellow with a purple-tipped keel, 10–16 mm long; calyx bell-shaped. Fruit oblong and curved, *very inflated*, with a groove along the back, 15–35 mm long. C.

1. *Lotus berthelotii*	4. *Lotus eremiticus*	7. *Lotus tenellus*
2. *Lotus maculatus*	5. *Lotus campylocladus*	8. *Lotus leptophyllus*
3. *Lotus pyranthus*	6. *Lotus tenellus* (habit)	9. *Lotus arinagensis*

Lotus ornithopodioides CUERNECILLO A short, spreading, hairy annual to 50 cm with stalked leaves that have oval to diamond-shaped leaflets, the lower 2 triangular and heart-shaped at the base. Flowers yellow, 6–8 mm long, borne in *stalked, terminal heads of 2–5 on stalks slightly longer than the leaves* (at least in fruit); calyx 2-lipped; *teeth very unequal*. Fruit 20–40 mm long, *constricted between the seeds*. Damp habitats near water. C. **L. angustissimus** CUERNECILLO FINO, SLENDER BIRD'S FOOT TREFOIL is similar but with flowers in heads of 1–2(3) on stalks as long, or longer than the leaves (10–40 mm) and *calyx teeth nearly equal*. Fruit 15–25 mm, *much longer than the calyx*. C, T, P.

Lotus campylocladus CORAZONCILLO DE PINAR A woody-based, spreading perennial. Leaflets narrowly lanceolate, 4–10 mm with dense, adpressed hairs. Flower stalks short; corolla 10–15 mm, in 3–6-flowered clusters. *T, H(?), P.

Lotus mascaensis CORAZONCILLO DE MASCA A dwarf, silvery shrub. Leaflets linear, 12–15 mm with petioles 1–3 mm. Flower clusters terminal with 3–7 flowers on stalks 20–50 mm. Calyx with long white hairs. Locally frequent in the Valle de Masca. *T (west).

Lotus tenellus CORAZONCILLO DE COSTA A shrubby perennial. Leaves stalkless or virtually so; leaves with 5 pinnate leaflets with adpressed hairs. Flowers yellow, borne in umbels of 1–4; standard 9–14 mm, about as long as the keel; wing petals 8–13 mm. Fruits straight. *T. The form in Gran Canaria has been described as *L. leptophyllus*. *C.

Lotus arinagensis CORAZONCILLO DE ARINAGA A sprawling, whitish perennial with dense adpressed hairs. Lateral leaflets obovate to circular; central leaflets broadly elliptic. Flowers in clusters of 1–6; calyx 5–6 mm; keel without a long beak. Coastal rocks and sands. *C.

Lotus sessilifolius CORAZONCILLO A sprawling perennial with a woody stock. Leaves stalkless or virtually so; leaflets linear, 3–9 mm, with short, silky hairs. Flowers borne in terminal and axillary clusters of 3–5; calyx purplish at the base. Fruits linear. *T, G, H.

Lotus emeroides CORAZONCILLO GOMERO A robust, sprawling perennial with circular leaflets, flowers borne in clusters of 3–4; petals purple-tipped. Dry lowland slopes. *G.

Lotus lancerottensis HIERBA MUDA A sprawling perennial with a woody stock. Leaflets obovate, 3–5 mm, silky-hairy. Flowers yellow, borne in terminal and axillary clusters of 1–5 on stalks 20–40 mm. Fruits blackish. Locally common on roadsides. L, F.

Lotus spartioides CORAZONCILLO DE GRAN CANARIA A dwarf shrub with slender, sub-erect branches. Leaflets linear, 5–10 mm, with short, adpressed, whitish hairs. Flowers yellow, 2–5 per cluster; calyx with purple stripes. *C. **L. holosericeus** CORAZONCILLO PLATEADO is similar but covered entirely with long, spreading, silky hairs and flowers borne in clusters of 6–10. Scrub; locally common around San Bartolomé de Tirajana. *C.

(c) Spreading annuals to perennials; flowers pink.

Lotus glinoides MATA PARDA A prostrate, spreading annual with rather thick, blue-grey, trifoliate leaves with sparse to dense white hairs; leaflets broadest above the middle. Calyx with long, sparse, white hairs; flowers pink. Fruit narrowly cylindrical, reddish and shiny, often borne in divergently spreading pairs. Bare ground; locally common. L, F, C, T, G, H.

(d) *Dorycnium* group (trébol de risco): shrubs. Flowers white or pink; stamens 10: 9 fused and 1 free. Fruits swollen, exceeding the calyces; 1–many-seeded, splitting.

Lotus broussonetii (syn. *Dorycnium broussonetii*) A subshrub to 1.5 m. Leaflets obovate to elliptic, 20–40 mm, with sub-circular *stipules free* at the base. Flowers white with pinkish tips, borne in

1. *Lotus sessilifolius*
2. *Lotus sessilifolius* subsp. *villosissimus* (LEOPOLDO MORO)
3. *Lotus lancerottensis* (habit)
4. *Lotus lancerottensis*
5. *Lotus spartioides*
6. *Lotus spartioides*
7. *Lotus glinoides*

dense clusters of 5–8. Calyx campanulate, *hairy*, with *unequal* teeth. Rare; coastal scrub. *C, T. ***D. spectabile*** is similar but has a *hairless* calyx with *equal teeth* and *pink flowers*. Rare. *T (Güímar Valley and Monte del Agua).

Lotus eriophthalmus (syn. ***Dorycnium eriophthalmum***) A subshrub to 1.5 m, similar to the species above but with *stalkless stipules, united* at the base. Flowers borne in outfacing circular clusters; corolla white; calyx hairy, reddish. Fruits linear. Rare; shady or sheltered cliffs. *C, T, G, H, P.

Ornithopus BIRD'S FOOT

Annual, erect or prostrate herbs with pinnately divided leaves with numerous leaflets. Flowers with 10 stamens: 9 in a tube, 1 free. Fruits *slender, often curved* (falcate).

Ornithopus pinnatus SERRADILLA AMARILLA, ORANGE BIRD'S FOOT A spreading, prostrate, slightly hairy annual. Leaves *pinnately lobed* with (3)6–7(14) leaflets. Flowers *orange-yellow*, 4.5–5.5 mm, borne in heads of 2–5(7) *without a leafy bract* below (bract absent or papery). Fruit linear, 16–33 mm long, flattened and markedly constricted between the seeds. Grassy and bare, stony habitats. T, P. ***O. perpusillus*** SERRADILLA ROSA has pinnately lobed leaves with 4–14 leaflets and a terminal leaflet. Flowerheads with a pinnately divided bract below, corolla 3.5–5.5 mm, *cream* to white with red veins. Fruit *curved and spreading*, in clusters of usually 2–3, and constricted between the seeds, 13–30 mm. L, F, T.

Ornithopus compressus UÑAGATO A low, prostrate to ascending, very hairy annual to 75 cm with pinnately divided leaves with *(2)8–15(18) pairs* of leaflets. Flowers 5–7 mm, yellow, borne in heads of 1–5, *subtended by a leafy bract*. Fruit slender, flattened and curved, slightly constricted between the seeds, 18–42 mm. Grassy habitats. C, T, G, H, P.

Trifolium TRÉBOL, CLOVER

Annuals or herbaceous perennials with trifoliate leaves. Flowers numerous, short-stalked, typically clustered into rounded, congested heads; stamens 10: 9 in a tube and 1 free. Fruit 1–9-seeded, often enclosed in the calyx. Calyx shape and vein number are important.

(a) Calyx 5-veined; flowers typically *yellow* with a *boat-shaped* standard.

Trifolium campestre HOP TREFOIL An erect or spreading *hairy* annual 50 mm–20(30) cm. Leaves alternate, the terminal leaflet longer-stalked than the laterals. Flowers, *yellow*, corolla (4)5–7 mm long, borne in *rounded heads of >20, 8–15 mm across*. L, F, C, T, G, H, P.

Trifolium dubium LESSER HOP TREFOIL A virtually hairless annual 50 mm–20(30) cm. Leaves alternate, the terminal leaflet longer-stalked than the laterals. Flowers small and *yellow*, corolla 3–4.5 mm, borne in *rounded heads* of 5–20, 5–9 mm across. Disturbed habitats. F, C, T, G, H, P. ***T. micranthum*** is similar but with stalkless leaflets, flower stalks shorter or slightly exceeding their subtending leaves and flowers 3–8 per head, *minute*, 1.5–3 mm. P.

(b) Calyx with *>5 veins*; flowers typically whitish or pink (or red), borne in *elongated heads*.

Trifolium angustifolium NARROW-LEAVED CRIMSON CLOVER A somewhat hairy annual to 50 cm. Leaves with *linear–lanceolate*, pointed leaflets. Flowers *pink*, 10–13 mm, borne in stalked,

---

1. *Lotus eriophthalmus*
2. *Ornithopus pinnatus*
3. *Trifolium campestre*
4. *Trifolium dubium*
5. *Trifolium angustifolium*
6. *Trifolium arvense*
7. *Trifolium scabrum*
8. *Trifolium tomentosum*
9. *Trifolium repens*

cylindrical heads, opening from the top downwards; corolla not (or scarcely) exceeding the calyx; calyx lobes unequal (the lowermost longest) and very slender and sharp, hairless at the tips. Disturbed or sandy habitats. L, F, C, T, G, H, P.

Trifolium arvense HARE'S-FOOT CLOVER A softly hairy annual to 40 cm with narrow leaflets, scarcely toothed. Flowers pale pink or white, *borne in dense, elongated oval or cylindrical, stalked heads* to 25 mm long, the flowers usually *exceeded* by the calyx teeth; corolla just 3–6 mm, with hairless wings; calyx lobes more or less equal. Sandy and disturbed areas. L, F, C, T, G, H, P.

(c) Calyx with >5 veins; flowers typically whitish or pinkish, borne in *stalkless, rounded heads* in the leaf axils, often on spreading or prostrate stems.

Trifolium suffocatum SUFFOCATED CLOVER A distinctive *prostrate*, tufted, low annual <50 mm. Leaves with oval leaflets overtopping the *unstalked* rounded flowerheads which are congested at the base of the plant. Flowerheads terminal and densely crowded; corolla white, 3–4 mm and shorter than the calyx; calyx tube nearly cylindrical. Local on coastal sands and bare ground. C, T, H, P.

Trifolium scabrum ROUGH CLOVER A spreading, often *prostrate* and *rather downy* annual to 20 cm with *leaves that have prominent, paler veins* which are thickened at the leaf margins. Flowers 4–7 mm, rather inconspicuous, cream-white, turning pink with age, borne in small, unstalked heads to 10 mm across; calyx soon becomes stiff after flowering, with *starry, spreading, recurved, spiny teeth,* equalling or exceeding the corolla. Disturbed habitats, towns and gardens. L, F, C, T, G, H, P.

Trifolium cherleri A spreading, *densely long-hairy* annual. Flowers white or pink, the corolla *equalling or shorter* than the calyx; *flowerheads densely woolly hairy*. Grassy habitats. C, T, G, H, P.

Trifolium tomentosum WOOLLY TREFOIL A slender, creeping annual to 15 cm, rather similar to *T. resupinatum* (see p. 151), but with leaflets 4–12 mm and smaller flowers 3–6 mm long and *inflated, spherical fruiting heads clothed in soft, white hairs*, short-stalked or stalkless (calyx teeth obscured). L, F, C, T, G, H, P.

Trifolium glomeratum An annual to 40 cm with squat, pinkish flowerheads borne stalkless within the leaf axils; corolla exceeding the calyx, with a standard 4–7 mm; calyx with 10 veins and reflexed teeth in fruit. L, F, C, T, G, H, P.

(d) Calyx with >5 veins; flowers typically whitish or pink (or yellowish), borne in *distinctly stalked, often rounded heads* terminally or in the leaf axils.

Trifolium lappaceum A slender, ascending to erect, virtually hairless annual 10–30 cm. Leaves with shallowly toothed, egg-shaped leaflets. Flowers borne on stalks much longer than their subtending leaves, in heads 12–16 mm across; calyx bell-shaped, 20-veined, *hairless*, with teeth slightly longer than their tube, with a broad, 3-veined base; corolla just exceeding the calyx, white or pink. L, T, G, P.

Trifolium repens WHITE CLOVER A variable, creeping perennial to 50 cm with stems rooting at the nodes, often hairless. Leaves trifoliate with green, oval to elliptic leaflets, paler along the veins. Flowers normally white, sometimes pink or reddish, 7–12 mm long, borne in dense, long-stalked spherical heads; calyx tube longer than wide, the lobes triangular–lanceolate; flowers scented. Fruit linear and compressed between the seeds. Damp, grassy places. C, T.

Trifolium stellatum STAR CLOVER A short, erect, hairy annual to 20(30) cm with stems simple or branched from the base. Leaflets oval and slightly toothed with oval, *toothed* stipules with bright green veins. Flowers pink or yellowish, 12–18 mm, borne in solitary, large *spherical* heads to

25 mm across when in fruit; calyx equalling the corolla, densely white-hairy with slender, reddish, *spreading lobes* (star-like). Disturbed and grassy habitats. L, F, C, T, G, P.

Trifolium resupinatum REVERSED CLOVER A spreading or sprawling, more or less hairless annual to 30 cm. Leaves trifoliate with wedge-shaped leaflets 10–20(25) mm long. Flowers borne *upside down* in circular, flattened heads (superficially similar to *T. tomentosum*); pink to purple, 5–8 mm long, on rather short stalks; calyx conspicuously inflated and papery when in fruit, rather hairy. Moist, grassy habitats. C, T, G.

Trifolium fragiferum STRAWBERRY CLOVER A creeping perennial with stems to 30 cm, rooting at the nodes. Leaflets oval, without whitish marks. Flowers pink or purplish, 5–7 mm long, borne in rounded heads to 15 mm across; calyx *greatly swollen when in fruit to give the appearance of a grey-pink berry*. Damp and grassy habitats. T, C.

Trifolium subterraneum A spreading annual to 90 cm. Leaves alternate with hairless to slightly hairy stipules; leaflets obcordate. Flowers *few*, borne in obovoid heads *in the axils*; corolla *whitish*; several of the flowers sterile, reduced to a calyx reflexed when in fruit; calyx with equal or sub-equal thread-like teeth, the tube with 20 veins and a hairless throat, sometimes purplish. C, T, G, H, P.

(e) Calyx with ›5 veins; flowers typically whitish or pink (or yellowish), borne in terminal, short-stalked, rounded heads with (1)2 leafy bracts close below.

Trifolium pratense RED CLOVER A perennial to 60 cm, initially rosette forming, later spreading; rather hairy. Leaflets circular to oval, hairier below than above. Flowerheads large, *to 30 mm across*, spherical or ovoid, solitary or paired, and usually virtually *stalkless*; calyx tube 10-veined with triangular teeth that are bristle-pointed at the tips, the lowest tooth *much longer than the others*; corolla 12–18 mm, usually pink, rarely cream or white. Cool, grassy places. F, C, T.

Trifolium spumosum A spreading to erect annual to 30(70) cm, rather similar to *T. resupinatum*. Leaves trifoliate with wedge-shaped, toothed leaflets that are not strongly veined. Flowerheads more or less spherical, borne on short stalks; corolla 12–16 mm, pink, slightly exceeding the calyx; calyx *hairless* and inflated when in fruit, forming somewhat spiky fruiting heads. C.

Trifolium squamosum An erect or ascending, more or less hairy perennial to 40 cm with elliptic leaflets; *stipules long and spreading*. Terminal flowerheads becoming ovoid, short-stalked and subtended by a pair of leaves; flowers 7–9 mm, pink-white; calyx teeth unequal and each with 3 veins ›½ their length, exceeded by the corolla; calyx tube broadly cylindrical. Fruiting heads *resemble miniature teasels* due to spreading calyx teeth, broadly *cylindrical*, about 12 mm wide. Widespread though local, in coastal grassland. P. *T. squarrosum* is similar but larger with lax flowerheads, hairy calyces *constricted* at the tips and with hairs that have pimpled bases, and *large, ovoid* fruiting heads 20 mm across. C, T, G, P.

Trifolium hirtum HAIRY TREFOIL A short, spreading, *hairy*, branched annual to 35 cm. Leaflets wedge-shaped and finely toothed towards the tip. Stipules *long and straight* with a hairy tip. Flowers pink to purple, 12–17 mm, borne in large, *densely hairy* heads to 20 mm across with a pair of leaves immediately below; corolla greatly *exceeding* the calyx; calyx 20-veined and with long hairs. Dry, stony habitats. T.

Trifolium striatum An annual to 60 cm with alternate leaves with ovate, entire stipules, hairy on the underside; leaflets obovate to obcordate. Flowerheads short-stalked with a veined bract; ovoid, terminal or arising from the axils with 1–2 subtending stalked to stalkless leaves; corolla pinkish, shorter or equalling the calyx; calyx regular, with triangular, unequal teeth, hairy, 1-veined; throat open and hairy when in fruit. F, C, T, G, H, P.

Coronilla

Hairless shrubs with pinnately divided (rarely trifoliate) leaves. Flowers with 10 stamens: 9 in a tube, 1 free. Fruit a cylindrical, pod-like legume.

Coronilla valentina A blue-green, hairless dwarf shrub to 1.5 m, with pinnately divided leaves with 4–6 pairs and heart-shaped stipules. Flowers 8.5–12(14) mm, borne in lateral clusters of 4–8(13), *yellow*, and *strongly scented*; calyx bell-shaped. Fruit 9–40(65) mm long, with 1–4(10) swollen segments. Scrub. C.

Coronilla viminalis CORONA DE LA REINA A blue-green, hairless, spreading perennial with pinnately divided leaves with up to 10 pairs, and free, whitish stipules. Flowers borne in stalked, lateral clusters of up to 10, *white to pinkish-mauve;* calyx with 5 short teeth. Fruit to 88 mm, arched, slender, with up to 11 poorly defined segments. Cliff scrub; local. L, F, C.

1. *Coronilla viminalis*
2. *Coronilla viminalis*
3. *Hippocrepis multisiliquosa*
4. *Scorpiurus muricatus*
5. *Scorpiurus sulcatus* (fruits)

Hippocrepis

Herbaceous perennials with pinnately divided leaves. Flowers borne in axillary clusters; stamens 10: 9 fused and 1 free or partially so. Fruit divided into distinct, *horseshoe-like* 1-seeded sections, strongly compressed.

Hippocrepis multisiliquosa ARETE PERRO A variable annual similar to *H. biflora*, though variable. Stipules with 2 gland-like dots. Flowers larger, 4.5–7 mm long, solitary or in clusters of 3–6, each with hairless stalks, the cluster borne on a distinct stalk (peduncle). Fruit *curved downwards* (sometimes into an almost complete circle); hairless, 30–35 mm; *swellings on the exterior (convex) side of the curve*. L, F, C, G, H.

Scorpiurus

Annual herbs with simple, alternate leaves. Flowers yellow; stamens 10: 9 fused and 1 free. Fruits elongated, strongly curved, with swellings or spines. Some authors recognise numerous taxa, most of which probably belong to the 2 species described below.

Scorpiurus muricatus ROSQUILLA A low, hairy or hairless sprawling annual to 60 cm with simple, elliptic, entire leaves with 3 prominent veins; the upper leaves short-stalked or unstalked. Flowers yellow, sometimes flushed with red, 9–13 mm long, borne in long-stalked clusters of 1–5; *calyx tube shorter than its teeth* (teeth 2–2.5 mm). Fruit to 50 mm long, coiled and twisted (so appearing short), variably covered in short, robust hairs. Common in dry, rocky and sandy habitats and roadsides. L, F, C, T, G, H, P. *S. sulcatus* has a *calyx that has lower teeth shorter than their tube* (teeth 1.5–2.5 mm) – also considered a mere form of the previous species. *S. vermiculatus* is similar to the above, but with *solitary* flowers (rarely 2–3) 11–14 mm long, and fruits swollen, with *warts not hairs,* 5–7 mm across. Similar habitats. C, T, P.

ROSACEAE | ROSE FAMILY

Annual or perennial herbs, shrubs and trees. Leaves alternate, simple or compound, and with stipules (often soon-falling). Flowers with 4(5)(–10) sepals and (0)4–16 petals; stamens usually 2–4 x as many; carpels 1–many. Fruit highly variable, for example an achene, drupe follicle or capsule.

Agrimonia

Rhizomatous perennials with erect, hairy stems. Flowers yellow, in spike-like racemes; petals 5; stamens 5–20; carpels 2. Fruit with hooked bristles.

Agrimonia eupatoria A hairy perennial 40–90 cm with leaves pinnately divided into 3–5 elliptic, toothed pairs of leaflets, alternating with much smaller leaflets. Flowers yellow, borne in long, very narrow spike-like racemes; petals 3.5–5.5 mm. Fruit nodding, grooved, with forward-directed, hooked bristles; achenes 1–2. Grassy habitats. T.

Rosa ROSAL SILVESTRE, ROSE

Prickly shrubs, generally deciduous with pinnately divided, toothed leaves and stipules. Flowers terminal, usually with 5 sepals and petals; stamens numerous; styles separate or fused into a column. Fruit a nut, enclosed in a fleshy (often edible) structure, called a *hip*.

Rosa canina DOG ROSE A lax shrub to 3(4) m with arching stems and *stout, curved spines,* longer than the width of the base. Leaves with 5–7 leaflets to 40 mm long, hairless, green. Flowers borne in *small* clusters (usually up to 4), 30–50 mm across and pink or white, with *hairless stalks*; sepals with narrow, usually entire, projecting side lobes. Hips red and hairless, without sepals when ripe. Woods and barrancos. C, T, G, P. *R. micrantha* is similar but with double dentate leaflets and glandular stalks. Abandoned orchards. T, P.

Potentilla CINQUEFOIL

Herbs or shrubs with lobed or pinnately divided leaves. Flowers with (4)5(6) petals; styles feathery in fruit; stamens (5)10. Fruit a head of achenes.

Potentilla reptans CREEPING CINQUEFOIL A creeping perennial with a central rosette and stems *rooting at the nodes* with long-stalked, hairless leaves with 5(7) leaflets; stipules leafy and entire. Flowers solitary in the leaf axils, yellow, 15–25 mm across with 5 notched petals; petals 2 x the length of the calyx. Woods. G. *P. indica* is similar but with 3 leaflets and receptacle much enlarged, spongy, bright red. Woods. T.

Rubus BRAMBLE

A complex genus of typically woody, scrambling vines. Flowers with 5(–8) sepals and 5(–10) petals; stamens numerous. Fruit a head of 1–many 1-seeded drupes.

Rubus ulmifolius A spreading perennial with robust, thorny, arching or prostrate stems that are angled and grooved; hairy or hairless. Leaves typically with *3 small* leaflets, *leathery dark green above and shortly white-hairy beneath*, variously toothed and oval in shape. Inflorescence rather narrowly pyramidal, and leafy at the base, with hairy prickles; sepals deflexed after flowering, and white-hairy; petals 9–14 mm long, usually *pink*, with stamens *equalling or scarcely exceeding* the styles. Fruit black and shiny (known as a blackberry). Common in thickets and woods. C, T, G, H, P. *R. bollei* is similar, typically with *5 broad* leaflets on *long stalks* with many curved spines 3–5 mm long. Flowers *white* (rarely pink); sepals greyish. Local in thickets. *F(?), C, T, G, H, P. *R. palmensis* has 5 leaflets with inconspicuous, simple appressed hairs on the upper surface. Stems or petioles with few to many gland-tipped setae. Flowers pale pink. *C, T, G, P.

Prunus CHERRY

Spineless or spiny deciduous or evergreen trees or shrubs with large, simple leaves. Flowers solitary or in racemes; sepals and petals 5; stamens 15–30(–numerous); carpel 1. Fruit a drupe (often edible).

(a) Flowers in clusters of 1–3; fruit with a waxy bloom (a plum).

Prunus domestica CIRUELO, PLUM A shrub or tree to 8(12) m with dull brown bark and spiny branches in naturalised populations; *young twigs dull brown or grey*. Leaves 30–80 mm, *dull green and hairy, at least when young*. Flowers white, appearing with the leaves; petals 7–12 mm. Fruit a pendent, round, blue-black, red, green or yellow drupe 20–60 mm across (a plum). Planted. C, T, G, P.

---

1. *Rosa micrantha*
2. *Rosa micrantha* (fruits)
3. *Potentilla indica*
4. *Rubus ulmifolius*
5. *Rubus ulmifolius* (leaf underside)
6. *Rubus bollei* (pink-flowered form)
7. *Rubus bollei* (white-flowered form)
8. *Rubus bollei*
9. *Rubus bollei* (leaf underside)

(b) Flowers in clusters of 1–3; fruit characteristically velvety (a peach or almond).

Prunus persica MELOCOTONERO, PEACH A small tree to 6 m with familiar *velvety* (hairless in **var. nucipersica**; the nectarine) yellow, orange or red, *large* rounded fruits with solitary flowers that remain *mid-dark pink*; petals 10–20 mm. Cultivated for peaches. *Prunus armeniaca* DAMASCO, ALBARICOQUE has broader, almost round leaves on stalks to 40 mm long, flowers always pale *pink-white*; cultivated for apricots. Planted.

Prunus dulcis ALMENDRO, ALMOND A deciduous shrub or small tree to 8 m, densely branched and spiny. Leaves oblong to lanceolate, hairless and toothed, 40 mm–12 cm. Flowers appearing before the leaves, pink in bud, fading to white, borne in pairs; petals 15–20 mm. Fruit oval and laterally *strongly compressed*, grey-green to yellow and conspicuously *velvety*, becoming dry and splitting, 35–46(50) mm. Hillsides and cultivated. F, C, T, G, P.

(c) Flowers borne numerously in elongated clusters.

Prunus lusitanica subsp. *hixa* HIJA, PORTUGAL LAUREL A dense, evergreen shrub to 12 m with leathery, *shiny green* elliptic–oval leaves 60 mm–13 cm, with deep red young leaf stalks, and numerous whitish flowers borne in *long* erect racemes exceeding the leaves. Fruit oval, purplish black when ripe, 8–10 mm. Laurel forests. C, T, G, H, P.

Cydonia QUINCE

Small, spineless, deciduous trees with simple leaves. Flowers solitary; stamens 15–25; carpels 5. Fruit hard and pear-like.

Cydonia oblonga MEMBRILLO, QUINCE A shrub or small tree to 3(6) m with shoots hairy when young, later hairless. Leaves oval and entire, grey-hairy below. Flowers to 40–50 mm across, borne on short, hairy stalks; petals pink, exceeding the sepals. Fruit to 12 cm, many-seeded, yellow, pear-like and fragrant (a quince). T, H.

Pyrus PEAR

Small, spiny-branched deciduous trees with simple leaves. Flowers in clusters; stamens 20–30; carpels 2–5. Fruits large, fleshy and characteristically pear-like.

Pyrus communis PERAL, PEAR A tree to 10 m with an open crown. Leaves *hairy-margined, untoothed*. Flowers with white petals 13–15 mm. Fruit the cultivated pear, *large*, 50–60 mm across. Hills. C, T, P.

Eriobotrya LOQUAT

Small evergreen trees with large, leathery, entire leaves. Flowers borne in branched terminal clusters, petals 5; stamens 15–25; carpels 5. Fruits plum-like.

Eriobotrya japonica NÍSPERO, JAPANESE LOQUAT A dense, robust, small tree to 10 m. Young stems covered in red-brown felted hairs. Leaves elliptic, large, 10–30 cm long, dark green and shiny above, felted below; strongly veined. Flowers white, borne in terminal panicles 70 mm–17 cm long; petals 6–11 mm. Fruit ovoid, 30–60 mm, yellow when ripe with 2–3(7) large seeds, 10–20 mm. Various habitats, often near towns. L, C, T, G, P.

---

1. *Prunus persica*
2. *Prunus dulcis*
3. *Prunus lusitanica* subsp. *hixa*
4. *Cydonia oblonga*
5. *Sanguisorba megacarpa* (habit)
6. *Sanguisorba megacarpa*
7. *Bencomia caudata* (habit)
8. *Bencomia caudata*
9. *Bencomia caudata* (in fruit)

Sanguisorba

Herbs with pinnately divided leaves. Flowers inconspicuous, borne in dense, spherical, terminal heads; unisexual and cosexual mixed; petals absent; stamens 0–50. Fruit 1–3 achenes.

Sanguisorba megacarpa ALGÁFITA A small, leafy perennial. Leaves forming a basal rosette, pinnately divided with 9–12(25) pairs of elliptic, toothed leaflets. Flowers tiny, borne in egg-shaped heads 6–15 mm with the upper flowers female with reddish-pink styles, the lower male with yellow anthers; sepals bright green; petals absent. Fruit enclosed in a ridged receptacle. Grassy places and fallow land. L, F, C, T, G, H, P.

Bencomia ZUMAQUERO

Shrubs with pinnate leaves. Inflorescences unisexual; receptacle fleshy, enclosing the achenes.

Bencomia caudata A shrub to 3 m. Leaves with 7–11 ovate, toothed, short-stalked leaflets. Inflorescence *simple*. Fruits *pear-shaped* to sub-globose, 4–5 mm across. Forest cliffs and ravines; locally common. *C, T, P. *B. brachystachya* has narrower leaflets and erect, *branched* inflorescences. Mountain cliffs; rare. *C. *B. exstipulata* has stalkless leaflets and flattened fruits. Shaded crevices; very rare. *T, P. *B. sphaerocarpa* has leaves with many (13–25) leaflets and *globose* fruits. Forest cliffs; very rare. *H.

Marcetella

Dioecious shrubs with pinnate, hairless, glaucous leaves. Flowers borne in lax spikes; receptacle dry, compressed, 4-angled.

Marcetella moquiniana PALO DE SANGRE A shrub with upper stems bearing red, glandular hairs. Leaves borne in terminal rosettes; leaflets 7–15, crenate. Flowers borne in hanging spikes (male fewer). Fruits dry, winged, elm-like. Rare and local in craggy habitats. *C, T, G.

RHAMNACEAE | BUCKTHORN

Trees or shrubs with simple leaves with stipules. Flowers with 4–5 sepals and petals; petals small or absent, often hooded over the stamens; stamens 4–5. Fruit a fleshy black berry.

Rhamnus

Evergreen or deciduous shrubs with alternate, or almost opposite, toothed leaves and winter buds with scales.

Rhamnus alaternus MEDITERRANEAN BUCKTHORN An, erect, superficially holly-like evergreen shrub to 5 m, not spiny and more or less hairless. Leaves 10–60 x 10–35 mm, alternate, *leathery*, oval and shiny dark green, toothed or sparsely toothed, with 3–6 side veins. Flowers borne in dense, cylindrical racemes, yellowish, borne in small clusters in the leaf axils, with 5 sepals and lacking petals. Fruit a berry, red, ripening black, 4–6 mm with 2–3 seeds. Introduced. C.

---

1. *Bencomia brachystachya*
2. *Bencomia exstipulata*
3. *Marcetella moquiniana*
4. *Marcetella moquiniana* (female flowers)
5. *Marcetella moquiniana* (male flowers)
6. *Rhamnus glandulosa*
7. *Rhamnus glandulosa* (fruits)
8. *Rhamnus crenulata* (fruits)

Rhamnus glandulosa SANGUINO A small *tree* to 10 m. Leaves to 10 cm, ovate with *small, round glands in the vein axils*; margins *bluntly toothed*. Inflorescences exceeded by their leaves. Fruits black. Common in laurel woods. C, T, G, P. *R. integrifolia* MORALITO is similar but a smaller shrub with *entire* leaves *lacking glands*. Fruit almost 4-lobed, reddish-black. Midlands to sub-alpine cliffs. *T. *R. crenulata* ESPINO NEGRO is similar but with *toothed leaves* without glands. Inflorescences few-flowered. Hill scrub. *L, F, C, T, G, H, P.

ULMACEAE | ELM FAMILY

Trees with simple, toothed, deciduous leaves; leaves usually asymmetrical at the base. Flowers borne in small axillary clusters, cosexual or male, regular and inconspicuous; perianth bell-shaped, 4–5-lobed; stamens 4–5; styles 2. Fruit a 2-winged achene, notched at the tip.

Ulmus OLMO, ELM

Distinctive for having leaves asymmetrical at the base and unmistakable flowers and fruits. A complicated genus, mainly temperate.

Ulmus minor ELM A deciduous tree to 20 m with greyish bark and slender twigs, *white-downy* in their first year. Leaves alternate, oval-elliptic, bluntly toothed and *grey-hairy below*, with 12–16(18) pairs of veins, with scallop-toothed margins. Flowers borne in clusters before the emergence of the leaves; stamens purplish. Fruit a brown, broadly-winged nut 7–15(17) mm, notched at the tip. Rocky gorges and ravines. C, T, P.

MORACEAE | FIG FAMILY

A large, mainly tropical family of trees including the fig, mulberry and breadfruit. Flowers small and inconspicuous, crowded; male flowers with 4–5 stamens; styles 1–2. Fruit a mass of drupes surrounded by a succulent perianth.

Morus

Spineless trees with toothed to lobed leaves. Flowers unisexual, 4-parted, borne in short, dense spikes. Fruit fleshy.

Morus alba MORERA A smooth, slender tree. Leaves 60 mm–18 cm, oval, rounded or heart-shaped at the base, virtually hairless and *waxy to glossy*. Fruit compound and fleshy (a syncarp), 10–25 mm across, white, pink or purplish. Native to China; planted. C, P. *M. nigra* MORAL is similar but with rather thick, *matt* (not glossy) leaves. Widely planted. C, T, G, H, P.

Ficus HIGUERA, FIG

Trees with (normally) simple leaves. Flowers unisexual; male flowers with 3-parted perianth; female flowers with 5-parted perianth. Fruit succulent (a fig).

Ficus carica HIGUERA, FIG A semi-deciduous tree 4–5(10) m with greyish branches and large, palmately lobed leaves to 35 x 28 cm. Flowers minute, borne within a *syconium*. The aggregate fruit is a fig; pear-shaped and purplish when mature and edible, 50–80 mm. Widely planted and naturalised in various habitats including malpaís. L, F, C, T, G, H, P.

MORACEAE

Ficus elastica ÁRBOL DEL CAUCHO, RUBBER PLANT A shrub to tall tree 30–35 m with large, thick, leathery green or yellow-green leaves 10–35 cm long. Native to the tropics; commonly planted as an exotic ornamental in gardens, parks and on roadsides throughout.

Ficus benjamina Generally a short tree with a grey trunk, pointed, leathery leaves, and small, reddish fruits. Native to the tropics; planted as an ornamental on streets throughout.

1. *Morus nigra* (fruits)
2. *Ficus carica* (habit)
3. *Ficus carica*
4. *Ficus elastica*
5. *Ficus benjamina*

URTICACEAE | NETTLE FAMILY

Annual or perennial herbs with opposite or alternate, simple leaves. Flowers typically unisexual, greenish with perianth with 1 whorl of 4, without petals; male flowers with 4 stamens; female flowers with 1 style. Fruit an achene.

Urtica ORTIGA, NETTLES

Annual or perennial herbs with stinging hairs. Monoecious or dioecious; flowers usually inconspicuous and borne in dense clusters in the axils, forming elongated inflorescences; perianth of free tepals; stamens 4.

Urtica membranacea (syn. *U. dubia*) MEMBRANOUS NETTLE A bright green annual to 1.5 m. Leaves 20 mm–12 cm x 15–80 mm, with stalk of similar length to the blade. Racemes slender, typically equalling the leaves; lower racemes female, shorter than their stalks, upper racemes male and longer than their stalks, often coiling. Towns, ditches. F, C, T, G, H, P. *U. morifolia* ORTIGÓN is a similar, *woody-based perennial*. Leaves ovate, cordate at the base, coarsely toothed. Inflorescences slender, *exceeded* by the leaves. Laurel forests. C, T, G, H, P. *U. stachyoides* is an annual with *long, slender, green* inflorescences, far-exceeding the leaves. *C, T, H, P.

Urtica urens SMALL NETTLE A short annual 10–60 cm, with abundant stinging hairs but otherwise virtually hairless, similar to *U. membranacea* but with *very short, spreading clusters of flowers to 20 mm or less*, not in elongated racemes, the flowers upward-swept in fruit; male flowers few; female tepals hairy-margined and unequal (outer segments smallest). Waste and agricultural land. L, F, C, T, G, H, P.

Parietaria RATONERA MANSA, PELLITORY-OF-THE-WALL

Annuals or perennials with alternate leaves. Flowers borne in clusters in the leaf axils, mostly unisexual; perianth of equal tepals, eventually enclosing the fruit; stamens 3–4.

Parietaria judaica PELLITORY-OF-THE-WALL A short, tufted and *spreading, densely hairy perennial* with much-branched, reddish stems to 80 cm. Leaves oval, pointed, 10–50(70) mm. Plants monoecious; flowers borne in clusters of ≥ 3; perianth tubular, 3–3.5 mm. Achenes blackish. Walls and damp rocks. Common in towns. L, F, C, T, G, H, P. *P. mauritanica* is similar, but a sparsely hairy *annual* with oval, long-tapering pointed leaves, perianth 2–3 mm long when in fruit, and *olive or reddish* (not blackish) achenes. L, T, H. *P. debilis* is similar but with ovate leaves 10–30 mm long with stalks of similar length. Achenes olive green. L, F, C, T, G, H, P. *P. filamentosa* is a small *shrub* with ovate long-hairy leaves and stalkless lateral flower clusters. *T, G, P.

Forsskaolea

Small shrubs or perennials. Leaves alternate, toothed, spiny. Flowers unisexual; males with 1 stamen.

Forsskaolea angustifolia RATONERA A small subshrub with prickly-margined leaves, densely white-woolly beneath. Flowers borne in small lateral clusters, pinkish. Very common in dry, lowland areas. *L, F, C, T, G, H, P.

1. *Urtica morifolia*
2. *Urtica morifolia*
3. *Urtica morifolia* (inflorescence)
4. *Urtica stachyoides*
5. *Urtica urens*
6. *Parietaria judaica*
7. *Parietaria filamentosa*
8. *Forsskaolea angustifolia* (habit)
9. *Forsskaolea angustifolia*

Gesnouinia

Shrubs with unisexual flowers borne in panicles; male with 4 stamens, female with short styles.

Gesnouinia arborea ESTRELLADERA A small tree or shrub with entire, 3-veined, hairy leaves; stipules absent. Inflorescence a dense panicle. Seed an achene enclosed by the calyx. Laurel forests. *C, T, G, H, P.

FAGACEAE | OAK FAMILY

Evergreen or deciduous trees or shrubs with alternate, simple leaves. Monoecious; male flowers borne in catkins, stamens 4–20; female flowers 1–few with 3–9 styles. Fruit 1–3(6) nuts enclosed in cup-like cupule of fused scales.

Castanea SWEET CHESTNUT

Deciduous trees with erect catkins; male flowers with 10–20 stamens; female flowers 3 at the base of long catkins. Fruits 1–3 enclosed in spiny, splitting cupule.

1. *Gesnouinia arborea*
2. *Castanea sativa*
3. *Myrica faya*

Castanea sativa CASTAÑO, SWEET CHESTNUT A large deciduous tree to 35 m, trunk with grey-brown bark, often with spiralled fissures. Leaves 10–30 cm long, oblong-lanceolate, with pronounced veins and sharply toothed. Male flowers yellowish in long catkins (to 18 cm); female flowers few, on lower branches. Fruit a chestnut in a spiny, splitting husk. Ravine thickets and forest scrub. C, T, G, H, P.

Quercus OAK

Evergreen or deciduous trees or shrubs. Male flowers with 4–12 stamens; female flowers solitary or in clusters. Fruit a distinctive nut (acorn) within a cupule.

Quercus ilex ILEX ENCINA, HOLM OAK A large evergreen tree to 27 m with downy young branches and grey bark. Leaves 20–80(90) mm, leathery, oblong to lanceolate, untoothed or sometimes spiny-toothed, *downy*, with 7–15 pairs of *prominent veins beneath* when mature; stalks 3–10 mm. Fruit 15–35 mm, bitter-tasting, with dense scales on the cupule. Planted. L, C, T, G, P.

Quercus suber ALCORNOQUE, CORK OAK A large evergreen tree 10–15(25) m with *thick, corky and deeply ridged bark,* red beneath, and downy young branches. Leaves 25 mm–10 cm, oblong, dark green above and grey and downy beneath, toothed; midrib wavy (sinuous). Fruit 10–20 mm, ripening in the first year, the cupule with long and spreading scales. Planted. T, C.

MYRICACEAE

Dioecious trees or shrubs. Flowers borne in catkins; stamens 2–many; ovary superior.

Myrica

Leaves alternate, simple, aromatic. Fruit a berry.

Myrica faya (syn. *Morella faya*) FAYA An evergreen shrub or tree to 20 m with simple, alternate, leathery leaves 40 mm–12 cm with irregularly and bluntly toothed margins. Catkins branched, borne among the current year's growth. Fruit a reddish to blackish blackberry-like drupe. Very common in forest scrub; extinct in the easternmost islands. C, T, G, H, P.

JUGLANDACEAE

Trees with alternate, pinnately divided leaves. Flowers unisexual; the male in drooping catkins with 3–many stamens, the female in terminal spikes, each flower with 2 styles. Fruit a drupe or winged nut.

Juglans

Trees with pinnately divided leaves with 3–9 leaflets. Female flowers borne in racemes, erect in fruit. Fruit an unwinged nut.

Juglans regia NOGAL, WALNUT A large, deciduous tree to 24 m with pale, smooth bark, becoming fissured. Leaves alternate, pinnately divided with 7–9 elliptic, untoothed lobes to 15 cm. Male flowers borne in drooping catkins to 15 cm, female flowers in short, erect spikes. Fruit a large, edible nut >30 mm. Planted in rural areas. C, T.

CASUARINACEAE

Superficially conifer-like trees and shrubs native to Australia, Southeast Asia, Malesia, Papuasia, and the Pacific Islands, characterised by drooping horsetail-like twigs and cone-like fruiting structures.

Casuarina

Evergreen trees native to Australia. Leaves much reduced, borne on drooping greyish twigs. Flowers unisexual or cosexual, borne in catkin-like inflorescences.

Casuarina equisetifolia CASUARINA, PINO MARÍTIMO   A large, evergreen, *conifer-like* tree to 35 m with a straight trunk with light grey-brown bark and much-branched crown. Twigs needle-like, grey-green, 23–38 mm x 0.5–1 mm, jointed, with reduced, minute, tooth-like leaves in whorls of 2–8 per node. Flowers unisexual; male flowers borne in elongated spikes 7–40 mm; female flowers borne on short, cone-shaped spikes 10–24 mm. Fruit solitary, grey-brown, nut-like, borne in a woody, cone-like structure. Native to Australia; planted. F, C, T.

CUCURBITACEAE | CUCUMBER FAMILY

Herbs, often climbing or trailing with tendrils, and alternate leaves. Flowers unisexual, sepals and petals 5; stamens usually 3; stigmas 2–3 on 1 style. Fruit succulent and berry- or pod-like.

Citrullus

Prostrate herbs with *branched* tendrils. Flowers solitary; corolla 5-parted beyond the middle, bell-shaped. Fruit fleshy or dry.

Citrullus colocynthis COHOMBRILLO, DESERT GOURD   A bristly, grey, spreading perennial with branched tendrils. Leaves triangular in outline, deeply 3–7-lobed, the lobes again incised. Flowers solitary, bright yellow. Fruit *melon-like*, pale green with dark green mottled stripes, later all yellow; about the size of a grapefruit. Seeds smooth, without markings. Damp flats and ravine bottoms in otherwise arid places. L, F, C, T, P.

Ecballium

Herbaceous, bristly monoecious perennials without tendrils. Corolla yellowish; stamens 5. Fruit an explosive pod.

Ecballium elaterium PEPINILLO DEL DIABLO, SQUIRTING CUCUMBER   A spreading, very bristly perennial to 1.5 m without tendrils, with a tuberous rootstock. Leaves rough, with bristles, more or less triangular and long-stalked (to 13 cm). Flowers small, 20–50 mm across and pale yellow; male and female flowers borne separately. Fruit bristly and pod-like, 30–45 mm borne on long stalks, *exploding violently* from the point of attachment when ripe. Uncommon. C, T.

1. *Casuarina equisetifolia*
2. *Citrullus colocynthis* (immature fruit)
3. *Citrullus colocynthis* (mature fruit)
4. *Ecballium elaterium*
5. *Bryonia verrucosa*
6. *Gymnosporia cryptopetala*
7. *Gymnosporia cryptopetala* (fruit)
8. *Gymnosporia cassinoides*

Bryonia BRYONY

Climbing dioecious perennials with long, spiralling, unbranched tendrils. Flowers greenish white; stamens 5. Fruit a red berry.

Bryonia verrucosa VENENILLO A climber with large 5-angled to lobed leaves. Flowers 10 mm, greenish-white. Fruits greenish with pale stripes, later yellow-orange. Locally common in lower forest zones. *L, C, T, G, H, P.

CELASTRACEAE | SPINDLE FAMILY

Shrubs or trees (or woody climbers) with simple leaves. Flowers cosexual or unisexual with a nectar disk, with 4–5 sepals, petals and stamens; style 1. Fruit variable, often a succulent, 3–5-angled capsule with seeds that have a bright orange-red aril.

Gymnosporia

Shrubs and trees, often spiny, with leaves alternate or in fascicles. Flowers unisexual. Seed with an aril.

Gymnosporia cassinoides PERALILLO A small tree to 4 m with leathery, glossy, alternate ovate leaves with irregularly toothed margins. Fruit a 3-angled, pale greenish to brown capsule, splitting; seeds with a white aril forming a basal cup. Forests. *F, C, T, G, H, P.

Gymnosporia cryptopetala ARTISCO A stout, grey-stemmed, spiny shrub to 1.5 m with leathery, stalkless obovate leaves; leaves bunched, often few. Flowers small and inconspicuous. Fruit a fleshy, pink to red-tinged capsule. Seeds with a red-orange basal cup. Dry malpaís and cliffs. *L, F.

OXALIDACEAE | OXALIS FAMILY

Perennial herbs, often with a bulbous stock or rhizomes, with clover-like leaves. Flowers with 5 petals and sepals; stamens 10; styles 5. Fruit a capsule.

Oxalis TRÉBOL, TREBINA

Perennial herbs, often on disturbed ground, with clover-like leaves with 3 leaflets (ternate). Flowers regular with 5 petals and 5 sepals.

Oxalis pes-caprae TREBINA, BERMUDA BUTTERCUP A low, tufted perennial with numerous leaves arising from a bulbous stock; far spreading; leaves withering soon after flowering. Leaves trifoliate and clover-like. *Flowers bright yellow, tubular* with petals 13–26 mm, borne in loose umbels on long stalks to 30 cm. An abundant and highly invasive weed on disturbed land. L, F, C, T, G, H, P. *O. corniculata* PROCUMBENT YELLOW SORREL has yellow flowers but with *stems rooting at the nodes* and (often) dull-purple leaves. Petals 5–9 mm long. Cultivated land. L, F, C, T, G, H, P.

Oxalis latifolia A tufted perennial with bulblets formed at the end of short rhizomes and long-stalked, all basal, clover-like leaves with heart-shaped leaflets. *Flowers pink,* borne in broad, umbel-like clusters; on stalks to 35 cm; petals slightly hairy, 9–13 mm long. Waste ground and gardens. F, C, T, P.

HYPERICACEAE

Shrubs or herbs, often with numerous glands and simple, opposite or whorled leaves. Flowers regular, yellow, with 5 free petals and sepals; stamens numerous; styles 3–5. Fruit a capsule, or succulent and berry-like.

Hypericum

Herbs and shrubs easily recognised by their opposite leaves and flowers with 5 yellow petals and many stamens.

(a) Plants without glands along the leaves and sepals.

Hypericum grandifolium MALFURADA A hairless shrub to 1–2 m with reddish-brown stems. Leaves broadly ovate, virtually stalkless, 40–70 mm, without glandular margins. Inflorescences lax, 2–4-flowered; flowers large, 45 mm across; petals lanceolate; sepals with glandular margins. Fruit a dark brown capsule. Laurel and pine forests. C, T, G, H, P.

1. *Oxalis pes-caprae*
2. *Hypericum grandifolium*
3. *Hypericum glandulosum*
4. *Hypericum reflexum*

Hypericum canariense GRANADILLO A hairless shrub to 2.5 m with linear–lanceolate to narrowly elliptic leaves 20–70 mm long. Flowers yellow, 20 mm across, *borne numerously* in large, dense panicles. Sepals ovate, fused, without glandular margins. Fruit hard and brown. C, T, G, H, P.

(b) Plants with glandular margins to the leaves and/or sepals.

Hypericum glandulosum MALFURADA DE MONTE A small shrub to 1 m with hairless or hairy stems. Leaves ovate to elliptical, attenuated into short stalks, 40–50 mm with glandular margins. Inflorescences dense, many-flowered. Frequent on forest cliffs in the west. F, C, T, G, P. *H. reflexum* CRUZADILLA is similar but with leaves in an overlapping formation, stalkless and sub-clasping at the base. Lower basalt cliffs. *C, T, G.

Hypericum perforatum HIERBA DE SAN JUAN, PERFOLIATE ST JOHN'S-WORT An erect *perennial* to 50(80) cm with a cylindrical stem with *2 lines running down its length*. Leaves oval to linear, to 30 mm long, *scarcely stalked*, hairless and blunt with numerous blackish glands. Flowers bright yellow, with petals 9–15 mm, often with black dots along the edges; sepal margins entire. Thickets. T, G, P.

Hypericum perfoliatum An erect to spreading, normally hairless perennial to 70(80) cm with stems with 2 lines. Leaves blue-green, opposite, lanceolate, *clasping the stem at the base*, without wavy margins, 8–50 mm long; glands pale (some blackish). Flowers yellow with petals 8–12 mm and blunt sepals with black markings. Fruit 8–10 mm with raised orange warts. Thickets. C.

VIOLACEAE | VIOLET FAMILY

Herbs and shrubs. Flowers zygomorphic, solitary; sepals 5, separate; petals 5, the lowermost forming a lip, and extended behind into a spur; stamens 5; style 1. Fruit a 3-valved capsule.

Viola VIOLETA, VIOLETS

Herbs or shrubs with alternate stalked leaves with stipules at the base. Flowers with 5 sepals and 5 petals; stamens 5; carpels 3. Fruit a capsule. Early spring-flowering.

(a) Annual or perennial herbs. Stipules fringed with narrow teeth. Flowers typically mostly white or blue-violet, with *downward-pointing* lateral petals.

Viola odorata SWEET VIOLET A small perennial 80 mm–20 cm with *creeping, rooting stems above ground*, rounded, blunt leaves, and stipules broad and short-fringed. Flowers usually dark violet-purple (rarely white) and fragrant; spur 4–5(7) mm. Forests. C, T, G, P.

Viola riviniana A rosette-forming perennial with side shoots, to 15 cm. Leaves 20–40 mm; stipules with equal or shorter teeth. Sepal appendages *conspicuous*, 2–3 mm, concealing the top of the flower stalk; spur stout, blunt and whitish. Forests. C, T, G, P.

(b) Annual or perennial herbs. Stipules often leafy and divided but *not fringed*. Flowers white, yellow, violet or multi-coloured, with *spreading to upward-pointing* lateral petals.

1. *Viola riviniana*
2. *Viola anagae*
3. *Viola cheiranthifolia* (habit) (RACHEL GRAHAM)
4. *Viola cheiranthifolia* (RACHEL GRAHAM)
5. *Viola palmensis* (OLIVER WHITE)
6. *Salix canariensis*
7. *Salix canariensis* (inflorescence)

Viola arvensis An erect, branched annual to 20(40) cm with short, deflexed hairs. Leaves 20–50 mm, oblong–spathulate with wavy margins; stipules as long as the leaves and coarsely lobed with a terminal leaf-like segment. Flowers 10–15 mm across, the lower petal cream to yellow, the others cream to bluish-violet; sepals lanceolate, equalling or exceeding the petals. Spur equalling calyx. T, G, P.

Viola kitaibeliana A low, bristly-hairy annual 40 mm–25 cm, with tufted leaves, oval below, narrower above, and somewhat toothed. Stipules pinnately divided with a large terminal lobe. Flowers violet or white *with a yellowish centre* and with darker veins; spur short, 1.5–3 mm long. Grassy habitats. T.

Viola anagae VIOLETA DE ANAGA A small, rhizomatous perennial herb with sub-circular, small-toothed leaves. Flowers blue-violet with darker veins and a white spur; lateral sepals upward-swept with prominent white bristles. Capsules hairless. Anaga; rare. *T.

Viola cheiranthifolia VIOLETA DEL TEIDE A small, rhizomatous perennial herb with densely hairy ovate to spathulate, toothed or entire leaves. Flowers tricoloured with a short spur, predominantly mauve with white and yellow markings. Teide-Pico Viejo. *T. ***V. guaxarensis*** is similar but larger, with a glabrous flower spur. Las Cañadas. *T. ***V. palmensis*** PENSAMIENTO DE LA CUMBRE is similar but more robust with larger, pale mauve and yellow flowers with a longer, slender spur. High mountain slopes; rare. *P.

SALICACEAE | WILLOW FAMILY

Deciduous trees and shrubs usually with alternate, simple, toothed leaves. Flowers reduced, borne in catkins; male flowers with 1–many stamens; female flowers with 1(–2) short styles. Fruit a 2-valved capsule.

Populus CHOPO, ÁLAMO

Deciduous trees with buds that have unequal scales and oval or triangular, entire, toothed to lobed leaves. Flowers appear before the leaves in stalked catkins, each flower with a stalked, cup-shaped disk and subtended by a bract; stamens 4–many. Capsule 2–4-valved. Seeds numerous. Many non-native species cultivated.

Populus nigra A tree to 30 m with *a trunk that has prominent bosses* and a broad, uneven crown and cylindrical yellowish twigs, greyish when mature. Leaves 50 mm–10 cm, *diamond-shaped, long pointed*, with *minutely wavy-toothed* margins; leaves on the shorter shoots smaller and broader. Stamens 20–30. Fruiting catkins 10–15 cm long. Capsule 2-valved. Planted. C. ***P. alba*** is similar but with *irregularly 5-lobed leaves*, white beneath. Planted. F, C, T, G, H.

Salix

Trees or shrubs with winter buds that have 1 outer scale. Flowers borne before or after the leaves. Stamens 1–5(12). A complex group: hybrids are common and it is impossible to see all traits at any one time of year. There are many more species in cooler parts, and non-native cultivated taxa are not included here.

Salix canariensis SAUCE CANARIO, SAO A tall shrub or tree to 10 m. Leaves oblong to lanceolate, entire or crenate, hairy beneath. Catkins to 60 mm long. Laurel woods, streamsides and other damp places. Locally common; absent from the east. C, T, G, H, P.

EUPHORBIACEAE | EUPHORBIA FAMILY

A large family of herbs, shrubs, trees and lianas, often with a white latex. Monoecious or dioecious; perianth absent or 3(5)-lobed; male flowers with 1–many stamens; female flowers with 2–3 styles. Fruit a capsule; seed with a caruncle (a fleshy structure attached to the seed).

Euphorbia SPURGE

Leaves normally alternate and entire, with a sticky, milky latex. Flowers in small groups surrounded by a cup-shaped structure with 4–5 round or crescent-shaped glands, a solitary female flower and male flowers with 1–several stamens, the whole structure forming a cyathium. Often difficult to identify in the field.

(a) Trees to 15 m with terminal panicles, in laurel forests (Tenerife).

Euphorbia mellifera ADELFA DE MONTE A small, lax tree to 15 m with smooth, grey bark. Leaves crowded around the branches, thin but leathery, narrowly lanceolate, dark green with pale midrib. Flowers borne in dense panicles. Capsules large; seeds with a plate-like caruncle. Laurel forests; very rare, almost extinct as a wild plant. T, G, P.

(b) Cactus-like shrubs with spines, on rocky slopes (or planted).

Euphorbia canariensis CARDÓN A robust, succulent shrub to 2 m with square to 5-sided, *cactus-like* stems; spines paired, curved. Flowers greenish-red, solitary on short stalks. Capsules reddish-brown. Locally frequent in coastal regions in the west; rare and local in the east. Also widely planted. *F, C, T, G, H, P.

Euphorbia handiensis CARDÓN DE JANDÍA A small, mounded, succulent shrub to 80 cm, rather densely branched and *conspicuously spiny* (cactus-like); spines 20–30 mm. Flowers reddish. Capsules reddish-brown. A rare and endangered species confined to a few coastal rocky slopes. *F (Jandía).

Euphorbia ingens CANDELABRA TREE A large, imposing, cactus-like shrub to 9 m. Stems bright green and 4-angled with irregular, spiny margins. Native to South Africa, occasionally planted in landscaped areas or cultivated in parks and gardens throughout. *E. neriifolia* is somewhat similar in stature but with *large leaves* crowded around the upper branches. Widely planted.

Euphorbia grandicornis has long, robust, *3-branched* spines. Planted in gardens.

Euphorbia trigona has a compact habit with organ pipe-like stems, congested above. Planted in gardens.

(c) Spineless shrubs with leafless stems.

Euphorbia aphylla TOLDA A small, compact shrub to 50 cm. Stems slender, pencil-like, *leafless*. Flowers virtually stalkless, in small clusters around the stem tips. Fruits very small, brown to reddish; seeds small, brown with a plate-like caruncle. Very locally frequent on coastal rocky slopes in the west. *C, T, G.

Euphorbia tirucalli A shrub with dense, succulent, pencil-like stems. Planted as a hedge and in gardens on all the islands.

1. *Euphorbia mellifera* (habit) (RACHEL GRAHAM)
2. *Euphorbia mellifera*
3. *Euphorbia mellifera* (fruit)
4. *Euphorbia canariensis*
5. *Euphorbia canariensis* (in fruit)
6. *Euphorbia ingens*
7. *Euphorbia handiensis*
8. *Euphorbia neriifolia*
9. *Euphorbia grandicornis*
10. *Euphorbia trigona*
11. *Euphorbia aphylla*
12. *Euphorbia tirucalli*

(d) Spineless shrubs with at least some terminal leaves.

Euphorbia balsamifera TABAIBA DULCE A rounded shrub to 2 m with pale grey, often contorted stems. Leaves pale blue-green, oblong–spathulate. *Flowerheads solitary*, short-stalked to stalkless; floral glands oval to rounded, entire. Capsules solitary, globose; seeds brown, wrinkled, without a caruncle. Common to dominant on malpaís and lower slopes throughout. L, F, C, T, G, H, P.

Euphorbia regis-jubae HIGUERILLA, TABAIBA AMARGA A robust, much-branched, rather rounded shrub to 1.2(3) m with thick, succulent, spineless stems. Leaves 15 mm–10 cm long, light yellow-green and linear. Rays 4–10; bracts 7–15 mm, *bright pale yellow*; bracts <10 mm, free to the base and *persistent* (falling just before the fruits mature). Capsules 5.2–7.3 mm. Common to dominant on malpaís in much of the east. L, F, C.

Euphorbia atropurpurea TABAIBA MAJORERA A shrub with succulent brown stems with leaves *crowded towards the tips*; leaves glaucous, stalked, oblong–spathulate, blunt. Flowers *dark purple* (rarely yellow), 5–15-rayed. Capsules dark red-brown. Seeds wrinkled; caruncle stalked. Very locally common on lower slopes. *T. *E. bravoana* is similar but with leaves linear–lanceolate, and bracts <5 mm across. *G.

Euphorbia bourgaeana A shrub to 1.5 m with light brown stems. Leaves lanceolate to oblanceolate. Inflorescence a simple (occasionally compound), greenish-yellow umbel; floral bracts large, fused at the base; glands crescent-shaped. Capsules brown or sometimes reddish. Ravines and slopes; rare and local. *T. *E. lambii* is similar but taller with more slender branches and narrowly lanceolate leaves; *floral bracts large, fused over ½ their length; glands toothed*. Capsules light brown or yellowish. Recent authors consider them conspecifics. Cliffs in laurel forest. *G.

Euphorbia serrata A greyish or bluish perennial 20–60 cm tall with a woody stock. Leaves narrow-oblong with finely *toothed margins*, 20–70 mm long. Umbels with 3–5 rays, branched x 1–3(5) with lanceolate to rounded, *yellow* bracts. Glands oval-shaped, squared at 1 end. Seeds *pitted*. L, F, C, T.

Euphorbia lamarckii TABAIBA AMARGA, HIGUERILLA A shrub to 2 m with light brown stems. Leaves narrowly oblong. Inflorescence an umbel, simple (rarely compound) with 5–8 rays; floral bracts greenish-yellow, <10 mm, *persistent in fruit, free at the base*; floral glands crescent-shaped. Capsules light brown to red; seeds with a stalked caruncle. Common on lower slopes. *T, G, H, P. *E. berthelotii* is similar but smaller and stockier, densely branched with grey-red stems. *G.

(e) Erect or ascending annuals to perennials.

Euphorbia segetalis TABAIBILLA, HIGUERILLA A hairless annual or perennial, simple or with some branches at the base, to 80 cm (1 m). Leaves alternate and *narrow* (2–4 mm), linear to linear–lanceolate, 27–40 mm long. Rays 4–6(8), *bracts diamond-shaped* and yellow-green, glands with 2–4 horns. Capsule rough and glandular; seeds pale grey. Open habitats. L, T, P.

Euphorbia paralias HIGUERILLA DE PLAYA, SEA SPURGE A clump-forming, hairless, stiffly erect, fleshy perennial to 60(80) cm. Leaves grey-green, regularly and closely set around the stem, *overlapping*, oval, *broadest towards the base* and concave above; *midrib obscure below*, 8–20 mm long. Umbels with 3–6 rays, bracts oval and concave. Glands kidney-shaped with long horns. Capsule rough along the back, 4–5.5 mm. Seeds pale grey and *smooth*. Coastal dunes. L, F, C, T, G.

1. *Euphorbia balsamifera* (habit)
2. *Euphorbia balsamifera* (fruit)
3. *Euphorbia balsamifera* (cyathium)
4. *Euphorbia balsamifera* (cyathium)
5. *Euphorbia regis-jubae* (habit)
6. *Euphorbia regis-jubae*
7. *Euphorbia atropurpurea*
8. *Euphorbia atropurpurea* (RACHAEL GRAHAM)
9. *Euphorbia atropurpurea* forma *lutea*

Euphorbia terracina SANALOTODO A hairless, *succulent* perennial 40 mm–90 cm with erect to ascending stems and non-flowering lateral branches. Leaves oblong to linear–lanceolate, minutely toothed, regularly and closely set around the stem and overlapping but *flat*, 4–60 mm long. Umbels with 2–5(6) rays with as many oblong to diamond-shaped, green bracts. Glands with 2, long, slender horns. Capsule smooth; seeds pale grey and smooth with a boat-shaped, fleshy structure (a caruncle) attached. Open habitats. L, F, C, T, G, H, P.

Euphorbia helioscopia SUN SPURGE A short, erect, hairless annual, normally with a single stem, to 30(50) cm. Leaves oval or spoon-shaped, broadest above the middle and *toothed* in the upper ½, 4–35(60) mm long. *Umbel 5-rayed* (small plants with 2–3 rays), *with 5 distinctive bracts* at the base, yellowish and similar in shape to the leaves. Glands oval and untoothed. Capsule smooth and unwinged. Seeds brown and wrinkled. A common ephemeral on bare soil. L, F, C, T, G, H, P.

Euphorbia peplus PETTY SPURGE An erect, hairless annual 20 mm–20(40) cm, branched at the base. Leaves green, oval to rounded, untoothed and short-stalked, 4–30 mm. Umbels with *3 main rays* (rarely 2–5), with 3 triangular–oval to spoon-shaped, *green* and unstalked bracts. Glands kidney-shaped with long, slender horns. Capsule smooth but with *2 winged keels*; seeds pale grey and *pitted*. Disturbed ground. L, F, C, T, G, H, P.

Euphorbia exigua DWARF SPURGE A very small, hairless, grey-green annual, 20 mm–20(30) cm, branched from the base. Leaves *very narrowly* lanceolate, untoothed and unstalked, 2–25 mm long. Rays 2–5(7), branched x 1–3(5), with narrowly triangular bracts. Glands crescent-shaped with 2 horns. Capsule shallowly grooved, 1.5 mm. Seeds wrinkled and grey. Very common on cultivated, grassy and fallow habitats. L, F, C, T, P.

(f) Spreading prostrate annuals (*Chamaesyce* group: often treated as a separate genus).

Euphorbia prostrata (syn. ***Chamaesyce prostrata***) CHIRRIGÜELA POSTRADA A spreading herb, similar to *C. peplis* but often hairy, and with *up to 10 branches at the base* and stems (which are hairy above), *leaves slightly serrated on the margins,* and *capsules hairy on the keels*. A North American weed naturalised locally in cobblestone streets and ruderal habitats. L, F, C, T, G, H, P. ***E. maculata*** (syn. ***Chamaesyce maculata***) is similar, with up to 8 branches and *leaves often with a dark central spot* and with a capsule either virtually hairless, or more often *entirely covered with closely adpressed hairs*. A North American weed naturalised in ruderal places. C, T, P. ***E. chamaesyce*** (syn. ***Chamaesyce canescens***) is similar but with up to 25 branches, leaves often without a dark spot, and with capsules with *spreading* (not adpressed) hairs. L, F, C, T, G, P.

Euphorbia peplis (syn. ***Chamaesyce peplis***) PURPLE SPURGE A *prostrate*, hairless annual with *4* (sometimes 3 or 5) main branches at the base. Stems red or purple, *leaves fleshy*, grey-green and small to 11 mm, opposite, oblong, with a single rounded lobe at the base. Flowers tiny, greenish with semi-circular red-brown glands, borne laterally or in clusters but not in umbels. Capsule nearly smooth and purplish. Shingle and sand. L, F, C, T, P. ***E. serpens* (syn. *Chamaesyce serpens*)** is very similar (often confused with *E. peplis*) but with up to 16 branches and leaves symmetrical rather than with rounded lobes at the base. Native to tropical America; introduced with ornamental plants, now a weed in gardens and ruderal places. L, F, C, T, G, H.

1. *Euphorbia bravoana*
2. *Euphorbia bravoana* (in fruit)
3. *Euphorbia lambii*
4. *Euphorbia lamarckii*
5. *Euphorbia lamarckii*
6. *Euphorbia berthelotii*
7. *Euphorbia segetalis*

Mercurialis MERCURY

Annual or perennial herbs (or shrubs) with opposite leaves. Flowers unisexual, green and inconspicuous; sepals 3; stamens 8–25; carpels 2(3–4). Fruit a 2(3–4)-parted capsule.

Mercurialis annua ORTIGA MANSA, ANNUAL MERCURY A dioecious, branched, erect *short annual* 30(–50) cm, more or less hairless. Leaves 20–70 mm long, opposite, oval to elliptic, toothed and long-stalked, shiny *light* green, with minute hairs along the margins (<0.4 mm); stipules *small*, just 1.5–3.5 mm. Male flowers borne on *unbranched greenish spikes without bracts*, female flowers few, borne in lateral clusters. Fruit 2.4–2.6 mm, 2-lobed and bristly. A common weed on disturbed ground. L, F, C, T, G, H, P. *M. canariensis* is a similar but *larger* understorey herb to 1 m with long stipules (4–8 mm); male flowers borne on *branched* spikes *with* bracts. Local. *L, T (north), H, P.

Ricinus CASTOR OIL PLANT

Shrubs without a milky latex, with *palmately lobed leaves*. Flowers unisexual; perianth with 3–5 tepals; *stamens conspicuous in fascicles* (branched); ovary 3-parted. Fruit a capsule.

Ricinus communis TÁRTAGO, CASTOR OIL PLANT A robust annual or shrub to 5(7) m, flushed red, bronze or purple. Leaves shiny, large, 10–36(60) cm across and palmate, with 5–9 coarsely toothed lobes. Flowers borne in large terminal panicles with the male below with yellowish stamens, and the female above and with bright red stigmas. Fruit a 3-parted, spiny capsule 18–20 mm across; seeds bean-like, 10–15 mm long. Native to tropical Africa; naturalised. L, F, C, T, G, H, P.

LINACEAE | FLAX FAMILY

Annuals or perennials with simple, opposite or alternate leaves. Flowers in branched inflorescences; sepals and petals 4–5, free; stamens 4; styles 4–5. Fruit a 8–10-valved capsule.

Linum FLAX

Hairless annual or perennial herbs or shrubs. Flowers 5-parted with white or blue petals. Capsule 10-valved, often short-beaked.

Linum strictum LINO SILVESTRE, UPRIGHT YELLOW FLAX A short, erect annual 10–45 cm with narrowly lanceolate leaves with inrolled, very rough margins. Flowers small with yellow petals 6–12 mm long that exceed the long-pointed sepals, borne in branched, spreading clusters or short lateral clusters in a rigid, open, corymb-like inflorescence. Coastal sands and other dry places. L, F, C, T, G, H, P.

Linum bienne LINO BRAVO, PALE FLAX An annual or perennial herb with slender, erect to spreading stems, often branched below (not in small specimens), 10–60 cm. Leaves alternate, linear and long-pointed, mostly 3-veined, 0.5–1.5 mm wide. Flowers *pale blue*, borne on slender stalks in loose clusters or singly, petals 8–12 mm, exceeding the oval, long-pointed, papery-margined sepals (4–6 mm long). Capsule 4–6 mm. C, T, G, H, P. *L. usitatissimum* LINO, CULTIVATED FLAX is a similar, *usually an unbranched annual* to 85 cm with *larger, darker blue* or white flowers with petals 12–20 mm long. Capsule 6–9 mm. Disturbed habitats; probably introduced. L, C, T, G, P.

---

1. *Euphorbia paralias*
2. *Euphorbia terracina*
3. *Euphorbia serpens*
4. *Mercurialis annua* (inset: detail)
5. *Ricinus communis* (flowers)
6. *Ricinis communis* (fruits)
7. *Linum strictum*

Radiola

Annuals with opposite leaves. Flowers 4-parted with white petals. Capsule 8-valved.

Radiola linoides LINILLO A much-branched, *extremely slender* and small annual to 10 cm with 1-veined leaves. Flowers tiny with 4 sepals and petals to 1 mm. Capsule 0.7–1 mm. Open, sandy ground. C, T, H, P.

GERANIACEAE GERANIUM FAMILY

Herbs with alternate palmately or pinnately lobed leaves. Flowers borne in cymes, umbels, or solitary, usually more or less regular with 5 sepals and 5 petals; stamens (3)5 or 10; style 1. Fruit with 5, 1-seeded portions united into a prominent beak.

Geranium

Annuals or perennials with simple, palmately lobed leaves. Flowers regular with 10 stamens; style with 5 branches. Fruit beaked.

Geranium molle PATAGALLO BLANDO, DOVE'S-FOOT CRANE'S-BILL A short, sprawling annual to 40 cm with stems branched from the base, *grey-green and softly hairy*. Basal leaves long-stalked, rounded or kidney-shaped, divided into 5–7(9) wedge-shaped, 3-lobed segments (½–⅔ to the base); upper leaves more deeply divided and short-stalked or unstalked. Flowers pink-purple; petals 4–6 mm, *deeply notched*, borne in lax clusters; outer stamens lacking anthers; flower stalks with short and long hairs. Grassy habitats. L, F, C, T, G, H, P. *G. rotundifolium* PATAGALLO REDONDO is similar but with leaves shallowly 5–9-lobed (<½ to the base) and bright pink flowers with *unnotched or slightly notched* petals, rounded at the tips, 5–7 mm long. Similar habitats. L, F, C, T, G, H, P.

Geranium dissectum PATAGALLO CORTADO, CUT-LEAVED CRANE'S-BILL A spreading, hairy annual to 60 cm with ascending flowering stems. Leaves circular in outline but *deeply dissected into 5–7 lobes almost to the base,* with sub-lobes. Flowers bright pink with shallowly notched petals 4.5–6 mm; *flower stalks <15 mm long*; sepals spreading and with pointed tips. Fruit ridged and hairy. Common in a range of habitats, especially damp, disturbed, grassy places. C, T, G, H, P.

Geranium robertianum HERB ROBERT A hairy, *very aromatic* annual or biennial to 50 cm; usually *strongly flushed with red* or purple. Leaves palmate, the lower leaves with 3–5 pinnately lobed segments. Flowers pink (sometimes white); petals slightly notched or rounded, 8–10(14) mm long; pollen orange. Fruit hairy and ridged. Cool, damp shady places. C, T, G, H, P. *G. purpureum* PATAGALLO PÚRPURA, LITTLE ROBIN is similar but less flushed with red, flowers purplish-pink and smaller; petals 5–9 mm; pollen yellow. Similar habitats. L, C, T, G, H, P.

Geranium reuteri (syn. *G. canariense*) PATAGALLO CANARIO A robust perennial with a woody stock or short stem and leaves arranged in a large rosette; leaves broadly ovate, deeply dissected, the end-middle lobe stalkless. Flowers pink, borne few in branched inflorescences; flowers 20–30 mm across; anthers reddish. Frequent in laurel forests. *C, T, G, H, P.

---

1. *Geranium rotundifolium*
2. *Geranium rotundifolium* (leaf)
3. *Geranium purpureum*
4. *Geranium purpureum* (fruits)
5. *Geranium reuteri*
6. *Erodium malacoides*
7. *Erodium malacoides* (leaf)
8. *Erodium hesperium*
9. *Erodium cicutarium*

Erodium ALFINELEJO, ALFILERILLO, STORK'S BILL

Annuals or perennials like *Geranium* but generally with pinnately lobed leaves and often slightly zygomorphic flowers with 2 petals larger than the other 3; stamens 5. Fruit beaked; beak length important. Mericarps important to observe in closely related species, for which a hand lens is required.

(a) Some (or all) leaves *shallowly lobed* (in at least some leaves to <½ the width of the blade).

Erodium chium THREE-LOBED STORK'S-BILL A robust, hairy perennial or biennial to 40 cm. Leaves oval, lowermost divided into 3 toothed, blunt lobes. Flowers pink-purple, 10–18 mm across, borne in 2–8-flowered clusters on non-glandular flower stalks; 2 petals slightly larger than the remaining 3; sepals with hairs *not glandular*. Fruit with short white hairs, the beak 20–40 mm long. Grassy and rocky places. L, F, C, T, G, H, P.

Erodium malacoides MALLOW-LEAVED STORK'S-BILL An erect to sprawling, glandular-hairy biennial to 40 cm. Leaves oblong, heart-shaped at the base, those below toothed, sometimes 3–several-lobed, covered in shiny glands. Flowers purplish pink, 11–18 mm across, borne in 3–7-flowered clusters with at least 3 bracts at the base, borne on glandular-hairy stalks; *sepals with glandular hairs*. Fruit beak 22–30 mm. Mericarp with a pit (foveola) with *glandular* hairs. Very common in towns and on disturbed ground. L, F, C, T, G, H, P. The following are similar and difficult to differentiate: ***E. laciniatum*** has leaves more deeply cut (pinnatisect) and not markedly glandular-hairy beneath, and flowers in clusters of 4–9 with just 2 bracts at the base. Fruit beak *longer*, 45–70 mm. Mericarp with a *pit with hairs along the border*. Coastal sands. L, F, C, T, G, H. ***E. hesperium*** has lower leaves with broad lobes and upper leaves with narrow segments. Fruit beak 40–60 mm; mericarp pit *without* a groove. L, F. ***E. neuradifolium*** also has upper leaves that are more deeply cut but with fruit beak *shorter,* to just 28 mm. Mericarp with a *non-glandular* pit. L, F, C, T.

(b) Mature leaves 1–2-*pinnately divided* into discrete leaflets.

Erodium cicutarium COMMON STORK'S-BILL A variable, erect or prostrate, hairy (sometimes sticky), annual to 60 cm. Leaves *deeply pinnately divided* without smaller lobes between the larger ones. *Stipules pointed* and whitish. Flowers purplish, pink or white, (7)10–18 mm across with 3–7(12) in a cluster; petals 4–9 mm, the upper 2 petals normally larger and with a blackish patch; sepals with inconspicuous netted veins and a bristly point. Bracts brownish. Fruit hairy, with a beak 15–40 mm. Mericarp with a pit containing a groove, all one colour, dark. Common in open and disturbed habitats. L, F, C, T, G, H, P. ***E. touchyanum*** is a similar, densely hairy desert annual, flowering after rain. Sepals with *conspicuously* netted veins, the point *without* bristles; petals 5–7 mm. Dunes, malpaís. L, F. ***E. salzmannii*** has mericarps with a pit containing a *whitish* groove, and erect-spreading hairs and blackish tubercles. L, H.

Erodium moschatum MUSK STORK'S-BILL A spreading annual to 60 cm, similar to *E. cicutarium* but always stickily hairy and smelling faintly of musk, leaflets only shallowly lobed (<½ to the midrib). *Stipules blunt*. Flowers larger, to 28 mm across, violet or pinkish purple. Fruit with beak 20–45 mm long. Cultivated and waste ground. L, F, C, T, G, H, P.

Erodium botrys MEDITERRANEAN STORK'S-BILL A short, hairy annual to 50 cm with an obvious stem above ground. Leaves bristly, to 50 mm across, oval and deeply pinnately lobed and toothed, at least on the upper leaves; often purple-veined. Flowers to 30 mm across, bluish with darker veins, borne in clusters of up to 4; bracts brown. Fruit with a long beak, 60 mm–(9)11 cm. Grassy habitats. L, F, C, T, G, H, P.

Pelargonium

A diverse genus of perennials or shrubs. Leaves alternate, opposite or in rosettes. Stamens 10.

Pelargonium capitatum MALVAROSA A densely hairy (non-glandular), fleshy shrub to 80 cm with erect or ascending stems. Leaves 20–80 mm with heart-shaped bases and broad lobes. Flowers with petals 15–20 mm, pink, unequal. An invasive garden escape. L, T.

LYTHRACEAE — LOOSETRIFE FAMILY

Annual or perennial herbs with leaves simple and opposite or in whorls of 3. Flowers cosexual, regular, usually with (4)6 sepals and *6 petals* (0–5) often pink or purple; stamens (2)6–12; style 1. Fruit a 2-valved capsule.

Lythrum ARROYUELO, LOOSESTRIFE

Herbs with tubular or bell-shaped calyx with 4–6 teeth and 4–6 petals <8 mm long.

Lythrum junceum A hairless perennial to 70 cm with much-branched, sparse stems. Leaves mostly alternate, elliptic and stalkless. *Flowers small*, borne 1(2) in each leaf axil, purple, rarely white, solitary; petals 6, 5–6 mm long; *stamens 12, some or all protruding*. Damp habitats. F, C, T, G, P.

Lythrum hyssopifolia GRASS POLY An erect, hairless annual to 25 cm, similar to *L. junceum*, with linear–lanceolate, rough-margined leaves. Flowers pink, borne 1(2) in each leaf axil with 4–6 *stamens, not protruding;* petals 2–3 mm. Seasonally flooded areas and damp places. L, C, T, G, H, P.

1. *Erodium touchyanum*
2. *Erodium moschatum*
3. *Pelargonium capitatum*
4. *Pelargonium capitatum* (fruits)
5. *Lythrum hyssopifolia*

Punica POMEGRANATE

Fruit-bearing deciduous shrubs or small trees, best known for the pomegranate. Flowers with 4–6(9) sepals and petals and as many or 2 x as many stamens. Fruit a capsule or berry. Recently established as part of the family Lythraceae.

Punica granatum GRANADO, POMEGRANATE A deciduous shrub or small tree to 5 m with spiny, 4-angled young stems. Leaves opposite, shiny, bright green, roughly oblong, untoothed and virtually unstalked. Flowers to 40 mm across, with 5–9 scarlet, crumpled petals and a fleshy calyx (hypanthium), borne in clusters of 1–3 near the ends of the branches. Fruit spherical, to 90 mm across. Widely planted. F, C, T, G, H, P.

ONAGRACEAE | WILLOWHERB FAMILY

Annual or perennial herbs with simple, alternate or opposite leaves. Flowers mostly regular, with 2–4 free sepals and petals and 2, 4 or 8(10–12) stamens; style 1. Fruit a capsule or berry splitting lengthways with distinctive cottony seeds, or a 1–2-seeded nut.

Oenothera

Annual or perennial herbs with alternate leaves. Flowers regular, large, borne in leafy spikes; sepals and petals 4; stamens 8, in 2 whorls; stigma deeply 4-lobed. Fruit elongated; seeds small and numerous. An American genus cultivated for ornament.

Oenothera rosea YERBAVINO A herbaceous annual to 1 m. Lower leaves entire or toothed, 20–50 mm, with short bristles; upper leaves entire to finely toothed. Flowers pink, opening in the evening, with petals to 10 mm. Capsules to 10 mm. Widely naturalised. L, F, C, T, P.

Oenothera parodiana An erect or decumbent annual with loose rosettes, to 50 cm. Leaves narrowly elliptic to oblanceolate, 30–50 mm, short-stalked below, stalkless above. Flowers small, yellow, petals 11 mm long, turning reddish-pink with age. Stigma 4-lobed. Naturalised. P. Several similar, related species have also naturalised across the islands.

Epilobium ADELFILLA, WILLOWHERB

Perennial herbs with opposite lower leaves. Flowers pink or purple with 4 sepals and 4 petals; stamens 8; ovary 4-parted. *Fruit a linear capsule splitting into 4 valves to reveal seeds with long plumes of hairs.*

Epilobium hirsutum GREAT WILLOWHERB A robust, densely and softly hairy perennial to 1.8 m with spreading non-glandular and glandular hairs. Leaves opposite, lanceolate, unstalked and *partially clasping the stem below*; markedly toothed. Flowers with bright pink, notched petals 10–16(18) mm long, borne in a leafy raceme; stigma 4-lobed. Damp places and near rivers. T, G, P.

Epilobium parviflorum SMALL-FLOWERED HAIRY WILLOWHERB is similar to (and hybridising with) the previous species but *smaller* in all parts (to 75 cm), with leaves not clasping the stem and *small, pale pink flowers with petals 5–9 mm long*. Local in damp places. C, T, G, P.

1. Oenothera rosea
2. Oenothera parodiana
3. Eucalyptus globulus
4. Pistacia atlantica
5. Schinus molle (flowers)
6. Schinus molle (fruits)
7. Schinus terebinthifolia
8. Rhus coriaria

Epilobium palustre MARSH WILLOWHERB A perennial to 60 cm with sparse, adpressed, non-glandular hairs (some glandular hairs above) *rounded stems without lines or ridges, and with untoothed, virtually stalkless leaves* (leaf stalks short, <4 mm). Flowers pale pink to white, borne in lax, coarsely hairy racemes; petals 4–7 mm; *stigmas club-shaped*. Marshes. T.

Epilobium tetragonum SQUARE-STALKED WILLOWHERB An erect perennial to 75 cm, similar to the previous species but with inflorescences with dense, white, adpressed hairs (no glandular hairs), distinguished by its *clearly 4-ridged, often winged stems*. Petals pink-purple, 5–7 mm. Fruits 65–80 mm (10 cm) long. C, T, G, P.

MYRTACEAE | MYRTLE FAMILY

A mainly tropical family of shrubs and trees with normally opposite, simple leaves. Flowers with 4–5 sepals and petals and numerous stamens; style 1. Fruit a many-seeded capsule or berry.

Myrtus MYRTLE

Trees or shrubs with opposite leaves. Flowers with 5 free sepals and petals; stamens numerous. Fruit a berry.

Myrtus communis MIRTO, COMMON MYRTLE An erect, much-branched evergreen shrub to 1.5 m, glandular-hairy when young. Leaves opposite, shiny deep green, lanceolate and pointed, *aromatic* when crushed, 20–50 mm long. Flowers white, to 30 mm with rounded petals and numerous, conspicuous protruding stamens. Berry blue-mauve then bluish-black when ripe, 7–10 mm long. Scrub. Very rare in the wild. C, T, G.

Eucalyptus EUCALIPTO

Evergreen trees native to Australasia with peeling bark. Leaves of saplings broad and erect; leaves on mature trees narrow and pendent. Flowers borne in clusters, with >100, prominent stamens. Fruit a woody capsule concealed by a fleshy cap (transformed sepals) when in bud. Planted for timber production.

(a) Flowers white or cream, borne on *stalked inflorescences*.

Eucalyptus camaldulensis RIVER RED GUM A tree to 15(50) m with smooth, white, peeling bark. Juvenile leaves oval-lanceolate, grey-green, mature leaves much narrower, 80 mm–25(30) cm long. Flowers yellowish-white, borne in clusters of 5–12, and *distinctly stalked; stalks 6–15(20) mm long*. Fruit *hemispherical*; longer than wide, to 6 mm and with a broad, raised rim. Planted for timber. F, C, T, G, H.

(b) Flowers white or cream and stalkless or virtually stalkless.

Eucalyptus globulus BLUE GUM A large tree to 30(40) m with smooth, peeling bark. Juvenile leaves oval–lanceolate with a heart-shaped base; mature leaves narrower and tapered, 10–15(40) cm long, grey, and aromatic when crushed. Flowers white or pink, to 35 mm, and *unstalked*. Fruit rounded and somewhat tapered towards the base, *large* to 10–15(18) mm across; wider than long. Native to Tasmania, widely planted. C, T, G.

ANACARDIACEAE

A large, mostly tropical family of shrubs, trees and lianas. Leaves alternate, simple or pinnately divided. Flowers with 5 sepals, 5 petals and 5–10 stamens; styles 3. Fruit a small, 1-seeded drupe.

Pistacia MASTIC

Shrubs and trees (mostly) with alternate leaves. Dioecious; flowers unisexual; stamens 3–5. Fruit a drupe.

Pistacia lentiscus LENTISCO, MASTIC A small, evergreen, dark green tree or shrub 6–8 m. Leaves dark green, pinnately divided *without* an end leaflet; leaflets 4–14, 10–50 mm, oval, leathery and untoothed, borne on winged stalks. Individual flowers rather inconspicuous, borne in dense *spike-like* clusters, the male with dark red anthers, the female greenish. Fruit 3.5–5 mm, spherical, red then black and shiny. Hill scrub; uncommon. L, F, C, T, G.

Pistacia atlantica ALMÁCIGO A deciduous tree to 7 m with a rounded crown. Leaves with 2–5 pairs of oblong to lanceolate leaflets 25–80 mm long, dark green above, paler beneath, with narrowly winged stalks. Fruits red, borne in panicles, 5–8 mm. Hill scrub; uncommon. F, C, T, G, P.

Schinus

Resinous trees native to South America with pinnately divided leaves with stalkless leaflets. Flowers with 4–5 sepals and petals and 8–10 stamens. Fruit berry-like.

Schinus molle ESPECIERO, CALIFORNIAN PEPPER TREE A small evergreen tree to 15(25) m with slender, pendent branches. Leaves pinnately divided with 11–47 linear–lanceolate, toothed leaflets 15–60 mm, hairy when young. Flowers 3–5 mm, yellow-white, borne in small, much-branched, pendent inflorescences; sepals and petals 5 each. Fruit 6–8 mm, pink, spherical. Commonly planted. L, F, C, T, G. *S. terebinthifolia* TURBINTO has leaves with 5–15 broader leaflets 30–60 mm and numerous red fruits 4–5 mm; similar in general appearance to *Pistacia* but with greenish-yellow (not red-purple) flowers with 10 stamens (not 3–5). Planted. F, C, T, H.

Rhus (including *Searsia*) SUMACH

Shrubs or small trees with pinnately divided leaves and thick twigs. Flowers often tiny and densely clustered, with 5 sepals, 5 petals and 5 stamens. *Searsia* is closely related and often still treated under *Rhus*.

Rhus coriaria ZUMAQUE, SUMACH A softly hairy shrub or tree 1–4(5) m with more or less evergreen leaves and densely downy shoots. Leaves pinnately divided with 7–21(25), toothed, green leaflets, *the stalk between the leaflets slightly winged*, at least towards the end; leaf stalk 20–30 mm. Flowers small and whitish, borne in very dense hairy panicles to 17–25 cm long. Fruit 4–6 mm, spherical and woolly, ageing brown-purple. Hill scrub. C, T, G, H, P.

Searsia albida (syn. *Rhus albida*) ZUMAQUE BLANCO An irregularly branched, low shrub 1–2.5 m with whitish branches. Leaves trifoliate with oblong–spathulate, grey-white, sometimes slightly blunt-toothed leaflets. Flowers with yellowish petals. Desert slopes; rare. *Fruit a shiny red, fleshy, berry* (drupe). F.

SAPINDACEAE

A large family of perennials, lianas and trees (genera very distinct from each other). Leaves pinnately or palmately lobed or divided, stalked; stipules absent. Flowers unisexual or cosexual with 4–5 sepals and (0)4–5 petals; stamens 5–9; styles 1–2. Fruit variably dry or fleshy.

Cardiospermum

Herbaceous annuals or perennials, sometimes woody below, often with lobed leaves and small, soon-falling stipules. Flowers unisexual and zygomorphic; sepals 4(5); petals 4; stamens 8; style 1. Fruit a capsule.

Cardiospermum grandiflorum FAROLITO A vigorous, creeping vine; young parts covered in stiff hairs. Leaves compound with tendrils at the forks; leaflets 20–80 mm with coarse-toothed margins. Flowers small, white, with stalks bearing tendrils. *Fruit a conspicuous, balloon-shaped, papery capsule,* 45–65 mm long with a seed in each of the 3 locules. Occasionally naturalised. C, T, G, P.

RUTACEAE

A large and widely distributed family of aromatic herbs, shrubs and trees. Leaves opposite or alternate with translucent glands. Flowers with 4–5 sepals and petals, free; stamens 2 x as many; style 1. Fruit a berry, capsule or drupe.

Ruta RUDA, RUE

Strong-smelling perennials and shrubs. Flowers yellowish with 4–5 sepals and petals, often lobed; stamens (8)10; ovary (4)5-parted. Fruit a capsule.

(a) Petal margins conspicuously fringed or toothed.

Ruta chalepensis FRINGED RUE A hairless shrub 20–60 cm. Leaves rather long-stalked, divided into narrowly oblong–lanceolate segments 1.5–6 mm wide. Flowers borne in lax inflorescences; petals oblong, *fringed with cilia,* the cilia not as long as the petal width; sepals hairless, triangular–oval. Capsule hairless, with pointed segments. Locally frequent in town and waste places. L, F, C, T, G, H, P. **R. graveolens** COMMON RUE is a similar, grey-blue shrub, with dense branches 14–45 cm. Leaves broad, with oval–lanceolate leaf segments. Flowers with *toothed* petals. Fruits 3.5–9.1 mm. C, P.

(b) Petal margins entire, sometimes wavy-edged (not conspicuously fringed or toothed).

Ruta pinnata A tall, lax shrub to 1.5 m. Leaves pale green, pinnate with linear to rhomboid, entire leaflets. Flowers yellow with small, cupped petals; margins entire, sometimes crenate. Fruits orange-brown, 6 mm. Rather rare; sporadic in rocky ravines. *T, P. **R. microcarpa** is similar but smaller, to 80 cm, much-branched and with leaflets sparsely toothed. Fruits small and yellowish. Very rare; rocks and cliffs. *G.

---

1. *Searsia albida*
2. *Searsia albida* (in fruit)
3. *Ruta chalepensis*
4. *Ruta pinnata* (habit)
5. *Ruta pinnata*
6. *Citrus* × *sinensis*
7. *Cneorum pulverulentum*
8. *Cneorum pulverulentum* (fruits)

(c) Petals erect, cupped.

Ruta oreojasme A small, spreading to ascending shrub. Leaves blue, pinnate, with blunt lobes. Flowers large, yellow, cupped with *erect*, keeled petals. Fruits rough, light brown. Locally frequent in cliffs and ravines. *C.

Citrus

Small evergreen trees with glossy leaves. Flowers solitary to few, often fragrant; petals (4)5(–8); stamens numerous — 16–20(100). Fruit large and edible. Relationships are complicated by a long history of cultivation and interbreeding.

The following are all widely grown: ***Citrus* × *limon*** LEMON An evergreen tree to 4 m with a rounded crown, flowering and fruiting throughout the year. Leaves elliptic–lanceolate and shallowly toothed. Flowers male or cosexual, with >4 x as many stamens as petals, petals white, often streaked purple. Fruit a lemon, yellow and warty when ripe. ***C. × aurantium*** SEVILLE ORANGE A small tree to 10 m with flexible spines. Leaves with a rounded base and *broadly winged stalks*. Fruit an orange. ***C. × sinensis*** SWEET ORANGE is similar but with narrowly *winged leaf stalks*. ***C. × paradisi*** GRAPEFRUIT is similar to *C. × sinensis* but with a distinct, broadly winged stalk, and large, yellow fruits. ***C. × aurantiifolia*** LIME has leaves resembling those of an orange tree and small, *spherical* fruits 25–50 mm across, harvested green (yellowish when mature).

Cneorum

A small genus of plants with alternate, entire, leathery leaves and regular, cosexual flowers; sepals, petals, stamens and carpels all 3(4). Fruit made up of 3(4) mericarps.

Cneorum pulverulentum (syn. *Neochamaelea pulverulenta*) ORIJAMA, LEÑA BUENA A blue-grey-hairy shrub to 1.5 m. Leaves broadly linear. Flowers yellow with four gently reflexed petals borne solitary in the leaf axis of the upper branches. Fruit a 3–4-parted reddish-purple fruit. Local. *C, T, G, H.

SIMAROUBACEAE

A predominantly tropical family of plants with alternate, pinnately divided leaves. Flowers small, usually unisexual with (3)5(7) sepals and petals and 2 x as many stamens; ovary superior, surrounded by a disk with 2–5 fused or free carpels. Fruit an aggregate of winged achenes.

Ailanthus

Trees with flowers with 5–6 sepals fused to the middle, the male flowers with 10 stamens; ovary superior surrounded by a disk. Fruit winged.

Ailanthus altissima ÁRBOL DEL CIELO, TREE OF HEAVEN A rapidly growing tree to 20(30) m with smooth bark and large, pinnately divided leaves with 5–12 pairs of oval–lanceolate leaflets 40 mm –17 cm long and terminal panicles of small, strong-smelling, greenish-yellow, 5(6)-parted flowers, unisexual; petals 2.2–4.5 mm. Fruit in clusters of 3-winged carpels, 25–50 mm, reddish brown. Native to China; planted and escaped. C, T, H, P.

MELIACEAE

A family of mostly trees and shrubs characterised by alternate, usually pinnately divided leaves without stipules. Flowers cosexual (or cryptically unisexual) borne in panicles, cymes, spikes, or clusters; stamens 3–10–numerous. Fruit a berry.

Melia

Trees with 2-pinnately divided leaves, usually toothed. Flowers with 5 sepals and 5 petals; stamens 10–12, in 1–2 whorls. Fruits berry-like.

Melia azedarach ÁRBOL DEL PARAÍSO, INDIAN BEAD TREE A deciduous tree to 15 m with furrowed bark. Leaves 20–40(60) cm, alternate and 2-pinnately divided; leaflets elliptic and toothed or lobed, 50–70 mm. Flowers lilac and scented, borne in panicles; petals 5, each 8–12 mm. Fruit a berry 8–15(25) mm across, yellow and *long-persisting* (even when not in leaf). Native to India and China; planted. C, T, H, P.

CYTINACEAE

A small family of obligate root parasites of other shrubs. Flowers unisexual and ant-pollinated. Seeds numerous, in a pulp, dust-like when dry and windborne.

Cytinus

Parasitic plants without chlorophyll or obvious leaves or stems. Perianth 4-lobed; stamens 8–10. Fruit a capsule; seeds minute. Host-specific races are probably in the process of forming cryptic species.

Cytinus hypocistis subsp. *subexsertum* MELERA DE JAGUARZO A yellow parasite of various *Cistus* spp., with underground stems 30 mm–16 cm (stemless above ground), and yellow, orange or red oval–oblong scale leaves. *Flowers bright yellow*, in dense clusters of 4–14(20), subtended by 2 bracteoles the same colour as the scale leaves. Rather rare and local; scrubby slopes. C, T, G, H, P.

Cytinus ruber MELERA DE JARA Similar to the above species, but with about 20 flowers with *crimson* (not red-orange) scale leaves and bracts contrasting a *white to pale pink* (not yellow) perianth 10–23 mm that slightly exceeds the bracteoles. Parasitic on pink-flowered *Cistus* spp. Rather rare and local; where *Cistus* is abundant. T, P.

MALVACEAE | MALLOW FAMILY

A large family of herbs or shrubs with star-shaped (stellate) hairs, and alternate, often palmately lobed leaves with stipules. Flowers often conspicuous, usually with both a *calyx and epicalyx* (an important character); sepals and petals 5 each; stamens numerous; styles 1 or 5–many. Fruit a nut with 1–3 seeds, or capsule splitting into nutlets.

Gossypium ALGODÓN, COTTON

Herbs and shrubs (to small trees). Leaves usually 3–9-palmately lobed (sometimes entire). Flowers solitary, borne in the leaf axils; epicalyx segments 3, large, leafy; corolla usually yellow with a crimson centre. Fruit a capsule; seeds covered in a mass of cotton.

Gossypium herbaceum A shrubby, rather woody annual to 1.5 m, usually hairy. Leaves 20–50 mm with hairy margins, palmately (3)5(7)-lobed; lobes oblong, elliptic or oval. Flowers erect, borne on stalks 7–15 mm; epicalyx segments 10–20 mm with heart-shaped bases; calyx cup-shaped, black-dotted; petals 25–35 mm, broad, yellow with a crimson centre. Fruit a capsule 25–30 mm long, beaked, 3–5-parted, woody and splitting widely when mature revealing *cottony mass* concealing the 5–7 seeds per chamber. Native to sub-Saharan Africa and Arabia; planted. F, G, H.

Malva MALVA, MALLOW

Annuals or perennials with epicalyx of 3(–10) segments, free or fused, and white-pink petals; carpels numerous; fruit splitting into nutlets. The traditional distinction between *Lavatera* and *Malva* is based on fusion or non-fusion of the epicalyx, but this character is now established to be artificial, therefore the genera are now merged. A difficult genus; observation of mature fruits often necessary.

(a) Herbaceous annual, biennial or perennial herbs.

Malva multiflora (syn. ***M. pseudolavatera, Lavatera cretica***) SMALL TREE MALLOW An erect or ascending, *non-woody* annual or biennial to 1.5 m, with lower leaves circular to heart-shaped, shallowly 3–7-lobed, the upper leaves more deeply 5-lobed. Flowers *lilac*, borne in the axils of the leaves in clusters of 2–8, on unequal stalks shorter than the leaves. Petals 10–20 mm long, epicalyx segments to 6 mm long, and free *almost* to the base (but fused there), shorter than the long-pointed sepals; *sepals just shorter than or as long as sepals in fruit*. Fruit smooth or slightly ridged. Waste places. L, F, C, T, G, H, P.

Malva parviflora A hairy or hairless annual to 50 cm. Leaves rounded to heart-shaped with 3–7, shallow, rounded and toothed lobes. Flowers borne in clusters of 2–4, pale mauve or lilac, *small* – the petals <5 mm long, *short-stalked* with linear–lanceolate epicalyx lobes; *sepals hairless at margins*. *Fruit with enlarged, spreading, papery sepals*, fruit strongly netted and hairy or not; with stalks <10 mm. L, F, C, T, G, H, P.

Malva neglecta DWARF MALLOW An erect or ascending annual to 60 cm with leaves less deeply lobed, petals *pale lilac* and *2–3 x the length* of the sepals (9–13 mm); epicalyx segments linear–lanceolate. *Fruit smooth* and hairless; *fruit stalks remaining erect*. Waste places. L, F, C, T, H. **M. nicaeensis** is similar but with broader epicalyx segments, leaves not distinctly heart-shaped at the base, pale mauve petals 10–12 mm. Fruit *netted; fruit stalks recurved*. Similar habitats. L, F, C, T, G.

Malva alcea GREATER MUSK-MALLOW A perennial to 1.2 m with *star-like* (not simple) hairs on all vegetative parts. Flowers pale pink; epicalyx segments 7–8 mm, *<3 x as long as wide*. Nutlets hairless or hairy. Dry and waste habitats. T, H.

---

1. *Melia azedarach*
2. *Cytinus hypocistis* subsp. *subexsertum* (GEORGE VANN)
3. *Cytinus ruber*
4. *Gossypium herbaceum*
5. *Malva acerifolia*
6. *Malva multiflora*
7. *Malva parviflora*
8. *Malva phoenicea*

(b) Woody perennials, or shrubs.

Malva arborea (syn. *Lavatera arborea*) TREE MALLOW A robust, *woody* biennial or perennial to 3 m, downy above with star-shaped hairs. Leaves large, circular, palmately lobed with 5–7 lobes to 20(22) cm long, velvety. Flowers borne in the leaf axils in clusters of 2–7, forming a long, terminal inflorescence; petals 14–20 mm, *pink-purple with darker veins*. Epicalyx segments to 10 mm long, *exceeding the sepals* and greatly enlarged in fruit and fused at the base. Fruit hairy or not, and sharply angled. Native to the Mediterranean; naturalised in waste places. L, F, C, T, G, H, P.

Malva acerifolia MALVARRISCO A tall shrub to 2.5 m. Leaves palmately lobed, long-stalked; lobes irregularly toothed. Flowers, few, borne in terminal and lateral clusters, sometimes solitary; petals whitish to mauve with dark bases. Locally frequent on cliffs in the west; very rare on remote sea cliffs in the east. *L, F, C, T, G, P. *M. phoenicea* HIGUERETA is similar in form but with leaves that have narrower segments and *orange-pink flowers*. On cliffs, very rare and local. *T.

Alcea HOLLYHOCK

Tall, erect biennials or perennials with *very large* flowers borne in tall, wand-like inflorescences; epicalyx with 6–7 lobes fused at the base and shorter than their calyx. Fruit splitting into 1-seeded nutlets.

Alcea rosea MALVA REAL, HOLLYHOCK A tall perennial to 3 m with stiff, erect, unbranched stems, and large, flat, palmate, bluntly toothed leaves. Flowers large and showy, usually pink but variable in colour, particularly in cultivated forms; petals 25–50 mm. Nutlets winged. Planted; possibly naturalised. F, C, T.

Brachychiton

Trees and shrubs native to Australia. Monoecious; the unisexual flowers with a bell-shaped, lobed perianth; stamens typically 10–30 and the same number of sterile stamens (staminodes); carpels 5.

Brachychiton populneus ÁRBOL BOTELLA, BOTTLE TREE A tree to 18 m (often much less) with a distended trunk, and pointed, simple or broad-lobed leaves. Flowers *bell-shaped*, variably greenish, yellowish or pinkish, with small red markings or entirely red within, and with usually 5 or 6 unequal, pointed lobes. Fruit a splitting pod. Occasionally planted along roadsides. F.

NEURADACEAE

A small family of hairy, prostrate annuals. Flowers solitary; sepals and petals 5 each; stamens and styles 10 each. Fruiting carpel dry, disk-like, convex.

Neurada

Unmistakable for its white-hairy, prostrate habit and disk-like fruiting heads.

Neurada procumbens PATA CAMELLO A prostrate, white-hairy annual to 14(32) cm, much-branched. Leaves 6–25 mm, oval, irregularly lobed. Flowers greenish, borne solitary in the leaf axils; petals 5, 2–4.3 mm long; stamens and styles 10 each. Fruiting carpel distinctly disk-like, 8–15 mm across, spiny above and 10-valved. Sand flats; dunes. C (south).

THYMELAEACEAE

Hairless shrubs with simple, untoothed, alternate leaves. Flowers in clusters or racemes, regular and cosexual; calyx a tube, petal-like; true petals absent; stamens 8, fused to the surface of the calyx-tubes; style solitary. Fruit a drupe, berry or nut.

Daphne

Shrubs with simple leaves. Flowers with 4 petal-like lobes; stamens 8. Fruit fleshy, enclosed in a persistent calyx.

Daphne gnidium TORVISCO An erect, lax, virtually hairless shrub to 2 m with branches bare beneath; superficially rather *Euphorbia*-like when not in flower, but without a white latex when cut. Leaves 20–30 mm, pale green, leathery, linear and pointed. Flowers cream-white, 5–6.5 mm long, borne in dense panicles in late winter to spring. Berry deep red, ageing black, 7–8 mm. Woods. C, T, P.

1. *Brachychiton populneus*
2. *Neurada procumbens*
3. *Daphne gnidium*

CISTACEAE | ROCK ROSE FAMILY

Leaves normally opposite, stipules often present. Flowers often showy, cosexual, solitary or in lax, terminal clusters; sepals 3–5, petals 5, free, stamens numerous. Fruit a capsule with 3–5 valves.

Cistus ROCK ROSE

Shrubs with opposite leaves without stipules. Flowers solitary, often showy; stamens 50–150; carpels 5(6–12). Fruit a capsule.

(a) Flowers white.

Cistus monspeliensis JAGUARZO, NARROW-LEAVED ROCK ROSE A slightly sticky bush to 1.8 m, lax below, compact above. Leaves mid green, *narrow*, linear to lanceolate, scarcely tapered at the base and unstalked, 15–45(70) mm. Flowers small, white; petals 9–14 mm; sepals 5, the outer 2 wedge-shaped at the base. Capsule 4 mm. Cliff-tops, open hill slopes; locally common. C, T, G, H, P. *C. grancanariae* is very similar but with oblong–elliptic, greyish leaves with simple and starry hairs and entire, flattish margins. Flowers to 28 mm across, borne in clusters. *C.

Cistus ladanifer JARA PRINGOSA, GUM ROCK ROSE An aromatic, *sticky*, lax, erect shrub to 2(4) m. Leaves linear–lanceolate, dark green (paler beneath), 40–80 mm long, 3-veined in the lower ⅓, and scarcely stalked. Flowers solitary, *large* (50–80 mm across), white, often with a crimson blotch at the base of each petal; petals 30–55 mm; *sepals 3*. Capsule 10–15 mm. Native to the western Mediterranean; locally naturalised in pine forests. C, T.

(b) Flowers pink. Many closely related, similar species.

Cistus symphytifolius JARA, AMAGANTE DE PINAR A variable shrub to 1 m. Leaves broadly lanceolate to ovate, variably hairy, wrinkled beneath with prominent veins. Flowers pink, to 50 cm across; stigmas capitate. Capsule brown, sparsely hairy. Very common in the west on mountain slopes and as an understorey in pine forests. *T, P. The following are similar: *C. osbeckiifolius* has lanceolate to elliptic, small, *densely hairy leaves with silvery margins*. Flowers pink. Capsules densely hairy. *T (Las Cañadas and Tágara). *C. ocreatus* has ovate–elliptic, hairy, sage-green leaves, *small* pink flowers and *hairless capsules*. Humid pine forests. *C. *C. chinamadensis* has *glaucous, sparsely hairy leaves*. Capsules ovoid and hairy. Regional forms include: subsp. *ombriosus* (*H), subsp. *gomerae* (*G) and subsp. *chinamadensis* (*T). *C. horrens* is a small shrub with *long-hairy leaves*. *C.

Tuberaria (syn. *Xolantha*)

Annuals or perennials with basal rosettes and erect flowering stems. Flowers yellow, sepals 5, the outer 2 smaller. Capsule 3-valved. Most floras classify under *Tuberaria* rather than *Xolantha*.

Tuberaria guttata HIERBA TURMERA, ANNUAL ROCK ROSE A very variable, hairy, low annual to 42 cm with a basal leaf rosette, dying when mature, and a normally unbranched flowering stem. Leaves elliptic to oval, often with downturned margins, 16–73 mm. Flowers yellow, with petals with or without a dark brown or purple spot at the base, 3–9 mm; flower stalks longer than the sepals at the point of flowering. Scrub. L, F, C, T, G, H, P.

---

1. *Cistus monspeliensis*
2. *Cistus grancanarieae* (inset: leaf detail)
3. *Cistus symphytifolius*
4. *Cistus osbeckiifolius*
5. *Cistus ocreatus*
6. *Cistus chinamadensis*
7. *Cistus horrens*
8. *Cistus horrens* (indumentum)

Tuberaria lignosa An ascending to erect *perennial* with a branched to 57 cm woody stock and with persistent plantain-like leaf-rosettes. Leaves oval to elliptic, gradually tapered towards the base, 3-veined and white-hairy beneath, 36–65 mm (10 cm). Flowers yellow and unspotted; petals 10–15 mm. Local in scrubby habitats. T, P.

Helianthemum JARILLA

Dwarf shrubs or herbs with opposite leaves. Flowers borne in 1-sided clusters; sepals 5, the outer 2 smaller; stamens numerous, style long and S-shaped. Fruit a 3-valved capsule. Many similar, closely related species on the islands, the distinctiveness of which is unclear in some cases.

(a) Shrubs and woody-based perennials 20–75 cm (many similar species).

Helianthemum broussonetii An erect shrub 25–75 cm. Leaves oblong–lanceolate, 20–25 mm, *shortly and very densely silvery-hairy* (hairs starry); stipules short and linear to thread-like, 1–4 mm. Sepals 4-veined, densely white-hairy; petals yellow, *without a dark spot* at base. *T, P. **H. teneriffae** has ovate–oblong *woolly-hairy* leaves <30 mm, stipules equalling the leaf stalks, and inflorescences of 4–12 flowers 20 mm across with *somewhat bristly* sepals; petals *with* a dark spot at the base. *T. **H. bystropogophyllum** is similar to the previous species but with *short stipules* (<½ the leaf stalk) and leaves with *jagged-scalloped* margins. *C. **H. tholiforme** is like the former, also with short stipules, only with *entire leaf margins*. *C. On La Palma, the following are similar: **H. linii** is like *H. tholiforme* but rather larger, with broader, longer stipules to 11 mm; **H. henriquezii** has simple (not starry) leaf hairs; **H. cirae** has linear, glaucous, long-stalked *hairless* leaves, and arching branches of numerous flowers.

Helianthemum juliae An erect, woody-based perennial 30–60 cm. Leaves stalkless, with inrolled margins; those below densely soft-hairy and linear–spathulate; those above narrower and less hairy. Flowers borne in loose cymes of 5–12; yellow, 25–30 mm across. A rare endemic of Cañadas del Teide. *T.

(b) Small shrublets, usually <20 cm.

Helianthemum canariense TURMERA, RAMA CRÍA A *small*, spreading subshrub, typically to just 20 cm with prostrate to ascending stems, much-branched below. Leaves *small, <10 mm*, softly white-grey-hairy, with downturned margins. Flowers *small*, c. 8 mm across, yellow *with conspicuous yellowish sepals with dark reddish prominent ribs*, enlarging in fruit. Locally very common; rocky hillsides. L, F, C, T, G, H, P. *H. thymiphyllum* is scarcely distinguishable (possibly a mere form of the above), with small *bright green*, oblanceolate–obovate, scarcely hairy leaves. Mainly on sea slopes and cliffs, sometimes inland; rare and local. *L, F. *H. gonzalezferreri* is a similar subshrub with ovate to lanceolate–elliptic leaves <20 mm, and flowers borne *numerously* and rather densely in groups of 7–20; petals broad, unspotted; sepals membranous with starry hairs; style 2.5 mm, straight or slightly curved. Very rare; high on remote sea cliffs. *L. *H. bramwelliorum* is a very similar, very low (5–20 cm tall) subshrub with longer, oblong–lanceolate leaves. Flowers *few* (4–10; often only one open); petals with a faint orange spot. Rare and local on lower cliff ledges. *L.

(c) Annuals.

Helianthemum salicifolium WILLOW-LEAVED ROCK ROSE A low, hairy, branched, erect or spreading *annual* to 30 cm. Leaves 5–25 mm, oval–lanceolate, flat and short-stalked. Flowers yellow, borne in lax clusters; petals 2–7 mm long and narrow, shorter to slightly longer than the sepals; bracts large and leafy; *flower stalks spreading in fruit*, upturned at the apex. Sandy and rocky slopes. C. *H. ledifolium* is *hairier* and with *petals 6–8 mm, shorter than the sepals*, and *flower stalks erect* in fruit. Rocky ledges. L, F.

1. *Helianthemum broussonetii* (habit)
2. *Helianthemum broussonetii*
3. *Helianthemum juliae* (habit)
4. *Helianthemum juliae*
5. *Helianthemum canariense* (habit)
6. *Helianthemum canariense*
7. *Helianthemum thymiphyllum*
8. *Helianthemum gonzalezferreri*
9. *Helianthemum bramwelliorum*
10. *Helianthemum ledifolium*

TROPAEOLACEAE | NASTURTIUM FAMILY

Herbaceous annuals and perennials, often with a rather succulent habit and hairless, alternate leaves. Flowers showy, with 5 sepals and petals and 8 stamens; stigmas 3. Fruit 3-parted, breaking into succulent segments.

Tropaeolum

Climbing or scrambling annuals with peltate leaves. Flowers zygomorphic with 5 free petals and sepals; stamens 8. Fruit 3-parted.

Tropaeolum majus MARAÑUELA, NASTURTIUM   A vigorous, creeping, hairless annual to 2 m with circular, blue-green, shallowly 5–6-lobed, upward-facing, *peltate leaves* (stalk attached to the centre of the blade). Flowers large, solitary, long-spurred, yellow or orange-red, 25–60 mm across, borne in the leaf axils; spur 25–40 mm. Native to South America; planted and escaping. L, F, C, T, G, H, P.

RESEDACEAE | MIGNONETTE FAMILY

Annual or perennial herbs with alternate, simple or pinnately divided leaves. Flowers borne in long spikes; flowers with 4–6(8) free sepals and the same number of often yellow, distinctively shaped petals; stamens 7–numerous (25); style 0. Fruit a capsule.

Reseda

Herbs or shrubs with flowers with 4–8 sepals and petals, 10–25 stamens and *bottle-shaped fruits*.

(a) Annual or biennial herbs.

Reseda crystallina SONAJERA   An annual or biennial to 55 cm, hairless, covered in minute, translucent vesicles. Leaves *trifoliate* (the lowermost sometimes with more lobes). Flowers borne in short, compact racemes; sepals 6, linear–oblong, persistent. Capsule a prism terminating in 3 short horns. Very common in a range of habitats in the east. *L, F, C.

Reseda luteola GUALDA, WELD   A tall, erect biennial to 1.3 m. All *leaves unlobed*, lanceolate and with wavy margins. Flowers greenish-yellow with 4 sepals and petals, borne in long, slender spikes. Fruit 3–6 mm, rounded with 3 pointed lobes. Local in disturbed areas and sandy waste places. L, F, C, T, H, P.

Reseda lutea WILD MIGNONETTE   An ascending, leafy perennial to 75 cm, bushy and sometimes woody at the base. Leaves stalked and pinnately lobed with 1–4 pairs of leaflets. Flowers pale *greenish-yellow*, borne on short flower stalks (<6 mm), each with 6 petals and 5–6 sepals, borne in many long, narrow spikes; petals entire to 3-lobed; stamens 15–20. Fruit 7–20 mm, oblong and erect, 3-parted. Seeds smooth. Waste places. F.

(b) Shrubs.

Reseda scoparia GUALDÓN   A rather sparse-branched *shrub* with numerous stiffly erect to ascending stems. Leaves *linear, entire,* grey. Flowers whitish with ochre stamens. Capsules smooth. Dry, rocky hillsides. *C, T, G, P.

1. *Tropaeolum majus*
2. *Reseda crystallina* (coastal form in fruit)
3. *Reseda crystallina*
4. *Reseda luteola* (habit)
5. *Reseda luteola*
6. *Oligomeris linifolia*
7. *Carica papaya* (habit)
8. *Carica papaya*

Oligomeris

Annual to perennial herbs with 4-parted sepals, 2 petals and 3–8 stamens. Ovary 4-angled.

Oligomeris linifolia ROMERILLO PARDO A small annual to 10(40) cm with bunched, grey, linear leaves 10–50 mm. Flowers virtually stalkless, greenish-white, minute; sepals 4-parted; petals slightly shorter; stamens 3. Capsule bead-like, toothed. Dry, stony slopes. L, F, C, T.

CAPPARACEAE | CAPER FAMILY

A small family of herbs and shrubs. Flowers with 4 petals, 4 sepals and 6–*numerous stamens*. Fruit a berry with numerous seeds.

Capparis CAPER

Shrubs with spiny stipules and simple leaves. Flowers with petals much shorter or longer than the sepals. Fruit a succulent berry (caper).

Capparis spinosa ALCAPARRA, CAPER A spreading or pendent shrub, often sprouting directly out of old walls. Leaves alternate, fleshy, grey-green and oval to circular, blunt or slightly notched at the tip with curved spines at the base of the stalks, *very sparsely hairy*; stipules *weak or virtually absent*. Flowers conspicuous, with white petals 20–35 mm, and numerous fine, violet stamens. Fruit a large fleshy berry 20–30 mm. Planted. H.

CARICACEAE

A tropical family of short-lived shrubs to trees with milky latex. Leaves alternate; stipules absent. Flowers unisexual or cosexual; calyx with 5 lobes; stamens 10. Fruit a pulpy berry.

Carica papaya PAPAYA A small tree with a slender, straight trunk to 5 m; leaves congested above, long-stalked, palmately lobed (7 segments). Plants dioecious; flowers lateral; fruit a berry 15–45 cm, soft and orange when ripe, edible. Grown for its fruit (a papaya), throughout.

BRASSICACEAE | CABBAGE FAMILY

A large family of annual and perennial herbs (or shrubs). Leaves alternate and simple or pinnately divided. Flowers often in racemes, characteristically with 4 sepals and 4 petals forming a cross; stamens usually 6 (2 shorter). Fruit dry, sometimes non-splitting; shape of the ripe fruit important for identification, classified as a *silicula* (broad and variously shaped), or a *siliqua* (long and thin).

Alyssum

Annual or perennial early-flowering herbs with branched or star-like hairs (rarely mixed with unbranched hairs). Flowers yellow; sepals erect-spreading; filaments of 2 kinds: long and winged, or short with an appendage. Fruit a silicula. Seeds 1–2(6) in each chamber, often winged. A taxonomically difficult group to identify in the field. Many local endemics exist that are beyond the scope of this book, so just a handful are included here.

Alyssum simplex A small annual with few prostrate to ascending stems 50 mm–20 cm. Leaves elliptic or broadest above the middle, often withered below, covered in adpressed, 6–8-rayed white hairs. Flowers small, borne in crowded heads, eventually elongating in fruit; sepals *soon-falling*; petals 2.5 mm, often slightly notched; fruits circular, with dense 8-rayed hairs. Seeds narrowly winged. Bare habitats. T.

Isatis WOAD

Erect annuals or perennials with simple leaves. Flowers small and yellow. Fruit winged, 1-seeded, pendent and non-splitting.

Isatis tinctoria HIERBA PASTEL, WOAD A large, erect, much-branched biennial to 1.5 m with grey-green leaves; leaves arrow-shaped and clasping the stem above. Flowers yellow, borne in dense and much-branched racemes. Fruit a *pendent*, oblong–elliptic silicula 10–25 mm. Introduced. H.

Capsella SHEPHERD'S PURSE

Annuals or perennials with simple basal leaves. Petals normally white. Fruits triangular–heart-shaped.

Capsella bursa-pastoris ZURRÓN DE PASTOR, SHEPHERD'S PURSE A distinctive, sparsely hairy annual to 40 cm with variable leaf shape and scentless white flowers borne in a long raceme; petals 2–3 mm. Fruit 5–9 mm and *heart-shaped* with straight to slightly convex sides. Common on disturbed ground. L, F, C, T, G, H, P.

Sisymbrium QUEMÓN, ROCKET

Annuals or perennials with simple, entire to deeply lobed leaves, the lowermost often withering in bloom. Petals yellow. Fruit a beakless siliqua, much longer than broad; seeds in 1 row under each valve.

(a) Fruits adpressed to the stems.

Sisymbrium officinale QUEMÓN, HEDGE MUSTARD A bushy, more or less hairless annual or biennial to 1 m with grey-green dense, oval, deeply lobed leaves with a large terminal lobe. Flowers yellow, borne on slender, branched inflorescences *without bracts*, small; petals 3–4.2 mm. Fruits 10–20 mm, cylindrical, straight, *closely and densely appressed along slender stems*. Common near buildings. Common on disturbed ground. L, F, C, T, G, H, P.

(b) Fruits erect to spreading.

Sisymbrium erysimoides QUEMONCILLO, FRENCH ROCKET An annual with erect stems to 60 cm and variously entire to (often deeply) divided leaves. Flowers yellow, rather crowded, with *small* petals ≤3.5 mm, not exceeding the sepals or stamens. Fruits long, spreading to erect-spreading, 20–50 mm with *robust* stalks about 1 mm wide (>⅓ the width of the fruit). Dry, waste habitats. L, F, C, T, G, H.

Sisymbrium irio AGONAL, QUEMONCILLO, LONDON ROCKET An erect, much-branched annual to 60 cm with variously (some deeply) pinnately lobed leaves, the end lobe pointed. Flowers pale yellow, petals 2.5–3.5 mm, equal to or longer than the sepals. *Fruits 30–60 mm long, overtopping the open flowers*, erect to spreading, with *slender stalks* (0.3–0.6 mm wide, <⅓ the width of the fruit). Disturbed ground. L, F, C, T, G, H, P. ***S. orientale*** EASTERN ROCKET is similar, grey-hairy, and with short-stalked uppermost leaves with few (0–2) lateral lobes, the middle lobe linear–lanceolate. Flowers with petals >6 mm (much longer than sepals and stamens). Fruits 40 mm–10(12) cm, slender, and erect to spreading. C, T, P.

Erysimum WALLFLOWER

Annuals, herbaceous perennials or shrubs with erect, often slightly winged stems and leaves entire, covered in branched hairs. Petals yellow or red. Fruit a flattened to 4-angled siliqua.

Erysimum scoparium ALHELÍ DEL TEIDE A woody-based perennial. Leaves narrow, upward-swept, silvery-grey, <70mm with forked hairs on both sides; margins sometimes weakly toothed. Flowers whitish-pink to purple; petals 4; sepals reddish-mauve. Stony slopes. Forms with smaller leaves and longer styles in Gran Canaria are sometimes classified as *E. albescens*. *C, T, P. *E. virescens* ALHELÍ MONTUNO has a shrubby habit, *narrowly lanceolate* leaves and remotely *toothed*, lanceolate leaves. *C, T, G, H, P.

Moricandia

Hairless plants with simple, fleshy leaves. Petals *violet*. Fruit linear, 4-angled.

Moricandia arvensis COLLEJÓN, VIOLET CABBAGE A hairless, blue-grey annual or perennial to 65 cm with slightly fleshy leaves, shallow-lobed below, clasping the stem with heart-shaped bases above. Flowers violet-purple borne in racemes of 10–20; petals 21–29 mm. Fruit linear, 30–60 mm and 4-angled. Dry slopes and roadsides; local. F, C.

Malcolmia ALHELÍ DE MAR

Annuals with branched hairs. Petals normally *pink or violet*. Styles absent, stigmas 2-lobed. Fruits linear with seeds in 1 row under each valve, not winged.

Malcolmia maritima An annual 10–35 cm, with 3–4-forked hairs. Leaves dark green, egg-shaped to oblong, wedge-shaped at the base, entire or toothed. Flowers pink to violet, borne in inflorescences without bracts; fruiting stalks 0.5–1 mm across when mature (*thinner* than the fruit itself); sepals 6–10 mm; petals 12–25 mm. Fruit a siliqua 35–80 mm. Introduced. L, F, C, T.

Malcolmia littorea A densely white-hairy perennial herb 10–40 cm tall, *woody at the base* and with numerous non-flowering shoots. Leaves 10–30 cm, lobed or not, *more or less stalkless*. Flowers 5–20 per cluster, pink-purple, with petals 15–18 mm. Fruit 30–60 mm and not beaded. Sandy and rocky maritime habitats. L, F.

Matthiola STOCK

Grey-leaved plants with simple basal leaves. Flowers with deeply 2-lobed stigmas; petals white or purple. Fruit linear with seeds in 1 row under each valve, broadly winged.

Matthiola bolleana ALHELÍ CANARIO A woody-based annual (or short-lived perennial) to 35 cm with short, starry hairs. Leaves wavy-toothed to lobed, blunt at the tips. Flowers fragrant; *petals wavy*, often recurved, pink-purplish, whitish in the centre; sepals 8–9.5 mm. Fruit cylindrical, 20–50 mm with parallel veins, often curved; densely short-hairy. Seeds short-winged. The following regional

---

1. *Sisymbrium officinale*
2. *Sisymbrium erysimoides*
3. *Erysimum scoparium*
4. *Erysimum virescens*
5. *Moricandia arvensis*
6. *Matthiola bolleana* subsp. *bolleana*
7. *Matthiola bolleana* subsp. *viridis* (habit)
8. *Matthiola bolleana* subsp. *viridis*
9. *Matthiola bolleana* subsp. *morocera*

forms are recognised: **subsp. *bolleana*** has *contorted* fruits (*L, F), **subsp. *viridis*** has elongated fruits with conspicuous *lateral horns* at the apex (L, F) and **subsp. *morocera*** (*C, T). Arid coastal habitats. L, F, C, T.

Matthiola incana ALHELÍ, HOARY STOCK   A stout, bushy perennial to 80 cm with a woody stock and numerous non-flowering shoots, stems and leaves grey-green and hairy. Leaves lanceolate, usually untoothed and unlobed. Flowers pink, purple or white, the petals 20–30 mm long. Fruit long and thin, 45 mm–15 cm, laterally compressed; hairy but not glandular. Scrub; a garden escape. L, C, T, G, P.

Matthiola parviflora ALHELÍ MENUDO   An annual to 20 cm with deeply pinnately lobed lower leaves. Flowers small, with *brownish purple petals 6–10 mm (the limb just 2.5–4 mm);* sepals 4–6 mm. Fruit narrowly *cylindrical* 50 mm–11 cm long and *slightly constricted at intervals*; horns 1–1.5 mm, straight and pointed. Various habitats. L, F, C, T, G, H, P.

Lobularia PANIQUESO, SWEET ALISON

Annuals or perennials with narrow, untoothed leaves. Petals white or purplish. *Fruit a disk-like siliqua.*

Lobularia canariensis   A grey-downy, spreading woody-based perennial. Leaves entire, linear–lanceolate, blunt. Flowers white, borne in dense rounded racemes; petals 2.5–4.5 mm long; sepals green (sometimes with darker tips). Fruit small, sub-circular, flat. Seeds borne singly in a locule. Rocky slopes and cliffs; common. Numerous endemic forms described (including **subsp. *intermedia*** from Tenerife westwards, and **subsp. *marginata*** on the eastern islands, the only subspecies shared with the mainland coast of SW Morocco and Western Sahara). L, F, C, T, G, H, P. *L. libyca* CAMOSILLA is a similar *annual* with smaller flowers (petals 1.5–2 mm long) and seeds winged along the entire perimeter. Seeds borne 4–5 per locule. Cliffs and scree; locally common. L, F, C, T, G, H, P. *L. maritima* is a predominantly Mediterranean species that has been widely recorded on the islands, possibly naturalised; it is very similar to *L. canariensis*. C, T, H, P.

Lepidium PEPPERWORT

Leaves simple to 2–3-pinnately divided. Flowers white or reddish. Fruits flattened and strongly keeled or winged.

Lepidium didymum (syn. *Coronopus didymus*) SERVELLINA, LESSER SWINECRESS   A small, spreading or ascending biennial to 40 cm, strong-smelling when crushed. Leaves divided, feathery, at first in a rosette, later on spreading stems. Flowers inconspicuous, petals 0.5 mm, shorter than the sepals, or absent. Racemes elongated in fruit, flower stalks longer than the fruit. *Fruit dumb-bell shaped,* 1.2–1.7 mm, veined and with a notch at the apex, style absent. Sandy waste places. L, F, C, T, G, H, P. *L. coronopus* (syn. *Coronopus squamatus*) SERVELLINA VERRUGOSA is similar, but with flowers with white petals to 2.5 mm across, and *ridged, kidney (not dumb-bell)-shaped* fruits 2–3 mm. F, T, C.

Diplotaxis WALL ROCKET

Annuals or perennials with (normally) deeply lobed leaves. Petals yellow. Fruits linear and flattened with seeds in 2 rows.

Diplotaxis tenuifolia RÚCULA   A perennial 20–80 cm (1 m). Leaves somewhat *fleshy*, lobed with 4–8 segments, hairless or with some hairs along the margins, strong-smelling when crushed.

1. *Matthiola incana*
2. *Matthiola parviflora*
3. *Lobularia canariensis* subsp. *marginata*
4. *Lobularia canariensis* subsp. *marginata*
5. *Lobularia canariensis* subsp. *intermedia*
6. *Lobularia libyca*
7. *Sinapis arvensis*
8. *Hirschfeldia incana*
9. *Cakile maritima*

Flowers lemon yellow; petals 8–10 mm. Fruit a slender, sometimes curved siliqua 20–60 mm borne on long, erect to spreading stalks. Thickets and roadside scrub. L, F. ***D. muralis*** **WALL ROCKET** is similar but a sparsely hairy *annual* to 60 cm with *thin* green leaves; petals 6–8(10) mm. Fruit 15–40 mm. L, T.

Sinapis MOSTAZA, MUSTARD

Annuals with crimped to lobed leaves. Flowers with spreading sepals and yellow sepals. Fruit a siliqua, splitting lengthways, with a distinct beak; valves with 3(7) veins.

Sinapis alba MOSTAZA, WHITE MUSTARD A tall, normally bristly annual (sometimes hairless) to 70 cm. Leaves *all stalked* and pinnately lobed. Flower pale yellow. Fruit 20–40 mm long, the beak flattened and sword-like, as long as, or exceeding the valves (10–30 mm). Cultivated and waste areas. L, F, C, T, G, H, P. ***S. arvensis*** JÉBANA, CHARLOCK is similar but with lanceolate *unstalked* upper leaves, and a larger fruit 25–45 mm long, with a *conical beak shorter than the 3–7 valves* (7–16 mm). Similar habitats. L, F, C, T, G, H, P.

Hirschfeldia HOARY MUSTARD

Annuals or weak perennials with lobed to divided leaves. Petals yellow; sepals erect. Fruit a siliqua with a short, swollen, club-shaped beak; valves with 1–3 strong veins.

Hirschfeldia incana RELINCHÓN, HOARY MUSTARD A tall, lax, erect annual to 1.2 m. Lower leaves stalked and pinnately divided with an oblong end-lobe and up to 9 pairs of lateral lobes; the uppermost unlobed. Flowers pale yellow, borne in crowded terminal racemes; petals 6–9 mm. Fruit 6–17 mm, closely appressed to the stem, peg-shaped with a swollen, 1-seeded upper segment, and a flattened, 2–6-seeded lower segment. Disturbed, sandy ground; frequent. L, F, C, T, G, H, P.

Cakile SEA ROCKET

Hairless annuals with succulent and blue-grey leaves. Petals white, pink or violet. Fruit a 2-parted siliqua.

Cakile maritima RÁBANO MARINO, SEA ROCKET A variable short, rather succulent, spreading, hairless annual to 50 cm. Leaves grey-green, irregularly pinnately lobed, the lobes narrow and untoothed, or undivided. Flowers lilac to white, borne in racemes that elongate significantly in fruit; petals 4–10 mm long. Fruit 7–25 mm, brown and succulent; bipartite, *the lower segment with an arrow-shaped base*, the upper oval and 4-angled. Common on maritime sands in the east. L, F, C, T.

Rapistrum

Annuals or perennials with toothed to lobes leaves. Petals yellow. Fruit a 2-parted siliqua: the lower slender with 0–1(3) seeds, the upper spherical, with 1 seed, wrinkled and with a persistent style.

Rapistrum rugosum RELINCHÓN BASTARDO, BASTARD CABBAGE An annual to 80 cm, bristly-hairy below and hairless above. Lower leaves pinnately divided, often toothed and stalked. Petals *lemon yellow*, 6–8 mm long. Fruit 4–10 mm, the upper segment ovoid–spherical *and abruptly contracted into the beak*; the lower segment cylindrical or swollen. Ruderal habitats. L, F, C, T, G, H, P.

Crambe COL DE RISCO, SEAKALE

Shrubs or bushy perennials with lobed leaves. Flowers with white petals and sepals erect to spreading. Fruit a 2-parted siliqua: the lower section stalk-like and seedless, the upper spherical and 1-seeded.

(a) Leaves <20 cm, rough.

Crambe strigosa A small shrub to 1.5 m. Leaves ovate, *thick and rough*, with irregularly jagged, wavy margins; the stalks of adult leaves often with a pair of sub-opposite, clasping auricles. Flowers small, borne in intricately diffuse panicles; petals white, much exceeding the calyx. Fruit weakly 4-lined. Common on cliffs and banks in laurel forests. *T, G, P. *C. scaberrima* is somewhat similar but with broadly ovate, flat leaves narrowing abruptly into short, winged stalks or stalkless. Flowers large, with petals 3 x the calyx, borne in panicles on robust stems with spreading branches. Local on cliffs and slopes. *T (northwest to southwest). *C. feuilleei* is an island endemic similar to *C. strigosa*. *H. *C. pritzelii* is like *C. strigosa* but with rough, *spiny* stems, with leaves very rough beneath. *C (Los Tilos de Moya).

1. *Rapistrum rugosum*
2. *Crambe strigosa*
3. *Crambe scaberrima* (habit)
4. *Crambe scaberrima*
5. *Crambe arborea* var. *indivisa*
6. *Crambe arborea* var. *indivisa* (left) and var. *arborea* (right)

Crambe arborea COLDERRISCO DE GÜÍMAR A shrub to 2 m with ridged stems. Leaves rather shiny, grass-green, ovate, coarsely lobed (**var.** *indivisa*), or sometimes into linear segments (**var.** *arborea*); *rough*. Flowers borne in inflorescences with slender branches; petals 2 x the calyx. Fruits 4-ribbed. Rocky cliffs; rare and local. *T (south: Güímar and Arafo Valley).

Crambe gomerae A small shrub with lanceolate or ovate, rough, stalked leaves, somewhat lobed towards the base; *stalks narrowly winged*. Flowers borne at the ends of the panicle branches. Fruit 4-ribbed. Cliffs and ravines; rare. *G.

Crambe tamadabensis A small shrub with a woody stock and loose rosettes of lanceolate to narrowly ovate–lanceolate, stalked, rough leaves with sharp teeth; leaves ageing red and yellow. Inflorescences *erect* with spreading branches, sparse, hairless. Rocky crevices; rare. *C.

Crambe scoparia A slender cliff plant, robust at the base; stems thick and woody when mature. Leaves linear to narrowly lanceolate with toothed to weakly lobed margins. Flowers borne in inflorescences with a few, long and pendulous branches. Fruit 4-ribbed. Mountain cliffs; rare. *C (west).

(b) Leaves large, up to 50 cm, not conspicuously rough.

Crambe santosii A *tall* shrub to 4 m with *large leaves*, to 50 cm; leaves rather thin, not particularly rough, stalked and with jagged edges, rather soft to the touch. *G, P.

(c) Leaves <20 cm, hairless.

Crambe sventenii COLINO A small shrub with lyre-shaped, coarsely lobed, smooth and glaucous leaves. Flowers borne in panicles with suberect branches. Fruit strongly 2-winged. Boulder crevices on Montaña del Cardón and adjacent peaks; rare. *F.

Crambe laevigata A small shrub with ovate or ovate–lanceolate, rather regularly and coarsely *toothed*, hairless, slightly glaucous leaves. Flowers few borne in lax panicles. Rocky cliffs; rare and local (Masca and adjacent ravines). *T.

Raphanus RADISH

Annuals or perennials with a peppery smell; leaves shallowly lobed. Petals white, mauve or yellow; sepals erect. Fruit a cylindrical siliqua, elongated into a seedless beak, constricted between the seeds.

Raphanus raphanistrum JARAMAGO, WILD RADISH A variable short to tall, bristly annual to 80 cm, erect and branched. Flowers white to pale yellow or mauve, *with lilac or reddish veins*, borne in branched racemes; petals 10–25 mm. Fruit 20–60 mm long, *jointed and beaded*. Common on arable land, and in grassy habitats. L, F, C, T, G, H, P.

Eruca

Annuals with deeply lobed leaves. Flowers with erect sepals; petals white or yellow with purple veins. Fruits with seeds in 2 rows under each 1-veined valve.

Eruca vesicaria (syn. *E. sativa*) HEDIONDO, RÚCULA A bristly annual to 1 m with stalked, lobed leaves with a large terminal lobe. Flowers with erect, purple sepals and white or pale yellow, purple-veined petals 15–20 mm. Fruit a small, *unbeaded siliqua 10–18 mm long with a flattened, sword-shaped beak*. Disturbed habitats. L, F, C, T, G, H.

1. *Crambe sventenii* (habit)
2. *Crambe sventenii*
3. *Crambe sventenii* (flowers)
4. *Raphanus raphanistrum*
5. *Erucastrum canariense*
6. *Erucastrum canariense* (habit)
7. *Carrichtera annua*
8. *Notoceras bicorne*
9. *Notoceras bicorne* (in fruit)

Erucastrum HAIRY ROCKET

Annuals or perennials with deeply lobed leaves. Petals yellow. Fruits linear, constricted between the seeds; seeds in 1 row under each 1-veined valve.

Erucastrum canariense ALCARCÁN A small annual with densely bristly entire to toothed leaves; those above *stalkless*. Flowers pale yellow, borne in lax racemes; petals 8–9 mm. Fruit a stalked, conical silicula. Common to abundant in disturbed habitats after winter rain, especially in the east. *L, F, C, T, G, P. *E. cardaminoides* is similar, with lower leaves segmented almost to the middle, and the upper leaves *stalked,* virtually *hairless*. Flowers small; petals just 6 mm. Locally common on disturbed lower ground. *L, F, C, T, G, P.

Carrichtera

A monotypic genus of annuals with pinnately divided leaves. Flowers pale yellow; filaments free. Fruit a 2-parted, pendent siliqua.

Carrichtera annua CUCHARILLA A small, rigid, bristly annual to 40 cm with linear 2–3-pinnately divided leaves. Flowers borne singly in elongated, leaf-opposed spike-like inflorescences; petals pale yellow, exceeding the hairy sepals. Fruit with 2 transverse segments: the lower ellipsoid with 2 chambers and 2 boat-shaped, 3-veined valves; the upper strongly flattened and sterile; seeds 3–4 per chamber. Bare, dry ground; common in the east. L, F, C, T, G.

Notoceras

Annual herbs with entire leaves. Flowers yellow, borne in racemes. Fruit a 4-sided, short siliqua.

Notoceras bicorne PATA GALLINA A small, spreading to ascending annual to 30 cm with entire, narrowly lanceolate to linear leaves to 25(30) mm with white adpressed hairs. Flowers small and yellow, borne in compact racemes, elongating in fruit; petals 2 mm; sepals 1.2–1.6 mm. Fruit erect, adpressed to the stem, to 10(12) mm long. Seeds oblong, rough, 1.2 mm. Deserts and bare ground. L, F, C, T, G, H, P.

Descurainia

Small shrubs with densely clustered 1–3-pinnate leaves. Flowers borne in dense to lax racemes, yellow. Fruit a siliqua, more-or-less 4-angled. Seeds usually winged.

Descurainia bourgaeana PAJONERA DE CUMBRE A shrub to about 1 m, branched and woody at the base, hairy and leafy. Leaves virtually stalkless, 2-pinnatisect with linear–lanceolate segments, often toothed at the tip. Fruiting stalks spreading; siliquas 16-seeded. Seeds brown. Dry slopes; common in Las Cañadas, rare in La Palma. *T, P. The following are similar: *D. gonzalesii* has leaves pinnate to almost entire; lobes linear, acute. Rare and local. *T (around Las Cañadas del Teide). *D. lemsii* has *silvery-grey* leaves with narrow, acute lobes and erect, fruiting stalks carrying many (28–32)-seeded siliquas. High pine forest slopes; locally common. *T.

Descurainia preauxiana A small, branched shrub 60–80 cm with leaves borne in rosette-like clusters towards the stem tips, divided into 6–10 pairs of linear to thread-like lobes, sparsely glandular. Siliquas sub-erect with 18–27 seeds. Local on exposed, dry slopes. *C (south and central).

Descurainia artemisioides A small shrub to 50 cm. Leaves stalkless below, 2-pinnatisect with 6–10 pairs of primary leaflets; lobes broad and blunt. Siliquas 20-seeded; seeds brown. Shaded cliffs and ravines; rare. *C (northwest).

1. *Descurainia bourgaeana* (habit) (RACHEL GRAHAM)
2. *Descurainia bourgaeana*
3. *Descurainia gonzalesii*
4. *Descurainia lemsii*
5. *Descurainia preauxiana* (habit)
6. *Descurainia preauxiana*

1. *Descurainia gilva*
2. *Descurainia gilva*
3. *Descurainia millefolia* (habit)
4. *Descurainia millefolia*
5. *Parolinia ornata*
6. *Parolinia filifolia*
7. *Parolinia intermedia*

Descurainia gilva A small shrub to 40 cm. Leaves stalkless, more or less erect, 2-pinnatisect, grey-felted. Flowers with petals 3–4 mm; sepals 2.5 mm. Siliquas 16–24-seeded; fruiting stalks erect-ascending. Seeds brown. High pine forest slopes; common very locally. *P.

Descurainia millefolia A shrub to 1 m. Leaves in loose rosettes towards the stem tips, 3-pinnatisect with toothed lobes. Flowers borne in congested racemes. Fruiting stalks ascending, curved. Siliquas with 10–20 seeds; seeds reddish. Cliff scrub, dry slopes and pine forests; locally common. *T, G, P.

Parolinia DAMA

Erect shrubs with starry hairs. Leaves linear to narrowly lanceolate, entire. Flowers pink or white. Fruit a siliqua, usually with a forked appendage.

(a) Gran Canarian endemics.

Parolinia ornata An erect, silvery shrub to 1.5 m. Leaves slender, linear–lanceolate, to 10 cm. Flowers pale pink, borne in clusters of 10–30, *small*, 15 mm across; sepals erect, 5–6 mm. Siliqua 20–25(30) mm with 5–9 seeds per valve; appendages deeply forked. Exposed slopes; locally common. *C (south and west). *P. platypetala* is similar but with larger flowers 15–20 mm across with petals that have *crimped edges*. Rare. *C (Guayadeque). *P. filifolia* is similar, with linear to *thread-like* leaves, few (1–6) flowers per cluster, and a short siliqua, 5–20 mm, with 2–5 seeds per valve. Hillside scrub; rare. *C. *P. glabriuscula* is similar, also with linear leaves; flowers 4–25 per cluster; siliqua short, recurved 10–18 mm, with a *simple* appendage ending in a point. Very rare. *C (Bandama).

(b) Species from Tenerife westwards.

Parolinia intermedia An erect, compact, silvery shrub with linear to oblanceolate leaves. Flowers pale pink, borne in elongated racemes 80 mm–15 cm long; flowers 10–15 mm. Fruits 10–15 mm, 4-parted, with 3–5 seeds per valve. Dry slopes; locally frequent. *T (west of the island from Punta de Teno to Guaza).

Parolinia schizogynoides A compact dwarf shrub to 80 cm. Leaves linear, 30 mm. Racemes with up to 20 flowers 10 mm across; petals narrow. Siliqua 10 mm with 2–3(4) seeds per valve; appendages shortly bifid or trifid at the apex. *G (southwest). *P. aridanae* is similar but with *small* flowers <10 mm across and siliqua very short – just 5 mm – with 1–3 seeds. Very local (east). *P.

SANTALACEAE

Woody or herbaceous perennials, hemi-parasitic on the roots of surrounding vegetation. Flowers small, cosexual or unisexual; stamens 3–5; style 1. Fruit a berry or nut.

Thesium ESCOBILLA

Perennial hemi-parasitic herbs with simple or branched stems. Leaves alternate, stalkless, linear and entire. Flowers borne in racemes or panicles; perianth 5-lobed; stamens 5. Fruit a nut. Most species rare and local.

(a) Annuals in grassy habitats.

Thesium humile ROMERILLO A slender, ascending to erect, branched annual 70 mm–20 cm. Leaves linear, 10–35 mm. Flowers inconspicuous, greenish-white or yellowish, 1.25–1.5 mm; tepals triangular, 0.5 mm. Fruit 2.53 mm with an oblong, erect, net-veined nut, and persistent perianth ¼ x as long; fruiting stalk slightly swollen. Grassy habitats; local. L, F, C, T, G, P.

1. *Thesium retamoides*
2. *Thesium retamoides* (flowers)
3. *Thesium retamoides* (fruit)
4. *Thesium subsucculentum*
5. *Osyris lanceolata*
6. *Frankenia pulverulenta*
7. *Frankenia pulverulenta*

(b) Broom-like shrubs on dry slopes.

Thesium retamoides A large shrub to 2 m, much-branched with frosted (pruinous) stems and scale-like, triangular, pointed leaves. Flowers short-stalked, the lateral sterile, the middle fertile, 3–4 mm; lobes acute. Fruit minutely bristly. Rare. *T.

Thesium subsucculentum A shrub to 80 cm with dense, divergent, succulent and minutely bristly stems. Leaves soon-falling, 1.5 mm. Flowers *stalkless*, 2–2.5 mm with triangular lobes, minutely bristly outside. Fruit globose, 6 mm, with a persistent calyx. Rare. *T. *T. palmense* is similar, with striated minutely hairy stems, linear leaves, virtually stalkless flowers and a pearly fruit. *P.

Thesium canariense A shrub to 80 cm. Lower branches brown, woody; upper branches green, hairless and flexuous. Leaves scale-like, triangular, 2 mm. Flowers small, cream with acute lobes. Fruit white, 3–6 mm, 1-seeded. Hill scrub; very rare. *C (Guayadeque). *T. psilotocladum* is similar but with minutely bristly branches and larger flowers with blunt lobes. Fruit ovoid, light green, 2.5–4 mm. Dry rocky slopes of the Teno massif; possibly extinct. *T (Masca).

Osyris

Dioecious shrubs with angled stems and entire leaves. Flowers with 3–4 tepals and 3–4 stamens or 1 style. Fruit a berry.

Osyris lanceolata BAYÓN A shrub 1–2.5(3) m, with yellowish leaves 18–30(45) mm, leathery and often with pronounced pinnate veins and *small, papery bracts that are shed before fruiting*. Tepals 4 (sometimes 3), and flowers male or cosexual, borne on separate plants (androdioecious); stamens hairless. Fruit a red berry 7–9 mm. Hill scrub; locally common on some slopes. T, G, P.

FRANKENIACEAE | SEA HEATH FAMILY

Dwarf shrubs (sometimes annual herbs) with opposite, entire leaves and no stipules. Flowers usually cosexual with 5 partly fused sepals and 5 free petals; stamens usually 6; style 1, divided. Fruit a small capsule.

Frankenia TOMILLO MARINO

Woody-based subshrubs, easily distinguished by their distinctive flower structure and small, *Erica*-like leaves.

Frankenia pulverulenta ANNUAL SEA HEATH An *annual* with numerous *prostrate branches* 50 mm–17(30) cm, not particularly woody, often spreading in a circle, with oval leaves 5–6(8) mm long, hairless above, crispy-hairy below, often reddish. Flowers borne in the axils of the branches and upper leaves; stalkless, pink, the petals notched. Bare and rocky ground; common in the east. L, F, C, T, H.

Frankenia capitata A woody-based, mound or mat-forming shrublet. Leaves *linear*, white-encrusted with *revolute* margins. Flowers pink, borne in *dense, terminal* clusters; *calyx covered in white bristly hairs*. Bare, dry places and sand dunes; common in arid areas. L, F, C. *F. boissieri* is similar, with erect branches 90 mm–30(40) cm long with *oval-heart-shaped leaves* 4–7(10) mm long, not white-encrusted, *hairless above*, and purplish flowers borne in dense terminal clusters; calyx with white hairs. Saline maritime habitats; confused with the former species. L, F, C. *F. ericifolia* is similar to *F. capitata* but laxer in habit, with *flat, lanceolate to oblanceolate* leaves and flowers scattered along the stems (not all terminal); sepals velvety; petals pale rose, sometimes whitish. Similar habitats. L, F, C, T, G, H, P.

TAMARICACEAE | TAMARISK FAMILY

Deciduous shrubs or trees with alternate, small (often scale-like), stalkless leaves. Flowers typically in catkin-like racemes, with (4)5 sepals and petals; stamens 5; styles 3–4. Fruit a capsule; seeds with hairy tufts.

Tamarix TARAJAL, TAMARISK

Shrubs or small trees with a distinctive habit and simple, alternate, small, scale-like leaves. Flowers tiny, borne in catkin-like spikes. Fruit a capsule with hairy, wind-borne seeds.

Tamarix africana A feathery, more or less hairless small tree to 3(6) m with *blackish or purplish* bark. Leaves small, 1.5–3 mm, pointed, growing close to the stem. Flowers *stalkless*, white or pale pink, borne in almost stalkless racemes 30–70 mm long and 6–8 mm wide, borne on the previous year's wood often before the leaves. Petals 5, to 2–3.3 mm long and usually *persistent* in fruit. Bracts hairy. Valley beds. F, C, T.

Tamarix canariensis A shrub, similar to the above, but with *reddish-brown* bark. Leaves 1–3 mm. Flowers pink, *stalked,* borne in shorter racemes 15–45 mm long and 3–5 mm wide. Locally abundant in coastal regions and lowland valleys. L, F, C, T, G, H, P.

PLUMBAGINACEAE | THRIFT FAMILY

Perennial herbs with basal, untoothed, simple leaves without stipules. Flowers regular, borne in lax or tight clusters, 5-parted with papery calyx lobes that persist in fruit; petals fused in the lower part to form a tube; stamens and styles 5. Fruit a 1-seeded capsule.

Plumbago

Herbaceous perennials and shrubs with clasping leaves. Inflorescence a dense, terminal spike; corolla tubular; stamens 5, free; style 1 with 5 stigmas. Fruit dry, 1-seeded and 5-valved.

Plumbago auriculata EMBELESO A scrambling shrub to 6 m. Flowers sky-blue (rarely white), 40 mm long. Native to South Africa, planted, sometimes naturalised. C, T, G, P.

Limonium SIEMPREVIVA, SEA LAVENDER

Perennial herbs with simple, leathery leaves forming basal rosettes and flowers in *branching cymes*; flowers persisting and papery in fruit; stamens and styles 5. A highly complex genus. Coastal habitats.

(a) Rosette plants; main flowering stems *winged*. Leaves deeply sinuately or pinnately lobed.

Limonium imbricatum A rosette plant. Leaves pinnately lobed with broadly lanceolate to oval lobes; hairy. Flower stem and branches broadly winged; calyx deep mauve. Coastal rocks; rare. *T, P.

---

1. *Frankenia capitata*
2. *Frankenia ericifolia*
3. *Tamarix canariensis*
4. *Tamarix canariensis* (infloresences)
5. *Plumbago auriculata*
6. *Limonium imbricatum*
7. *Limonium macrophyllum*

Limonium brassicifolium A woody-based rosette with a stock to 20 cm. Leaves stalked, lobed with a large terminal segment to 15 cm, tapering. Flower stems broadly winged (wings to 15 mm). Calyx pale mauve. *G (north coast: **subsp.** *brassicifolium*), *H (**subsp.** *macropterum*).

(b) Perennials, shrublets or shrubs; main flowering stems *winged*. Leaves entire (or rarely shallowly lobed).

Limonium macrophyllum A small shrub with entire, stalkless to short-stalked *large* leaves to 30 cm; oblanceolate to narrowly obovate, hairless. Stalks expanded at the base. Flower stems narrowly winged; calyx bluish-purple, persistent; flowers small, white. Rare. *T (northeast cliffs).

Limonium arboreum A *large shrub* to 1.8 m with smooth, woody stems. Leaves ovate, hairless, glaucous and long-stalked with jagged margins. Flowers borne in densely branched inflorescences; stems narrowly winged below; calyx blue-mauve, persistent; corolla small and white. Rare and local; also planted. *T (north cliffs), *P.

Limonium sinuatum WINGED SEA LAVENDER A bristly perennial 10–40 cm. Leaves wavy-lobed, 40 mm–15 cm long and 8–30 mm wide, borne in a basal rosette. Stems with 3–4 undulate wings. Flowers cream; calyx 11.5–14 mm, *conspicuous, blue-purple and with a papery margin*. Very rare on disturbed habitats and coastal areas. L, F, C. ***L. lobatum*** is similar but an *annual* with *angular* (not winged) stems and pale yellowish flowers; *calyx pale blue,* deeply lobed. F, C.

(c) Shrublets; main flowering stems winged or *unwinged*. Leaves pinnately lobed.

Limonium spectabile A small subshrub. Leaves pinnate with linear–lanceolate lobes with ciliate margins, the terminal segment triangular. Flower stems virtually unwinged but peduncles narrowly winged; calyx mauve. Very rare. *T (Teno Massif).

(d) Shrublets or woody-based rosette plants; main flowering stems winged or *unwinged*. Leaves entire (or rarely shallowly lobed).

Limonium preauxii A small, shrubby perennial with loose rosettes of large, rhomboid, hairless and entire leaves; stalks equalling the blades, the blades heart-shaped at the base. Flower stems wiry, unwinged; flowers with deep mauve calyx. Rocky slopes. *C (south). ***L. sventenii*** has leaf bases graduating into their stalks. *C (west).

Limonium bourgaei A virtually stemless woody-based rosette plant to 40 cm. Leaves hairy, long-stalked, broadly ovate–rhomboidal, often slightly lobed at the base. Flowers borne on hairy branches; stems unwinged; flowers with deep mauve calyx. High north-facing sea cliffs. *L (Famara), F. ***L. puberulum*** is similar but smaller, just 10–15 cm, with short-stalked leaves attenuated at the base, with ciliate margins. Flower stalks to just 60 mm; calyx violet-blue; corolla white. Sea cliff edges and feet. *L, F(?).

Limonium pectinatum A dwarf shrub. Leaves spathulate, 20–40 mm, borne in dense rosettes. Flower stems unwinged, often with leafy bracts. Flowers borne densely in upward-facing branches; calyx pale mauve; corolla whitish. Common on coastal rocks and sands from C westwards. Various forms described including **var. *solandri***. *C, T, G, H, P.

1. *Limonium arboreum*
2. *Limonium lobatum*
3. *Limonium preauxii*
4. *Limonium bourgaei*
5. *Limonium bourgaei*
6. *Limonium puberulum*
7. *Limonium puberulum* (leaf)
8. *Limonium puberulum* (in bud)

Limonium bollei A small, hairless perennial with basal rosettes. Flower stems not winged. Saltmarshes; a rare islet endemic. *F (Lobos).

(e) Perennials, shrublets or shrubs; main flowering stems winged or *unwinged*. Leaves entire (or rarely shallowly lobed at base), or reduced to absent.

Limonium dendroides A *large* shrub to 3 m with long, woody stems. Leaves in rosettes, lanceolate, hairless, rather leathery. Flowers borne in panicles with unwinged branches; calyx pinkish. Inland cliffs. *G.

Limonium papillatum A small, often *prostrate* perennial with hairless, obovate or spatulate leaves in rosettes turning reddish with age. Flowers borne in *zig-zagging*, terminal, *papillose* inflorescences; calyx white; corolla pale mauve. Locally common on wind-eroded cliffs and coastal sands in the east. Var. *papillatum* is endemic on the easternmost islands and islets. *L, F.

Limonium tuberculatum An irregular, greyish shrub to 60 cm. Leaves hairless, *reduced*. Inflorescences short, with papillose branches bearing pink flowers. Coastal flats, saltmarshes, and rocks; rare. F, C.

Limoniastrum

Similar to *Limonium* but *large, spreading shrubs* with leaves alternate along the stems and with chalky glands. Flowers with 5(6) stamens.

Limoniastrum monopetalum SALITROSA, LIMONIASTRUM A fleshy, spreading shrub 50 cm (2 m) with silvery-green, fleshy, spoon-shaped leaves 20–60(90) mm long covered in white scales, sheathing the stem at the base. Flowers bright pink and conspicuous, later violet, to 16 mm, borne in loosely branched spikes; corolla with 5 spreading, oval petals. Introduced as an ornamental and an escape in maritime habitats. F, C, T.

POLYGONACEAE | DOCK FAMILY

Herbs or small shrubs without latex, with alternate leaves and stipules that form a membranous sheath around the stem. Flowers cosexual or unisexual, often small and greenish or reddish; tepals 3–6; stamens (3)6–9; stigmas 2–3; sessile or with styles. Fruit an achene.

Polygonum

Annuals or perennials with tap-roots. Flowers single or few (<6) in the leaf axils, exceeded by the leaves; stamens 8; stigmas 3, virtually sessile. Achene with 3 rounded angles.

Polygonum maritimum TREINTANUDOS DE MAR, SEA KNOTGRASS A prostrate or ascending, branched perennial 10–50 cm with a woody stock. Leaves narrowly elliptic, blue-grey and sessile, with downturned margins, 15–25 mm long and 6–9(16) mm wide. Stipules silvery, reddish at the base with *8–12 conspicuous, branched veins*. Flowers white or pink, 5-lobed and 3–4 mm across, solitary or 2–3 in the nodes. Common on maritime sands and shingle. L, F, C, P.

---

1. *Limonium pectinatum* var. *solandri* (habit)
2. *Limonium pectinatum* var. *solandri*
3. *Limonium bollei*
4. *Limonium dendroides* (habit)
5. *Limonium dendroides*
6. *Limonium papillatum*
7. *Limonium tuberculatum*

Polygonum equisetiforme An ascending perennial with *long, wiry branches*, and stipules above each leaf *shorter than the internodes* (appearing bare towards the tips); leaves 2–5(10) mm wide. Flowers borne in *lax spikes that are bare above* or with small, leafy bracts only. F, C.

Polygonum aviculare TREINTANUDOS A small, prostrate annual. Leaves elliptical to lanceolate, ›5mm wide. Flowers white or pink, arranged in lateral fascicles. Grassy habitats. L, F, C, T, G, H, P.

Rumex DOCK

Perennials with terminal or axillary racemes, or panicles with whorled flowers; stamens 6; styles 3. Valves (fruiting inner tepals) important. Achene 3-angled.

(a) Annual to perennial herbs with entire leaves. Valves small (‹5 mm), rather inconspicuous (to absent).

Rumex bucephalophorus VINAGRERILLA, HORNED DOCK A variable, reddish, erect annual to 30(50) cm, branched or not. Leaves normally small, 6–35(65) mm long, lanceolate, oval or spoon-shaped, stalked, and greyish green. Flowers borne on variously sized flower stalks, very small, bright red and in clusters of 2–3 in the leaf axils, forming a long, dense spike. *Valves triangular–oval or narrow, with 3–4 teeth*. The Canary Islands forms belong to subsp. ***canariensis***, present also in Madeira. Waste ground and sea cliffs. L, F, C, T, G, H, P.

Rumex acetosella ACEDERILLA, SHEEP'S SORREL A spreading to erect, *slender* perennial to 45(80) cm with *small leaves*; leaves *arrow-shaped with small, forward-directed lower lobes and a large, oval–lanceolate central lobe* 6–60 mm long. Flowers greenish or reddish, borne in simple or branched racemes, unisexual, the male and female flowers borne on separate plants. Fruit valves to 1.6 mm, equalling the achenes. Fallow land and forest clearings; locally common. L, F, C, T, G, P.

Rumex conglomeratus A hairless perennial with erect stems to 1 m. Leaves short-lived, oblong–lanceolate, 10–30 cm, rounded at the base. Flowers borne in terminal inflorescences, 10–20 per whorl. Valves just 2–3 mm long, to 1.5 mm wide, narrowly triangular, reddish-brown. Woodland fringes; common in the west. C, T, G, P.

Rumex crispus LABASA A hairless perennial with erect stems to 1.2 m. Leaves short-stalked, lanceolate, 10–25 cm; those above small, narrower. Valves greenish, 3–6 mm, heart-shaped and *conspicuously net-veined*. Woodland fringes and thickets; common in the west. L, C, T, G.

Rumex pulcher A hairless perennial with erect stems to 70 cm. Leaves to 15 cm, ‹4 x as long as wide, the base truncated or weakly heart-shaped. Flowers borne 10–20 in rather dense whorls; valves triangular–ovate, 4–5 mm, conspicuously net-veined, and with *2–6 spine-like teeth* on each margin. Woodland fringes and thickets; common in the west. C, T, G, H, P.

Rumex spinosus (syn. ***Emex spinosa***) CAÍL, EMEX A hairless, short, somewhat fleshy annual with sprawling stems to 50(60) cm. Leaves oval, heart-shaped at the base, 12–14 x 8(10) cm, long-stalked (to 25 cm). Male flowers stalked in terminal clusters, female flowers sessile at the base. *Fruit a spiny nut*, 3 mm. Maritime sands and disturbed ground in arid areas; locally common. L, F, C, T, G, H, P.

---

1. *Polygonum maritimum* (habit)
2. *Polygonum maritimum*
3. *Rumex bucephalophorus*
4. *Rumex acetosella*
5. *Rumex conglomeratus*
6. *Rumex crispus*
7. *Rumex pulcher*
8. *Rumex pulcher* (tepal valves)
9. *Rumex spinosus*

(b) Annual to perennial herbs with entire leaves. Valves large (5–23 mm), red, and conspicuous.

Rumex vesicarius VINAGRERA, BLADDER DOCK An annual with oval-triangular to rounded leaves. Flower stalks each with 2 flowers, one of which is smaller and concealed by the *conspicuous*, inflated, rounded valves 12–18(23) mm long, *flushed bright pink or crimson* with darker netted veins when in fruit. Rocky and desert fringe habitats. L, F, C, T, G, H, P.

Rumex maderensis ACEDERA DE MADEIRA A perennial herb with ovate-deltoid leaves. Flowers bright pinkish-red, borne in dense, elongated panicles; valves 5–6 mm, papery, sub-circular. Rocky shelves, forest clearings, humid malpaís, and roadsides. C, T, G, H, P.

(c) Annual to perennial herbs with lobed leaves.

Rumex bipinnatus A perennial herb with stalked, variably *lobed* leaves: entire with 2 basal lobes to pinnately divided. Fruits pendulous; valves heart-shaped, reddish then papery. Coastal dunes. L, F.

(d) Woody-based perennials and shrubs.

Rumex lunaria VINAGRERA, CALCOSA A sparse shrub with slightly fleshy leaves wider than long, truncated at the base, blunt-tipped. Flowers greenish to brownish-orange, borne in compound panicles; valves kidney-shaped to circular. Rocky slopes and malpaís, waste ground and cliffs; locally common. *L, F, C, T, G, H, P.

Rumex scutatus A woody-based, much-branched perennial to 40(65) cm with leaves 10–23(45) x 5–26(40) mm, variably triangular–heart-shaped with *diverging basal lobes*, rather thick and blue-grey. Flowers unisexual, reddish, borne in branched clusters. Fruits with oval to rounded valves, heart-shaped at the base, *as long as wide* (4.5–6.5 mm). Mainly in mountains.

Fallopia

A small genus of annuals or herbaceous perennials and woody vines, previously included in the genus *Polygonum*. Flowers with 8 stamens; styles 3. Achene 3-angled.

Fallopia convolvulus FALSA CORREGÜELA An annual, clockwise twining, climbing or prostrate vine to 1 m with angular stems. Leaves heart- or arrow-shaped, pointed and mealy beneath, 30–70 mm long. Flowers greenish or yellowish-white, borne in loose clusters in the leaf axils. Fruit a triangular nut borne on a short stalk 1–3 mm long. Ruderal and bare places. Throughout. *F. dumetorum* is similar but with more rounded stems and *fruits borne on stalks 4–8 mm long,* often deflexed. Fruit black, smooth and glossy. Hedges, scrub and degraded woodland. T, G, H, P.

Coccoloba

Shrubs, trees and lianas native to tropical and subtropical America. Leaves alternate, often large. Flowers in spikes. Fruit a 3-angled achene.

Coccoloba uvifera UVA DE PLAYA An exotic-looking tree with *large, circular* leaves, glaucous with prominent pinkish veins, heart-shaped at the base. Inflorescences 10–30 cm. Fruits borne in long, *pendent clusters*; achenes small, *grape-like*, 8–11 mm, pale green, ageing purplish-red. Widely planted.

---

1. *Rumex vesicarius* (habit)
2. *Rumex vesicarius*
3. *Rumex maderensis* (habit)
4. *Rumex maderensis* (tepal valves)
5. *Rumex maderensis*
6. *Rumex bipinnatus*
7. *Rumex lunaria*
8. *Coccoloba uvifera*

CARYOPHYLLACEAE | PINK FAMILY

Herbs with opposite leaves. Flowers cosexual, regular, with 4–5 free or fused sepals and 4–5 petals (absent in some species), typically 8–10 stamens and 2–5 styles. Fruit a many-seeded capsule, sometimes a berry or 1-seeded achene.

Gymnocarpos

Erect, woody under-shrubs with opposite, stalkless, *fleshy* leaves. Flowers stalkless; sepals and petals 5, free; stamens 5; stigmas 3. Fruit dry, non-splitting.

Gymnocarpos decandrus　MATOCOSTA MILENGRANA　A woody-based shrub, often growing in bare rock. Leaves greyish, cylindrical, fleshy, stalkless, hairless and pointed, 8–16 mm. Flowers stalkless, yellowish-green to purplish, rather small and inconspicuous; sepals 2–3 mm long, with whitish margins; petals pointed, 1–1.5 mm long; stamens opposite the sepals and slightly exceeding the petals. Fruit membranous, 1-seeded and enclosed by persistent sepals. Dry, rocky habitats close to the coast; local. L, F (south), C, T, G.

Arenaria　MORUJILLA, SANDWORT

Leaves opposite and stipules present. Flowers with 10 stamens and 3 styles. Capsule with 6(–10) teeth. Many mountain-dwelling species occur besides those described here.

Arenaria serpyllifolia　THYME-LEAVED SANDWORT　A variable, downy, grey-green prostrate or ascending annual with unstalked, oval leaves. Flowers small, to 8 mm across, white, with unnotched petals, *exceeded by their sepals*. Fields and dry habitats. L, T, P. **A. leptoclados** is similar (also widely treated as a subspecies) with conical, straight-sided capsules; sepals 2–3 mm long. L, F, C, T, G, H, P.

Rhodalsine

Small, slender annuals or perennials. Flowers with 0 or 5 petals, 10 or fewer stamens and 3(–5) styles. Capsule with 3(–5) teeth. Including species formerly classified under *Minuartia*.

Rhodalsine geniculata (syn. *Minuartia geniculata*)　ROMERILLO　A prostrate to ascending, woody-based and much-branched, glandular-hairy perennial to 30 cm. Leaves 6–10 mm, elliptic to linear with a prominent vein beneath. Flowers pink, borne in lateral and terminal cymes; sepals 3–4 mm, with wide, translucent margins; anthers yellow. Capsule splitting into 3 boat-shaped valves; seeds kidney-shaped, slightly wrinkled. Bare, lowland habitats; common in the east. L, F, C. The following are similar, often treated as mere forms. **R. platyphylla** is a small, often purplish, hairy perennial. Leaves *ovate to sub-circular*, 15 mm long. Flowers small, pale pink; petals *much shorter* than the sepals. *F. **R. webbii** is an *annual*. *L, F, T.

---

1. *Gymnocarpos decandrus* (habit)
2. *Gymnocarpos decandrus*
3. *Rhodalsine geniculata*
4. *Rhodalsine platyphylla*
5. *Paronychia canariensis* (habit)
6. *Paronychia canariensis*

Cerastium MORUJA, MOUSE-EAR

Annuals or perennials, often hairy. Flowers with 4–5 sepals and petals (or absent), 3–5(6) styles and 4–5 or 10 stamens. Fruit a capsule with 2 x as many teeth as styles. Numerous similar species in the area.

Cerastium glomeratum STICKY MOUSE-EAR A short, erect or ascending annual to 45 cm, covered in sticky glandular hairs. Leaves oval–elliptic, to 20 mm long, hairy. White flowers borne in *dense clusters*, the *sepals hairy to the tips*, petals more or less equalling the length of the sepals. Bare ground and dry, grassy places. L, F, C, T, G, H, P.

Cerastium fontanum A variable, non-sticky, short-lived, tufted or matted *perennial* to 40 cm, with both short, non-flowering basal shoots with stalkless leaves. Flowers white, borne in loose clusters, petals slightly shorter than or slightly longer than the sepals, and deeply notched. Grassy habitats on higher ground. C, T, P.

Paronychia

Perennial herbs with small opposite leaves and *conspicuous silvery stipules*. Flowers small, borne in dense clusters *surrounded by silvery bracts*; stamens 5; styles 1–2.

Paronychia argentea NEVADILLA DE PLATA A branched, *mat-forming perennial* with stems 50 mm–50 cm long. Leaves oval–lanceolate, greyish, in opposite pairs, bristle-tipped and almost hairless. Stipules membranous, and shorter than the leaves. Flowers borne in lateral and terminal clusters 10–15(25) mm across; with prominent *membranous, silvery bracts* (most conspicuous in mature or fruiting heads), concealing the flowers; sepals with bristle-tips 0.4–0.65 mm. C, H. **P. echinulata** is a similar, spreading *annual* with stems to 20 cm and flowers with bracts much *shorter* (the flowers not concealed by them). C, P.

Paronychia capitata NEVADILLA DE RISCO A grey-green, hairy, tufted or mat-forming, short-lived perennial with stems to 12(15) cm. Leaves stiff, oblong to linear–lanceolate, 3–6 mm, hairy, equalling the length of the stipules; stipules 4 per node. Flowerheads rather dense with bracts completely concealing the flowers (like in *P. argentea*) and *markedly unequal*, somewhat fleshy *sepals inwardly curved at the tips*. Rocky slopes. Canary Island plants have been considered as subsp. *canariensis*. *C.

Paronychia canariensis NEVADILLA CANARIA A brownish-green, *woody-based perennial* to 60 cm with *erect stems*. Leaves soft, short-hairy, lanceolate, pointed. Flowers white, borne in dense, *paniculate cymes* towards the tips of the branches; sepals with a prominent, finely hairy spine (awn); style 0.5–0.6 mm. Damp rocks, cliffs and roadsides; locally common in the west. *C, T, G, H, P.

Illecebrum

Similar to *Paronychia* but with white, spongy sepals that persist in fruit. Stamens 5; stigmas 2. Fruit a 1-seeded, 5-valved capsule.

Illecebrum verticillatum CORAL NECKLACE A distinctive, low, hairless annual superficially similar to *Paronychia*. Stems creeping, rooting at the base, reddish, to 20 cm. Leaves 2–5 mm, opposite and with small stipules. Flowers borne in 2 whorl-like clusters at each node, 4–6-flowered with silvery bracts. Damp, sandy places. C, T.

Pteranthus

Small, fleshy annuals with minute, inconspicuous flowers. Calyx with hooded lobes, tipped with spiny appendages; petals absent; stamens 4.

Pteranthus dichotomus HIERBA BARROCA A spreading to ascending annual 10–15 cm with stems 1–3 x divergently branching, jointed at the nodes. Leaves linear, fleshy, 8–18 mm. Flowers borne in leafy, corymbose panicles with an *expanded, leaf-like* peduncle to 9 x 6 mm; calyx closed, 4-parted, with a *spiny appendage*; petals absent; stamens 4, opposite the calyx lobes; stigma 2-parted. Fruit 1-seeded; seeds compressed, 1–2 mm. Sandy habitats and deserts. L, F.

Herniaria RUPTUREWORT

Similar to *Paronychia* but often with tiny leaves and flowers, and *inconspicuous bracts*. Stamens 5; stigmas 2. Fruit an achene.

Herniaria cinerea ESTERILLA A variable annual (with many forms, sometimes treated as species) with bright green to grey-green leaves 4–8(11) mm long that are clothed in *dense, straight, white, spreading hairs*, and flowers 1.3–1.6 mm with 2 stamens, borne in roundish clusters of 7–12 in the leaf axils; *calyx with stout hairs*. Bare habitats. L, F, C, T, G, H.

Herniaria fontanesii A yellowish *woody-based perennial herb* with stems to 15 mm across, much-branched and spreading. Leaves fleshy to rigid, oblong–spathulate to elliptic, sometimes with adpressed, velvety hairs. Flower clusters 3–7-flowered, leaf-opposed, densely congested at the ends of the branches; flowers 2–2.5 mm, vase-like, with soft and hooked hairs. Dry, sandy habitats; local. F, C, T. *H. canariensis* is similar in habit but with much smaller flowers (1.25–1.4 mm) with *equal to subequal sepals*, in conspicuous clusters. *T. *H. hartungii* is *prostrate* with few flowers; sepals small with short hairs. Rare. F.

Polycarpon ALLSEED

Small herbs with forking stems and opposite or whorled leaves with papery stipules. Flowers with *keeled and hooded sepals;* stamens (1)3–5; stigmas 3. Fruit a 3-valved capsule.

Polycarpon tetraphyllum JABONERA, FOUR-LEAVED ALLSEED A small (sometimes minute) hairless annual without a woody stock, up to 35 cm high (usually much shorter). Leaves mostly in *whorls of 4*, oval, green or purple. Flowers in branched clusters, white, tiny to 2 mm across with notched petals shorter than the sepals; stamens 3–4. Seeds brownish, with protuberances. Disturbed places and bare, sandy habitats. L, F, C, T, G, H, P.

Sagina MORUJILLA, PEARLWORT

Small, often tufted, moss-like herbs. Flowers minute with 4–5 sepals and petals, or petals absent; stamens 4, 5, 8 or 10; styles 4–5. Capsule 4–5-valved.

Sagina apetala ANNUAL PEARLWORT A small, annual herb with very slender, *sub-erect stems to 15 cm, all producing flowers*. Leaves linear, tapered at the tips. Flowers solitary, small with 4 oval sepals, often hooded; petals minute and falling early. Sandy places. L, F, C, T, G, H, P.

Sagina procumbens PROCUMBENT PEARLWORT is a similar, moss-like, bright green *mat-forming perennial* with a short, non-flowering main stem bearing a central, dense leaf rosette and numerous lateral stems to 20 cm ascending from rooting bases. Sepals and stamens 4(5). Urban waste places. C, T, H, P.

1. *Herniaria cinerea*
2. *Polycarpon tetraphyllum*
3. *Spergula fallax*
4. *Spergula fallax*
5. *Spergularia fimbriata*
6. *Spergularia fimbriata* (stipules)
7. *Spergularia fimbriata* (seeds)
8. *Spergularia bocconei*

Spergula

Slender annuals with opposite leaves appearing whorled. Petals almost as long as sepals or slightly longer, white, entire; stamens 5–10; stigmas 5. Fruit a 5-valved capsule.

Spergula fallax ROMERILLO A slender, spreading to ascending annual with hairless stems 12–30 cm, branched from the base. Leaves *channelled beneath*, borne in whorls at the nodes, 10–25 mm, linear, hairless, often with sterile shoots. Flowers white, solitary or few, borne on long, slender stalks; sepals 3–4.5 mm with white, papery margins; petals *shorter* than the sepals. Seeds blackish-brown and *smooth with a shiny white membranous wing*, almost as broad as the seed. Dry and semi-arid areas. L, F, C, T, G, H.

Spergula arvensis A slender annual, branched from the base 10–30(60) cm, sparsely glandular-hairy throughout. Leaves in interrupted whorls of 6, linear, slightly fleshy, *not* channelled beneath. Flowers white, borne on long, slender stems; sepals 3–5 mm, slightly *exceeded* by the petals; stamens 5–10. Capsule exceeding the sepals. Seeds *black, scarcely* compressed, *keeled* (or only narrowly winged, <1/10 the seed width). L, F, C, T, G, H, P. **S. pentandra** is similar but with 5 stamens and *strongly* compressed, *winged* seeds. C, T, H, P.

Spergularia SEA-SPURREY

Annuals or perennials with narrow, opposite (sometimes seemingly whorled) leaves and leafy tufts at each node; leaves fleshy or not. Petals purple, pink or white; stamens 5–10; stigmas 5. Capsule 5-valved.

(a) Petals normally *exceeding* the sepals; seeds winged or partially winged.

Spergularia fimbriata ROMERILLO A woody-based perennial 50 mm–40 cm. Leaves awned; stipules silvery. Flowers lilac-pink with petals exceeding the sepals. Seeds *black with a pale, fringed wing*. Dry, bare places on lower ground; common in the east. L, F, C, T.

Spergularia media GREATER SEA-SPURREY A short, almost hairless perennial to 40 cm with fleshy leaves ending in an abrupt point, rounded beneath. Stipules broadly triangular. Flowers 10–12(13) mm across with white or pink *petals equalling or exceeding the sepals*; sepals 4–6 mm; stamens 0–10. Capsule 7–9 mm, greatly exceeding the calyx; seeds dark brown, all (or mostly) winged. Sandy shores and saltmarshes. L, F, C, T, G.

(b) Petals *equalling or shorter* than the sepals; seeds winged, not winged, or both.

Spergularia marina LESSER SEA SPURREY An annual to 35 cm with very fleshy leaves, normally a slender (not woody) stock, and short stipules that form a sheath. Inflorescence sparingly branched; flowers 5–8 mm across, petals pink above and white below, *not exceeding the sepals*; sepals 2.5–4 mm; stamens 2–7(10). Capsule 3–6 mm, exceeding the calyx; seeds light brown, unwinged, winged, or mixed. Salt marshes. L, F, C, T, P.

(c) Petals equalling or shorter than the sepals; seeds not winged.

Spergularia bocconei ESPARCILLA MAYOR A slender annual or biennial to 20 cm, with *densely hairy inflorescences*. Leaves scattered (not in dense clusters); stipules triangular. Petals pink with a white base, equalling or shorter than the sepals which are 2–4 mm. Capsule 2(–4) mm, shorter than the calyx; seeds grey-brown, not winged. Ruderal habitats and fields. L, F, C, T, G, H, P.

Spergularia diandra A slender, ascending annual to biennial 30 mm–30 cm; stipules short, triangular. Flowers pale lilac; sepals 2–3 mm; petals lilac (rarely white), narrowly elliptical, equalling the sepals; stamens 2 or 3. Capsule 1.5–3 mm, globose, about equalling the sepals; seeds 0.6–0.7 mm, unwinged, dark brown to black, wrinkled or with papillae. L, F, C.

Agrostemma

Annuals with sepals fused into a tubular calyx with long, green, leaf-like teeth; stamens 10, styles 5. Fruit a capsule with 5 teeth.

Agrostemma githago NEGUILLÓN, CORNCOCKLE An annual to 70 cm (1 m) covered in adpressed, greyish hairs. Leaves narrowly lanceolate and pointed, 5–7 mm across. Flowers borne on long individual stalks; petals 20–35 mm, pale reddish-pink, shallowly notched and shorter than the long-pointed, linear sepals; sepals 15–18 mm with even longer teeth, 20–35 mm. An arable weed; rare and local. C.

Silene CAMPION

Herbs with flowers in branched inflorescences. Sepals fused into a tube, often with 5 teeth; petals separate; styles 3 or 5, protruding. Fruit a capsule.

(a) Annuals; corolla absent or concealed within the calyx.

Silene apetala MORTERILLO A small, erect annual 10–35 cm with linear to narrowly oval leaves. Lower flower stalks 3 x as long as the calyx; calyx (6)7–10 mm, becoming bell-shaped in fruit; corolla *absent* (or included in the calyx). Capsule 4–6(7.5) mm. Seeds dull blackish-brown. Various habitats. L, F, C, T, G, H, P. **Silene inaperta** is an erect, slender annual with stems 17–80 cm, erect, usually simple below and branched above, shortly hairy. Leaves lanceolate below, narrow and linear above; calyx 7.5–13.5 mm with short teeth, corolla absent or within the calyx. C, T.

(b) Annuals; corolla conspicuous.

Silene gallica CONEJERA, TARRILLO, SMALL-FLOWERED CATCHFLY A downy, erect annual with simple or branched stems 80 mm–45(60) cm, sticky above. Leaves oval and stalked beneath, narrower and unstalked further up the stem. Flowers small, yellowish-white or pinkish, borne in a more or less *1-sided inflorescence with alternating, short-stalked flowers*; calyx small, 6.5–10(12) mm, cylindrical to ovoid with long teeth, $\frac{1}{4}$ the length of the tube, *sticky-hairy and 10-veined*. Waste places. L, F, C, T, G, H, P.

Silene cretica CRETAN CATCHFLY A *slender*, erect annual 20–50 cm, hairy at the base, *hairless* above, and sticky. Basal leaves oval and broadest above the middle and stalked; upper leaves pointed and unstalked. Flowers bright pink, borne in lax, branched clusters with notched petals; *calyx distinctly ridged and hairless*, 11–16 mm. Disturbed ground; rare. G.

Silene conica An erect, hairy annual 10–25 cm with broadly linear leaves. Flowers few, borne on *short, stout* stalks 5–10 mm; calyx 30-veined, toothed, 13–18 mm; petals' claw not exserted, the limb (3)4–5(6) mm, pink to magenta. Capsule 7–12 mm, pear-shaped, included within the calyx. Seeds grey, minutely warty. C, T.

Silene rubella A glandular annual with stems 10–60 cm, lower leaves spathulate and pubescent, often with rolled-in margins; calyx 9–12(13), red-tinged with oval and blunt teeth, *long and narrow cone-shaped below, inflated towards the apex*; petals bright pink. Arable and cultivated land. L.

Silene gracilis CALABACILLA An annual with stems hairless above. Flowers white; calyx green with 10 veins, smooth and hairless or slightly bristly along the 10, faint, reddish-brown veins; flower stalks exceeding their bracts. Coastal sands. L, F.

1. *Silene apetala*
2. *Silene gallica*
3. *Silene gracilis*
4. *Silene berthelotiana*
5. *Silene vulgaris*
6. *Silene behen*
7. *Silene tridentata*
8. *Petrorhagia prolifera*
9. *Dicheranthus plocamoides*

(c) Perennials, often woody at the base.

Silene berthelotiana CANUTILLO DE BERTHELOT A woody-based perennial to 1 m with short-hairy leaves. Flowering stems tall and slender, hairy; flowers white to pale pink; calyx with distinct dark, brownish-purple veins; flower stalks *long*, to 25 mm. Pine forests. *T. *S. nocteolens* is similar in form, with hairless to hairy leaves. Flowers pinkish-white; calyx cylindrical, tapered abruptly at the tip, with reddish-brown veins. Pumice pyroclastics; very rare. *T. *S. tamaranae* is hairless or virtually so, and has *green* calyx veins. *C (west and centre). *S. lagunensis* is a woody-based perennial with hairless *leaves with prominently pointed tips*. Flowers held atop erect, slender stems; corolla white with notched lobes; calyx densely glandular-hairy. *T.

Silene vulgaris RILLA, BLADDER CAMPION A variable perennial to 80 cm with a branching, woody stock and several erect or ascending shoots, all flowering, usually (not always) hairless. Leaves elliptic to oval, the upper stalkless. Flowers white, with *strongly inflated, 20-veined calyx* 18–20 mm, persisting in fruit. Capsule with erect-spreading teeth. Common in a range of habitats. L, F, C, T, G, H, P. *S. behen* RILLA also has flowers with an inflated, hairless calyx, similar to the previous species, but is an *annual* with broad, bluish leaves. L, F, C, T.

Silene nutans A variable, erect perennial 15–80 cm, usually unbranched and sticky above, downy below. Leaves spathulate at the base, narrower and unstalked above. Flowers white, greenish or reddish below with *deeply lobed petals and inward-rolling lobes* in a loose, 1-sided and *drooping inflorescence*; calyx 9–12(14) mm, with teeth to 2 mm. C, T, G.

Silene sabinosae A woody-based perennial with lanceolate to sub-spathulate, hairless to papillose leaves. Flowers pinkish-white; calyx green with reddish-brown veins. Rare. *H.

Silene tridentata A perennial with virtually hairless stems. Flowers pinkish, borne in rather 1-sided inflorescences; calyx strongly contracted at the base with *long, conspicuous, forward-pointing bristles*. Seeds netted. L, F, C, T, G.

Petrorhagia

Herbs with an epicalyx of papery scales at the base of a single flower, or with several bracts at the base of a dense flowerhead; stamens 10; styles 2. Capsule with 4 teeth.

Petrorhagia prolifera (syn. *Kohlrauschia prolifera, P. nanteuilii*) CLAVELITO SILVESTRE, PROLIFEROUS PINK A lax, *hairless annual* 60 mm–50(70) cm. Leaves linear, greyish and rough-edged, fused at the base into *leaf sheaths that are about as long as wide*. Flowers small and pale to mid pink, borne in a dense cluster but opening 1 at a time, *surrounded by brown, papery bracts*; bracts equalling the calyx; calyx 10–13 mm. *Seeds minutely netted, not warted*. L, C, T, G, H, P. *P. dubia* (syn. *Kohlrauschia velutina, P. velutina*) KOHLRAUSCHIA is similar to the previous species but with calyx smaller, 8–14 mm, *all* inflorescence bracts short bristle-tipped (mucronate) and *seeds covered in sparse, cone-like warts*. C.

Dicheranthus

Succulent shrublets with inconspicuous flowers; stamens 3.

Dicheranthus plocamoides PATA DE GALLINA A small shrub with *linear, succulent, blue-grey leaves* with stipules. Flowers very small, pale greyish-pink, borne in dense clusters; *corolla absent*; styles 2–3-parted; stamens 3. Dry, rocky slopes; rather rare; western. *T, G.

Loeflingia

Low, much-branched annuals with linear, opposite leaves. Flowers stalkless with 5 petals; stamens 3 or 5; style 1. Fruit a 3-valved capsule.

Loeflingia hispanica LOEFLINGIA A low, glandular-hairy, reddish or purplish, and much branched annual with ascending branches to 15(20) cm and linear, long-pointed leaves fused at the base, 3–6(14) mm long. Flowers greenish, tiny to 3 mm across in branched spikes. Petals 5, *shorter than the capsules when mature*; stamens 3–5. Dry, sandy habitats in the east; local. L, F.

Polycarpaea

Shrublets or perennial herbs, ascending to erect. Leaves opposite (seemingly whorled), with papery stipules. Flowers borne in compact cymes; sepals and petals 5; stamens 5; styles 3-lobed. Fruit a 3-parted, splitting capsule.

(a) Plants hairless or virtually so.

Polycarpaea nivea SALADILLO BLANCO A small, prostrate, *silvery-white* short-hairy perennial with lanceolate to ovate, succulent leaves, blunt-tipped. Inflorescences small, silvery, lax to dense. Dunes, rocks and occasionally sea cliffs; common on most coasts. L, F, C, T, H, P. *P. robusta* is similar but a *shrub to 50 cm*. Rare. L.

Polycarpaea aristata A *sparsely branched*, prostate to ascending, subshrub with greyish, hairy, linear or narrowly spathulate leaves, bristle-tipped. Flowers borne in small, *lax* clusters. Malpaís and pine forests. *C, T, P.

Polycarpaea tenuis A *dense*, bluish-white-hairy, *mat-forming* perennial with linear to sub-spathulate, bristle-tipped leaves. Flowers borne in *dense*, lateral and terminal dull-reddish inflorescences. Rocky cliffs. *T, P.

(b) Plants hairless or virtually so.

Polycarpaea smithii LENGUA DE PÁJARO A hairless cliff plant with *long*, linear, succulent, glaucous leaves. Inflorescence large, dichotomously branched and diffuse. Forest cliffs. *G, H, P.

Polycarpaea divaricata PATA CONEJO A low shrublet with virtually hairless (to sparsely hairy), lanceolate or spathulate leaves and large, much-branched terminal inflorescences. Roadsides, bare ground and cliffs; locally very common. L, F, C, T, G, H, P.

Polycarpaea latifolia PATACONEJO BASTA A low, woody-based subshrub with virtually hairless, *circular, fleshy* leaves; leaves rather flat, bristle-tipped. Flowers borne in lateral, small clusters and large, terminal clusters. Damp habitats in forests; locally common in the west. *F, C, T, G, H.

Polycarpaea carnosa PATACONEJO CARNOSA A very woody, cliff hanging plant with hairless, *circular*, glaucous and fleshy leaves, blunt-tipped. Flowers borne in compact, globular, terminal inflorescences. Coastal basalt cliffs. *C, T, G.

Polycarpaea filifolia PATACONEJO FINA A low subshrub with woody stems and upward-pointing, *linear, fleshy and hairless, green* leaves. Flowers *white*, starry, borne in lax terminal clusters. Rocky slopes and crevices. *C, T, G.

AMARANTHACEAE | AMARANTH FAMILY

Herbs and shrubs with alternate, simple leaves. Flowers unisexual, greenish with 2–5 tepals (female) and 3–5 bracteoles (male and female); stamens 1–5; styles 2–3. Fruit an achene or 1-seeded capsule. Often in maritime or waste habitats; many introduced species, particularly in eastern semi-arid areas. Many genera were traditionally described under the family Chenopodiaceae.

Bosea

Erect or scrambling shrubs, often intricately branched. Leaves alternate, entire. Flowers inconspicuous, often in panicles; tepals and stamens 5 each; stigmas 2–3. Fruit a large-seeded berry.

Bosea yervamora HEDIONDO A hairless shrub to 3 m with slender branches. Leaves to 70 mm, ovate to lanceolate, alternate and short-stalked. Flowers inconspicuous, greenish. Fruit a small, pinkish-red berry, borne numerously in conspicuous hanging, grape-like clusters. Thickets; locally common in the west, rarer further east. *F, C, T, G, H, P.

Amaranthus BLEDO

Annuals of waste land and disturbed habitats. Flowers inconspicuous with a brownish, papery perianth; tepals (2)3–5 (or absent). Fruit 1-seeded. A hand lens is required to examine the tepals. Most species are naturalised in dry or arid areas from Central and South America. Distributions are provisional; most are widespread, ephemeral weeds.

(a) Fruit non-splitting.

Amaranthus deflexus A slender, much-branched, spreading perennial to 40 cm. Leaves 30–50 mm, diamond- to oval-shaped, *ending in an acute point*, the margins finely wavy. Inflorescence brownish, dense, terminal, spike-like, interrupted and leafy below. Bracteoles as long as the tepals, oval and pointed; tepals 2–3, 1.2–1.5 mm. Fruits *inflated*, larger than the seeds, non-splitting. A cosmopolitan weed. L, F, C, T.

Amaranthus muricatus A hairless, spreading perennial to 60 cm. Leaves 20–50 mm, *narrow*, linear to lanceolate or narrowly oval (>3 x as long as wide), long-stalked. Flowers borne in long panicles, branched below; bracteoles as long as or *shorter* than the tepals and not spine-pointed but rather papery with green mid-veins; tepals 5, 2 mm long. Fruit non-splitting. A local weed. L, F, C, T, G.

Amaranthus viridis An annual to 80 cm, spreading to erect, *hairless*. Leaves diamond to almost triangular (<3 x as long as wide), with light or dark spots on the upper surface, often with wavy margins. Flowers borne in clusters in the axils, forming dense, more or less leafless and *weak* spikes towards the apex; lateral inflorescence branches spreading; bracteoles as long as the tepals; tepals 3(4–5), 1–1.5 mm, whitish with green mid-veins. Fruit contorted and non-splitting. Waste places. L, F, C, T, G, H, P.

1. *Polycarpaea nivea*
2. *Polycarpaea nivea* (coastal form in fruit)
3. *Polycarpaea nivea* (cliff-dwelling form)
4. *Polycarpaea smithii*
5. *Polycarpaea aristata*
6. *Polycarpaea carnosa*
7. *Polycarpaea divaricata*
8. *Polycarpaea latifolia*
9. *Bosea yervamora*
10. *Amaranthus deflexus*
11. *Amaranthus muricatus*

Amaranthus blitum (syn. *A. lividus*) An ascending to erect annual to 80 cm with diamond to oval-shaped leaves, often reddish and speckled. Flowers borne in clusters in the axils and forming a short, dense spike above; tepals 3, much longer than the bracteoles, unequal. Fruit compressed, non-splitting. L, F, C, T, G, H, P.

(b) Fruit splitting transversely. Flowers borne in terminal, virtually leafless spikes or panicles.

Amaranthus hybridus An erect annual, 20 cm–1 m, similar to *A. retroflexus* with broadly lance- to diamond-shaped, long-stalked leaves. Inflorescences rather sparse, compound; *tepals 3(5), tapering to a long point*, as long as the fruit; bracteoles in female flowers 4–6 mm, exceeding the tepals x 1.5. Fruit splitting transversely. Seasonally wet habitats. L, C, T, G, P. The following are similar: **A. hypochondriacus** has *broad, erect* panicles to 20 cm wide. An ornamental escape. C, T, G, P. **A. cruentus** is reddish or purplish, with dense, *longer* terminal inflorescences, often nodding above and with short branches at the base; bracteoles 2–4 mm (equalling the tepals) with a short spine-point (a mucro); most branches of the inflorescence laxly spreading. Casual. L, C, T, H, P.

1. *Amaranthus viridis*
2. *Achyranthes aspera*
3. *Atriplex semibaccata*
4. *Atriplex suberecta*
5. *Atriplex lindleyi*
6. *Atriplex semilunaris*

Amaranthus retroflexus PIGWEED An erect, robust, hairy annual to 1 m with few side branches, green (rarely red). Leaves oval to lance- or diamond-shaped and stalked. Flowers small and inconspicuous, greenish-*white* with 5 *linear, tapering tepals 2–3 mm with mid veins ending below the tips*; borne in compact, greenish-white spikes that are leafless towards the top, densely *hairy* below. Fruit capsule splitting when ripe. A North American native widely established on waste ground. L, C, T, H.

(c) Fruit splitting transversely. Flowers all borne in clusters in the leaf axils.

Amaranthus albus An erect annual 10–60 cm with rigid, whitish stems. Flowers borne in small clusters in the axils, exceeded by the leaves; bracteoles 2 x as long as the tepals, rigid and spine-like; tepals 3. Fruit splitting transversely. Casual. C, P.

Amaranthus blitoides A prostrate to ascending, often mat-forming annual. Leaves with white, narrow, whitish margins. Flowers borne in clusters in the axils, green or purplish, overtopped by the leaves; bracteoles exceeded by the tepals; tepals 4(5), unequal. Fruits splitting transversely. Disturbed habitats. L, C, T, P. ***A. graecizans*** is similar but ascending to erect to 70 cm with leaves not or scarcely margined and 3 tepals becoming papery when mature. Disturbed habitats. L, C, T.

Achyranthes

Perennial herbs with stems woody at the bases. Leaves opposite. Flowers small and numerous, borne in a lax inflorescence; tepals 5. Fruit an achene.

Achyranthes aspera MALPICA, SANGRADERA A lax, slender perennial to 70 cm with more or less hairy stems. Leaves oval–elliptic and pointed at the tips, 20–60 mm long. Flowers borne in *very slender, long terminal and lateral spikes*, lax below, crowded towards the apex and pointing downwards when fully open; each flower has 3 spine-like bracts and a 5-parted perianth. Fruit a pendulous achene. Usually in bare or scrubby coastal disturbed ground. Dry and disturbed habitats. F, C, T, G, H, P.

Atriplex SALADO, ORACHE

Herbs and shrubs, often greyish, with flat leaves and inconspicuous flowers. Flowers unisexual, the male flowers with 5 tepals, female flowers with no tepals but 2 bracteoles, enlarged in fruit. Often naturalised weeds.

(a) Annual to woody-based perennial herbs in waste places (or disturbed, coastal habitats).

Atriplex semibaccata A short, greyish, woody-based perennial to 40 cm with spreading-to-prostrate, herbaceous branches. Leaves 8–15 mm, oblong to oval–lanceolate, wedge-shaped at the base, wavy-toothed. Female flowers grouped around the male, in clusters in the axils, sometimes spike-like; fruiting bracts stalkless, diamond to broadly oval-shaped, fused in the lower half, *fleshy, reddish and berry-like* when mature. Bare, dry waste ground; common. L, F, C, T, G, H, P.

Atriplex suberecta A sparsely spreading, mealy, green to ash-coloured annual (or short-lived perennial) to 1 m with leaves 5–11 mm, oval to diamond–elliptic in outline, wedge-shaped at the base, with toothed margins, without basal lobes. Flowers 2–3 mm, clustered in the axils, dispersed all the way up the stem; fruiting bracts diamond to broadly oval, fused in the lower half, with toothed margins. Native to Australia; naturalised in dry areas, especially in the east. L, F, C, T.

Atriplex lindleyi A clumped, erect or spreading, weak, silvery annual with obovate to narrowly elliptic leaves to 40 mm with entire to toothed margins. Flowers borne in lateral or terminal clusters. Fruits with *conspicuous inflated bracteoles* with entire margins and no appendages. Native to

Australia; bare ground, uncommon. L. *A. semilunaris* is rather similar in habit: a clumped, erect, weak, silvery, nettle-like annual. Leaves rather stiff and fleshy with inrolled margins with remote teeth. Flowers borne in dense, terminal clusters. Native to Australia; common on roadsides and in towns in the east. L, F.

Atriplex prostrata (syn. *A. hastata*) SPEAR-LEAVED ORACHE A variable, tall, hairless and branched annual to 2 m, mealy when young. Leaves green-grey with spreading triangular basal lobes, the *largest lower leaves with straight-edged lower lobes at right angles to the leaf stalk*. Flowers reddish, borne in dense clusters in leafy panicles; bracteoles small, entire and triangular, 2–6 mm. Common on sandy shores and disturbed ground. L, T.

(b) Shrubs, usually in coastal habitats.

Atriplex halimus MATO SALADO, SHRUBBY ORACHE An erect to ascending, woody *shrub* to 2.5 m. Leaves small, to 30 mm long, alternate, *silvery-white*, narrowly oval, slightly diamond-shaped, or angled leaves. Flowers yellowish, borne in leafless branched terminal spikes; *bracteoles 1.5–3 mm, fused only at the base*. Sandy shores; common in the east. L, F, C. *A. glauca* is a similar whitish, shrubby perennial to 50 cm with *prostrate to ascending stems*. Leaves stalkless, 10 x 7 mm. Bracteoles *oval–diamond-shaped* and with protuberances on the surface. **Subsp.** *ifniensis* is the form on the islands. Dry, saline habitats. L, F, C, T, G, H.

Atriplex portulacoides (syn. *Halimione portulacoides*) FALSA VERDOLAGA, SEA PURSLANE A low shrub with spreading and often rooting branches, more or less mat-forming. Leaves upward-pointing, silvery, the *lowermost opposite*, narrowly oval and untoothed. Flowers small and green or reddish, borne in more or less leafless panicles; *bracteoles 2.5–5 mm, fused to just >½ their length*. Saltmarshes. L.

Beta ACELGA, BEET

Erect or ascending annuals or perennials with swollen roots and with alternate, not mealy leaves and cosexual flowers with 5 tepals.

Beta vulgaris SEA BEET A variable, hairless, *reddish* and rather fleshy annual or perennial to 1.5 m, but often prostrate. Leaves oval–lanceolate, dark green, leathery and shiny, and more or less untoothed, to 20 cm. Flowers small and green or purplish, borne at the top of a dense, long spike; stigmas mostly 2; lower bracts 2–20 mm; bracts above 10–20(35) mm or absent. Ruderal and maritime habitats. L, F, C, T, G, H, P. *B. macrocarpa* is similar but with flowering stems with flowers practically to the base, and *bracts to the top*. L, F, C, T, G, P.

Bassia

Small desert annuals and perennials with adpressed to spreading hairs (often concealing flowers and fruits). Leaves alternate, linear, often succulent. Flowers cosexual or unisexual; tepals 5; stamens 3–5, exserted. Fruit a capsule.

Bassia tomentosa (syn. *Chenoleoides tomentosa*) ALGOAERA A greyish-white-hairy, woody-based shrub to 40 cm. Leaves linear to lanceolate. Flowers minute, borne in the leaf axils, with 5, yellow exserted stamens and 2 stigmas. Very common to subdominant in bare, lowland habitats; less frequent further west. L, F, C, T, H, P.

1. *Atriplex halimus*
2. *Atriplex halimus* (in fruit)
3. *Atriplex glauca* subsp. *ifniensis*
4. *Atriplex portulacoides*
5. *Beta macrocarpa*
6. *Bassia tomentosa*
7. *Maireana brevifolia*
8. *Chenopodiastrum murale*
9. *Chenopodium nutans*

Maireana

Woody, succulent shrubs native to Australia. Fruiting perianth with horizontal wings (superficially like those of *Salsola*).

Maireana brevifolia MATO AZUL, COTTON BUSH An erect, succulent, virtually hairless shrub 20 cm–1 m with woody, ridged stems. Leaves alternate, fleshy, cylindrical to ovoid, 2–5 mm, narrowing into short stalks, grey-green or purplish. Flowers borne solitary in the leaf axils, green, cosexual; fruit to 3 mm, *prominent*, surrounded by a fruiting perianth, with 5 spreading, often overlapping winged bracts, pale green, pinkish when mature, *small*, 6–8 mm across. Native to Australia, naturalised; local. F, C.

Chenopodium (including *Chenopodiastrum*) CENIZO, GOOSEFOOT

Herbs or small shrubs with alternate, often mealy leaves and inconspicuous cosexual or female flowers; bracteoles absent, tepals 4–5; stamens 5. Leaves and seeds are important diagnostics. A complex genus; species described here are the most widespread and easily identifiable.

Chenopodium vulvaria STINKING GOOSEFOOT An erect annual to 40 cm, *not mealy* (except for perianth) and *unpleasant-smelling* when crushed. Leaves small (<25 mm), diamond-shaped with basal lobes but more or less entire. Tepals rounded. Seed *with faint furrows and zygomorphic thickenings*. Saline waste places. T, H.

Chenopodium album FAT HEN A tall, erect, *deep green* but *grey-mealy* plant to 1.5 m. Stems stiff, often marked with red. Leaves oval to diamond-shaped and markedly longer than broad (to 80 mm long) and *entire to coarsely and bluntly toothed*. Inflorescence usually leafless at the very top; tepals slightly keeled. Seeds smooth or faintly ridged. Very common on disturbed ground. L, F, C, T, G, H, P.

Chenopodiastrum murale (syn. *Chenopodium murale*) is similar to *Chenopodium album*, also mealy, occasionally reddish with leaves that have coarse, *irregular, forward-pointing teeth* (nettle-shaped). Flowers mealy, with blunt keels on the back of the *toothed tepals*; inflorescence leafy to the top. Seeds minutely pitted. Sandy, disturbed and maritime waste habitats. L, F, C, T, G, H, P. *C. coronopus* is similar but leaves deeply divided into linear lobes. Bare volcanic sandy soils, often along paths and roads, coastal cliffs. *C, T, H, P.

Chenopodium nutans (syn. *Einadia nutans*) A herbaceous perennial to 60 cm, sometimes woody at the base. Leaves 5–20(25) mm, lance to arrow-shaped, abruptly contracted into the stalks, mealy when young. Flowers tiny, borne in panicles; tepals 0.5–1.5(2) mm, fused below, mealy to hairless; stamens 0–2(3). *Fruit 3–5 mm, red, berry-like spherical*. Native to Australia, naturalised in bare places. C, T, P.

Dysphania

Very similar to *Chenopodium* but markedly *glandular*-hairy, aromatic and sticky.

Dysphania ambrosioides PASOTE An annual (or short-lived perennial) to 2 m, hairy, glandular and aromatic. Leaves lance-shaped, entire or toothed. Flowers borne in panicles with stalkless upper branches, usually with bracts; sepals free almost to ½ their length, rounded on the back. Seeds 0.5–0.8 mm across. Common on disturbed ground at low elevation. L, F, C, T, G, H, P.

Patellifolia

Annual or perennial herbs with alternate leaves. Perianth with 5 keeled lobes, fused at the base; stamens 5, tepal-opposed; ovary semi-inferior with 2 stigmas.

Patellifolia patellaris TEBETE A leafy, hairless annual herb with prostrate, creeping stems to 2 m (often smaller). Leaves alternate, stalked, *succulent* and entire, ovate–triangular. Flowers borne laterally in groups of 1–3; perianth 5-parted. Fruits globose, 5–6 mm, encased within the fruiting perianth. Bare ground and malpaís; very common. L, F, C, T, G, H, P.

Patellifolia procumbens MARMOHAY is a perennial, considered a mere form; it has less succulent, rather narrower ovate-rhomboidal, and darker green leaves. Similar habitats. L, F, C, T, G, H, P. *P. webbiana* is similar but with very narrow, almost *linear* leaves with two divergent basal lobes (hastate). Similar habitats. *C, T(?), G(?), H(?).

1. *Dysphania ambrosioides*
2. *Patellifolia patellaris* (habit)
3. *Patellifolia patellaris*
4. *Patellifolia procumbens* (habit)
5. *Patellifolia procumbens*
6. *Salsola tragus*

1. *Caroxylon vermiculatum*
2. *Caroxylon vermiculatum* (pale form)
3. *Caroxylon tetrandrum*
4. *Caroxylon tetrandrum* (in flower)
5. *Salsola divaricata* (foliage)
6. *Salsola divaricata* (in flower)
7. *Salsola divaricata*
8. *Arthrocaulon macrostachyum*
9. *Arthrocaulon macrostachyum* (in fruit)

Salsola (including Caroxylon) MATO, SALTWORT

Herbs or shrubs with fleshy, cylindrical leaves, and small cosexual flowers borne in the leaf axils. Tepals 5, often developing a transverse wing on the back in fruit.

(a) Plant herbaceous, with leaves and bracts sharp-pointed or spiny.

Salsola tragus PINCHO A robust annual with erect stems to 1 m; rigid, spiny stems, soft when young. Leaves linear, 10–40(60) mm, succulent with spines 10–15 mm. Flowers with tepals 2–2.5 mm, hairless. Fruits 6–8 mm. Coasts. F, C, T.

(b) Plant a shrub or subshrub, with leaves and bracts *not* spiny.

Caroxylon vermiculatum (syn. *Salsola vermiculata*) MATO, SOGAL A succulent shrub to 1 m with *very small, alternate, crowded leaves* 5–12(25) mm long, semi-cylindrical, oval at the base and semi-clasping the stem, *often pubescent*. Flowers borne in inflorescences with regularly arranged primary branches 20–40 cm long; flowers small and green or pink-tinged; stamens 5; fruiting perianth to 12 mm across, stigmas shorter than the style. Fruiting perianth with green to bright pinkish-red wings. Dry, saline habitats; abundant in the east. L, F, C, T. *C. tetrandrum* (syn *S. tetrandra*) is similar in form, with minutely white-woolly, blue-grey, *opposite* leaves. Stamens 4. Fruiting perianth without wings. Semi-desert and saline habitats in the east. L, F.

Salsola divaricata SALADO A succulent shrub to 2.5 m with *long, cylindrical, succulent, opposite, hairless* leaves. Fruiting perianth with enlarged, papery, pink wings tinged pale green. Coastal rocks and dunes; locally common. *L, F, C, T, G, H, P.

Arthrocaulon and Sarcocornia

Glabrous halophytic, rhizomatous shrubs, often forming mats. Flowers minute, in groups of 3 in a row.

Arthrocaulon macrostachyum (syn. *Arthrocnemum macrostachyum*) SALADO A succulent shrub to 1.5 m, with very woody, spreading to prostrate branches to 25 mm across at the base, bare below, glaucous (grey-yellow), often partly *yellowish or reddish*; leaves opposite, scale-like and fused in pairs, appearing as segments. Flowers tiny (1–2 mm), borne in groups of 3, each with 1–2 stamens; stigmas 3. Saltmarshes and coastal clays and sands. Locally common in the east. L, F.

Sarcocornia perennis (syn. *Arthrocnemum perenne*) A small, very succulent rather *short, spreading shrub to 30(70) cm with creeping, subterranean stems, at first dark green*, ageing red-brown or yellowish. Leaves opposite, scale-like and fused in pairs, appearing as segments. Flowers tiny, borne groups of 3, each with 2 stamens; flowers falling to leave 3 scars in the segment. Saltmarshes in the east; local. L, F.

Suaeda SEABLITE

Herbs or small shrubs in saline habitats with fleshy, alternate, linear leaves and minute flowers with 5 tepals and 2–3 bracteoles; flowers cosexual and female. Species rather similar.

(a) Mound- or mat-forming shrubs, woody-based below.

Suaeda vera MATOMORO, SHRUBBY SEABLITE A succulent, woody-based shrub to 1.2 m, leaves *densely crowded, blunt and fleshy*, alternate, 5–18 mm, cylindrical, often bright yellowish green, sometimes red-tinged; stalkless and with inconspicuous edges and a pointed tip. Flowers inconspicuous, 5-parted with 5 stamens and irregular fused stigmas. Coastal clays or sands, often in large patches; locally common except in the far west. L, F, C, T, G.

Suaeda mollis BRUSQUILLA A succulent, woody-based, mound-forming shrub. Leaves cylindrical to globose; small, red-flushed with age; stalked, *without edges*. Flowers inconspicuous, 5-parted with 5 stamens and 3–5 stigmas, borne 1–3 in the axils of the upper leaves; stigmas 3. Maritime and inland saline habitats; common in the east. L, F, C, T. *S. ifniensis* is very similar but with *stalkless* leaves with inconspicuous edges. Similar but coastal habitats; strongly eastern. L, F, T.

(b) Mound- or mat-forming annuals, scarcely woody-based below.

Suaeda spicata A succulent *annual* to 30(75) cm with *prostrate or ascending stems, scarcely woody below* bearing *upward-pointing*, alternate, fleshy leaves 3–25 mm long, ranging from blue-green to reddish or purple; *leaves slightly concave above in cross-section*. Flowers minute, green and borne in the leaf axils, tepals keeled, stigmas 2. Seeds <1.5 mm long. Saltmarshes, coastal habitats; rarely abundant. F, C. *Suaeda maritima* is a similar but perennial prostrate herb and flowers without keeled tepals. Seeds >1.5 mm long. **Var. *perennans*** is the form on Lanzarote.

Traganum

Succulent, woody shrubs with woolly nodes and fleshy, cylindrical leaves. Fruiting perianth wingless, with *2 horn-like teeth*.

Traganum moquinii BALANCÓN A robust, spreading, woody based, much-branched, succulent shrub to 1.5 m. Leaves 10 mm, succulent, yellowish with a whitish bloom, broadly cylindrical and bluntly pointed and angled. Flowers inconspicuous, dull yellow, borne solitary in the leaf axils, each with 2 secondary bracts. Fruiting perianth with *2 horn-like teeth*. Maritime dunes; locally common in the east. L, F, C, T, G.

AIZOACEAE

Herbaceous annuals or perennials with simple, opposite or alternate, succulent leaves. Flowers regular with numerous linear 'petals' (derived from staminodes, described as petals below); sepals 4–5(6); stamens 3–numerous. Fruit a capsule with (1)3–numerous seeds.

Aizoon

Annual or perennial herbs with a crystalline (*papillose*) surface. Leaves (mostly) alternate. Flowers with 4–5 tepals, often fused below into a short tube; ovary superior. Fruit a capsule with numerous seeds.

Aizoon canariense PATILLA A short, slightly succulent annual to 30 cm, sprawling when mature. Leaves 40 mm wide, opposite below, alternate above, oblong–lanceolate, blunt and minutely hairy. Flowers solitary and unstalked with 5 yellow tepals 1–3 mm long and yellow stamens. Semi-desert habitats. L, F, C, T, G, H, P. *A. hispanicum* (syn. *Aizoanthemopsis hispanica*) COSCO MACHO has leaves just 10 mm across and tepals yellow or whitish, 7–15 mm long. Rare in the area; bare sand and deserts. L, F, G.

---

1. *Sarcocornia perennis* (habit)
2. *Sarcocornia perennis*
3. *Suaeda vera* (habit)
4. *Suaeda vera*
5. *Suaeda vera* (flowers)
6. *Suaeda mollis* (habit)
7. *Suaeda mollis*
8. *Suaeda mollis* (flower)
9. *Suaeda ifniensis* (habit)
10. *Suaeda ifniensis*
11. *Suaeda ifniensis* (flower)
12. *Suaeda spicata*

Carpobrotus HOTTENTOT FIG

Fleshy perennials with leaves triangular in cross-section. Flowers with numerous petals and 8–20 stigmas. Fruit succulent, the seeds embedded in mucilage.

Carpobrotus edulis UÑA DE GATO, HOTTENTOT FIG A succulent, trailing, mat-forming perennial with woody stems to 2 m long. Leaves opposite, 3-angled and upwardly curving, finely toothed along the edge and tapered towards the apex, 40 mm–9(13) cm long. Flowers large, 8–10 cm across, solitary, virtually stalkless, bright pink or yellow with yellow stamens. Fleshy, edible fruit. Native of South Africa; naturalised. L, F, C, T, G, P.

Malephora

Succulent perennials native to Africa. Leaves opposite and 3-angled. Flowers showy with numerous (65) petals and (150) stamens. Fruit an 8–12-parted capsule.

Malephora crocea CLAVEL DE SOL A spreading succulent perennial with stems that have long internodes and *a white-grey bloom*. Leaves up to 50(60) mm long, opposite, cylindrical and bluntly 3-angled. Flowers showy, 50 mm across, with numerous (to 65) pinkish-orange or magenta petals and yellow-orange stamens. Planted in coastal areas, locally naturalised. L, F, C, T.

Mesembryanthemum

Succulent annuals or biennials *densely covered in crystal-like vesicles*. Flowers with numerous stamens. Capsule with 4(5) valves.

Mesembryanthemum crystallinum BARRILLA, ICE PLANT A spreading annual covered in glistening, frost-like *crystalline hairs*. Leaves stalked, flat, fleshy, oval, alternate and untoothed, to 11.5 cm long. Flowers solitary, 20–30 mm across, virtually stalkless, yellowish or whitish. Very common in disturbed sandy, rocky and saline environments, especially in the east. L, F, C, T, G, H, P. *M. nodiflorum* COSCO is similar but less crystalline and with *cylindrical-linear leaves* 10–25(30) mm long, red-tinged; flowers smaller, to 15 mm across. Similar habitats. L, F, C, T, G, H, P.

Mesembryanthemum cordifolium (syn. *Aptenia cordifolia*) A mat-forming perennial to 60 cm high with stems to 3 m, with slightly fleshy, bright green, oval leaves that are pointed at the apex and heart-shaped at the base, 13–56 mm long. Flowers bright magenta, 10–18 mm across. An ornamental native to South Africa, widely planted. L, F, C, T, G, H, P.

Drosanthemum

Crystalline, creeping, succulent perennials with showy (often pink), rather small flowers with numerous stamens; sepals 5; petals numerous; stigmas mostly 5. Capsule woody.

Drosanthemum floribundum A crystalline, creeping, all-green succulent perennial to 12 cm high with opposite, fleshy, stalkless leaves 5–16 mm long. Flowers 20–25 mm across, *bright pink* with many narrow petals and numerous anthers; perianth tube semi-globose. Native to South Africa but planted. C, T.

---

1. *Traganum moquinii*
2. *Aizoon canariense*
3. *Aizoon hispanicum*
4. *Malephora crocea* (orange form)
5. *Malephora crocea* (pink form)
6. *Mesembryanthemum crystallinum* (fruits)
7. *Mesembryanthemum crystallinum*
8. *Mesembryanthemum nodiflorum*
9. *Tetragonia tetragonoides*
10. *Commicarpus helenae*
11. *Commicarpus helenae* (fruits)
12. *Portulaca canariensis*

Tetragonia

Prostrate or ascending, leafy annuals or shrubs with alternate, flat leaves with long stalks. Flowers mostly solitary with 4–5 sepals, 0 petals; stigmas 3–8. Fruit woody, ridged with 1 seed in each of 3–8 cells.

Tetragonia tetragonoides ESPINACA DE NUEVA ZELANDA, NEW ZEALAND SPINACH A robust, trailing or scrambling, leafy plant to 1 m with triangular leaves 15 mm–11 cm long, superficially like a member of the Amaranthaceae. Flowers small, borne in the axils, virtually stalkless and yellow with 5 petals. Fruit horned. Native to New Zealand and Australia. L, C, T, G, H, P.

NYCTAGINACEAE

Trees, shrubs and woody climbers native to South America with opposite leaves. Flowers with 5 showy bracts, cosexual; stamens 5; style 1. Fruit surrounded by the perianth tube, forming a false fruit.

Bougainvillea

South American vigorous climbing shrubs with flowers that have showy bracts; stamens 8. Fruit a 5-lobed achene.

Bougainvillea glabra BOUGAINVILLEA A vigorous, woody, virtually hairless climber to 10 m. Leaves opposite or alternate, to 60 mm long, untoothed and stalked. Flowers in groups of 3, inconspicuous, whitish-yellow and funnel-shaped, surrounded by *large, flower-like crimson bracts*, 1 to each flower. Planted in towns. C, G.

Mirabilis

South American tuberous perennials typically with fragrant, deep-throated flowers; stamens 3–5. Fruit a 0–5-angled achene.

Mirabilis jalapa DON DIEGO DE NOCHE, MARVEL OF PERU A hairless or short-hairy perennial to 1.5 m with large, oval, untoothed leaves narrowing to a point; rather wrinkled, stalked below and short-stalked to sessile above. Flowers borne in terminal clusters, variably white, yellow, pink, red or variegated; tube 25–35 mm long, 5-lobed. Native to tropical America, widely planted and naturalised. L, F, C, T, G, H, P.

Commicarpus

Succulent perennial herbs or shrubs with opposite, entire leaves. Flowers cosexual with 5 lobes; stamens 2–5. Fruit a 10-ribbed, club-shaped anthocarp and large, wart-like glands.

Commicarpus helenae (syn. *Boerhavia helenae*) A straggling, branched perennial with hairless, whitish stems. Leaves oval to rounded with undulating margins, and stalks to 15 mm. Flowers borne 3–5 per node in very lax, rigid inflorescences, small, 6–6.5 mm long, pinkish red; flower stalks short and stout (1–4 mm). Fruit aggregated (an anthocarp) 5–7 mm, with a *ring of large, wart-like glands* around the apex. Dry habitats; rare. F.

PORTULACACEAE | BLINKS FAMILY

Annual or perennial herbs, often fleshy, with cosexual flowers with only 2 opposite sepals and 4–6 petals; stamens 3–14; styles 1–3(6). Fruit a capsule.

Portulaca VERDOLAGA, PURSLANE

Herbaceous annuals with 1–few stalkless flowers; stamens numerous and ovary ½ inferior. Fruit a many-seeded capsule.

Portulaca oleracea PURSLANE A fleshy, prostrate or ascending, patch-forming, leafy annual with stems to 30(50) cm, Leaves to 30 mm long, spoon-shaped and more or less opposite and densely crowded beneath the flowers. Flowers yellow with 4–5(6) petals 4–8 mm long that soon fall, revealing blunt sepals. Exposed habitats. L, F, C, T, G, H, P.

Portulaca canariensis A fleshy, prostrate annual to 30 cm, often strongly red-flushed. Leaves alternate. Flowers borne 2–3. Fruit a splitting capsule, revealing black seeds 1 mm across with a *regularly* tuberculate surface. Rocky (natural) habitats. L, F, C, T, G. *P. granulatostellulata* has minute seeds (<1 mm) without a *regularly* tuberculate surface. Disturbed (manmade) habitats. L, C, T, P.

CACTACEAE | CACTUS FAMILY

A family of (typically) spiny succulents from the Americas, often with showy flowers. Cultivated and naturalised in the region.

Lophocereus

Robust, slow-growing cacti with large, cylindrical stems.

Lophocereus marginatus FENCE POST CACTUS A robust cactus with columnar, spiny-ribbed, unbranched stems to 4 m tall and 18 cm across; surface matte, with a greyish bloom. Flowers pale yellow with many stamens. Native to Central America, planted in the east. L, F.

Hylocereus

Succulent shrubs from Tropical America. Stems succulent, 3-angled. Flowers opening at night, white with numerous tepals; stamens numerous. Fruit fleshy, many-seeded.

Hylocereus undatus PITAYA, DRAGON FRUIT A robust, cactus-like shrub to 2 m with yellowish-green stems with spiny, undulate margins. Flowers conspicuous, about 25–30 cm across, white, with greenish outer tepals. Fruit reddish with a white pulp and >200 seeds. Commonly planted as a hedge or for its edible fruit. F, C, T, G, P.

Lophocereus marginatus

CACTACEAE | 257

Opuntia PRICKLY PEAR

Woody cacti with jointed stems and soon-falling leaves. A confused genus with conflicting names and descriptions in the literature (morphological traits are often continuous, or not clear or useful in the field). Native to North and South America.

Opuntia ficus-indica (incl. *O. maxima*) TUNERA, PRICKLY PEAR A robust, blue-green cactus 5–6 m; woody and trunk-like at the base (rather tree-like) with *large*, flattened jointed stem segments 30–50 cm long and soon-falling, inconspicuous leaves. Bristles hooked and yellowish, *straight spines usually absent*. Flowers bright yellow. Fruit egg-shaped, 50–60(90) mm long, yellow or reddish when ripe; edible. L, F, C, T, G, H, P.

Opuntia tuna TUNERA INDIA A shrub to 1.5 m with blue-green stem segments 70 mm–40 cm long, with *1–10 conspicuous, stout, yellow, flattened spines of various lengths*. Flowers yellow. Fruits 50–75 mm, *purple-red*. Planted in hot, dry areas. L, F, C, T, G, H, P. *O. stricta* has a *spreading habit* with erect branches, *shorter* than the previous species, to 1 m high; stem joints 20–30 cm; spines cylindrical, normally absent. Fruits reddish. C.

Cylindropuntia

Woody cacti similar to *Opuntia* (and previously classified in this genus) but with *cylindrical stem joints* and papery sheaths around the spines. Native to South America.

Cylindropuntia imbricata (syn. *Opuntia imbricata*) A robust succulent 1–3(4) m high with *cylindrical* stem segments 20–50 cm long with prominent ribs and clusters of 2–12 cream-white spines 8–30 mm. Flowers *dark pink-magenta*. Fruit grey-green, 40 mm. Native to Mexico, naturalised. C.

Austrocylindropuntia

Woody cacti similar to *Cylindropuntia*, but with spines lacking papery sheaths. Native to South America.

Austrocylindropuntia subulata (syn. *A. exaltata*) A spreading, robust shrub to 3 m, similar to *Cylindropuntia imbricata* but with less prominently ribbed segments 35–50 cm long with prominent clusters of 1–4 spines to 70 mm long that lack papery sheaths; leaves to 20 mm, persistent on young stems. *Flowers pale pink*. Fruit green, 10 cm long. Native to Peru, naturalised; common in the east. L, F, C, T, G, P. *A. cylindrica* is similar, but with leaves to about 10 mm, soon-falling, and shorter spines. L, F, C, P.

---

1. *Hylocereus undatus* (habit)
2. *Hylocereus undatus* (fruit)
3. *Opuntia ficus-indica* (habit)
4. *Opuntia ficus-indica* (in fruit)
5. *Opuntia ficus-indica* (orange-flowered form)
6. *Opuntia tuna* (habit)
7. *Opuntia tuna*
8. *Austrocylindropuntia subulata*
9. *Austrocylindropuntia cylindrica*

BALSAMINACEAE | BALSAM FAMILY

Herbaceous, hairless annuals with simple leaves and translucent, fleshy stems. Flowers strongly zygomorphic; sepals 3, petal-like; stamens 5; style 1. Fruit a capsule that violently expels seeds when ripe.

Impatiens

Herbaceous annuals with succulent stems swollen at the nodes and opposite or alternate leaves. Flowers zygomorphic. Fruit an explosive capsule.

(a) Herbaceous annuals to perennials.

Impatiens walleriana PERIQUITO A spreading to upright annual to perennial, hairless, with alternate, stalked, ovate to elliptic leaves 25–55 mm long, sometimes variegated, with scalloped margins. Flowers to 40 mm across, white, pink, violet or various other bright shades. Planted. C, T.

(b) Woody-based shrubs.

Impatiens sodenii BALSAMINA A stout, bushy perennial with fleshy stems to 1 m. Leaves to 20 cm, short-stalked to stalkless, regularly fine-toothed. Flowers 50–80 mm across, white to pink with darker central markings. C, T.

SAPOTACEAE

Trees or shrubs with milky latex. Leaves alternate, simple. Flowers clustered in the leaf axils, often 5-parted. Fruit a berry.

Sideroxylon

Leaves stalked, thick, leathery, with prominent lateral veins. Flowers congested in leaf axils. Corolla 5-lobed.

Sideroxylon canariense MARMULÁN A shrub to small tree to 15 m with dark grey bark. Leaves elliptic to obovate, virtually hairless. Flowers white. Fruit black when ripe. Various habitats including roadsides, cliffs, barrancos; rare and local, especially in the east. *F, C, T, G, H, P.

PRIMULACEAE | PRIMULA FAMILY

Herbaceous annuals or perennials. Flowers regular with free to fused sepals; petals 5(–9), fused (sometimes only just, seemingly free); stamens usually 5; ovary with a single style. Fruit a capsule.

Heberdenia

Trees. Flowers borne laterally, cosexual, 5-parted; sepals fused below; stamens attached to the petal bases. Fruit a hard, 1-seeded berry.

Heberdenia excelsa ADERNO, SAQUITERO A tree to 10 m. Leaves entire, dark green, leathery, obovate to oblong, short-stalked. Flowers borne in lateral clusters, whitish- to pinkish-green,

1. *Impatiens sodenii*
2. *Sideroxylon canariense*
3. *Sideroxylon canariense*
4. *Heberdenia excelsa*
5. *Lysimachia arvensis* (blue form)
6. *Lysimachia arvensis* (red form)
7. *Lysimachia wildpretii*
8. *Pleiomeris canariensis* (habit)
9. *Pleiomeris canariensis*

starry, *strongly scented,* glandular. Fruit a hard, reddish-purple berry, 5 mm across. Laurel forests and shaded, north-facing cliffs. F, C, T, G, H, P.

Lysimachia

Annuals, perennials, or dwarf shrubs. Leaves flat, opposite (rarely alternate). Flowers often borne in *panicles or racemes* or solitary in the leaf axils; calyx with 5 teeth; corolla with a short tube and 5 lobes; stamens 5. Fruit a capsule with 5(7) valves; seeds numerous. Now includes *Anagallis*.

Lysimachia linum-stellatum LINO DE LAGARTIJA A small, erect annual to just 12 cm. Whitish flowers borne over most of the stem; minute, framed by 5 pointed sepals. Fruit a globose capsule, splitting into 5 parts. Disturbed ground. L, F, C, T, G, H, P.

Lysimachia arvensis **(syn.** *Anagallis arvensis***)** MURAJE, SCARLET PIMPERNEL A low, weedy annual with prostrate or ascending, 4-angled stems to 40 cm. Leaves often >10 mm, oval, opposite and gland-dotted; unstalked. Flowers *blue, red or orange*, 4–7(10) mm, long-stalked (35 mm) and becoming curved in fruit; *petals with minutely hairy margins*, sometimes toothed at the tip. Common in towns and disturbed places. L, F, C, T, G, H, P. **L. foemina (syn. A. foemina)** is often treated as a subspecies, but is genetically distinct, and has flowers that are always blue, with narrower petals, *without hairy margins*; corolla 5–8 mm. T.

1. *Samolus valerandi*
2. *Visnea mocanera* (habit)
3. *Visnea mocanera*
4. *Visnea mocanera*
5. *Arbutus canariensis*

Lysimachia wildpretii (syn. *Pelletiera wildpretii*) FALSO LINO DE LAGARTIJA A hairless annual to 16 cm, erect, usually branched from the base. Leaves opposite, stalkless, 4–14 mm, linear–elliptic, thin and net-veined. Flowers minute, white; corolla with 3 unequal segments; stamens 3–5. Tracksides; easily overlooked. *L, F, C, T, H, P.

Pleiomeris

Shrubs to small trees. Flowers unisexual; petals united along their lower margins to form a tube. Fruit a berry.

Pleiomeris canariensis DELFINO A shrub or tree to 15 m. Leaves oblong to elliptic with prominent lateral veins, to 20 cm. Flowers dull creamy-white, with 4–6 reflexed corolla lobes united at the base, *borne in small, almost stalkless lateral clusters along the stems*. Berries 5–7 mm, reddish-violet with *long, persistent styles*. Laurel forests. *C, T, G, P.

Samolus

Hairless perennials with leaves in basal rosettes. Flowers white with 5 sepals and corolla lobes. Capsule with 5 teeth.

Samolus valerandi JABONERA, BROOKWEED A hairless perennial herb with a rosette of leaves 10–50 mm long, and erect flowering stems 50 mm–70 cm. Leaves rather shiny, spoon-shaped and scarcely stalked below; stalkless above. Flowers small and white, cup-shaped, to just 3 mm across and with 5 petals. Fruit 3 mm. Damp rocks and seeps. L, F, C, T, G, P.

THEACEAE

Trees and shrubs with alternate, usually simple leaves. Flowers usually cosexual with 5 sepals and petals and many stamens. Fruit a capsule.

Visnea mocanera MOCÁN A small tree to 15 m. Leaves 30–40 mm, alternate, smooth and without glands, lanceolate with a regularly fine-toothed margin. Flowers 5-parted, creamy-white, with pointed petals; stamens many, creamy-yellow; calyx lobes blunt, hairy. Fruit a fleshy, splitting capsule, grey-brown to reddish. Cliffs, hillsides and forests; fairly frequent in the west, very rare and local in the east. F, C, T, G, H, P.

ERICACEAE | HEATHER FAMILY

Evergreen trees and shrubs with alternate, opposite or whorled leaves, or herbaceous perennials. Flowers often borne in clusters (rarely solitary); sepals and petals (3)4–5, petals fused; stamens 2 x as many as petals; style 1. Fruit a capsule, drupe or berry.

Arbutus

Evergreen shrubs with alternate leaves. Flowers with 5 petals, borne in terminal clusters; stamens 10. Fruit a warty berry.

Arbutus unedo MADROÑO, STRAWBERRY TREE An evergreen shrub or small tree 4–5(11) m. Bark dull brown and fissured; young twigs at least partially *downy*. Leaves 80 mm long, oblong–lanceolate, short-stalked and somewhat toothed, more or less hairless. Flowers scented; white, tinged green or pink, bell-shaped with recurved petal lobes, 7–8(11) mm borne in drooping panicles in *autumn*. Fruit a spherical berry 7–15(20) mm, ripening deep crimson; strawberry-like. Forests; local. C, T.

Arbutus canariensis MADROÑO CANARIO A tree or tree-like shrub with a reddish-brown trunk with bark peeling in strips. Leaves oblong–lanceolate, 15 cm, toothed, with glandular-hairy stalks to 20 mm. Flowers greenish-white or pinkish. Fruit a spherical, fleshy, warty berry, orange-red when ripe. Rather local in dry laurel and laurel-pine mixed forest fringes and along roadsides; strongly western. *C, T, G, H, P.

Erica HEATHER

Shrubs with whorled, narrow to linear leaves. Flowers bell-shaped to spherical, borne in spikes or panicles, generally 4-lobed; stamens 8. Fruit a dry capsule.

Erica canariensis BREZO A tall shrub or small tree to 15 m with densely hairy young twigs. Leaves *erect*, in groups of 3–4, to 4–9 mm long. Flowers pure *white*, broadly bell-shaped, 2–4 mm long with erect lobes, borne in dense terminal panicles; sepals 1.2–2 mm. Fruit a capsule, 2 mm. Common to abundant in laurel forest zones. C, T, G, H, P. *E. platycodon* TEJO has *spreading* leaves in groups of 3–4 in irregular rows; rolled margins almost conceal the undersides. Flowers *pinkish-red*. Similar habitats; far more local. *T, G.

RUBIACEAE | MADDER FAMILY

A large family of herbs with opposite or distinctly whorled leaves; stipules present between each pair of leaves, often leaf-like. Flowers typically small, funnel-shaped and tubular, borne in dense heads, branched cymes or panicles; sepals 0 or minute, petals 4–5; stamens 4–5; styles 1–2 (if 1, branched); ovary inferior. Fruit fleshy, dry or berry-like, 1–2-seeded.

Rubia

Scrambling perennials or shrubs with leaves in whorls of 4, equal and stalkless. Flowers with a 4-lobed, yellow corolla; the terminal cosexual, the laterals male. Fruit a pair of smooth nutlets.

Rubia occidens RAPASO, AZAICO, WILD MADDER A trailing or scrambling evergreen perennial to 7 m with a creeping rootstock. Stems square and rough with downturned bristles. Leaves in whorls of 8; dark green linear–lanceolate, gradually tapering to an apex, *25–80 mm long, 3–5 mm wide*, leathery and dark shiny green, 1-veined, margins rough. Flowers pale yellow-green, 3.5–8 mm, 5-lobed, forming dense, leafy panicles. Fruit 3–7 mm, black and fleshy when ripe. Laurel forest. C, T, G, H, P. *R. peregrina* WILD MADDER is similar but leaves in whorls of 4–6(8); oval to elliptic, *4–27 mm wide*. Wet cliffs (Jandía). F.

Rubia fruticosa TASAIGO A woody-based, shrubby climber with spines along the stems and leaf margins. Leaves elliptic to ovate, borne in whorls. Flowers small, pale greenish-yellow, borne in lateral and terminal racemes. Fruit a translucent-*white or black* berry, 4–6 mm. Laurel forests and *Euphorbia* scrub. L, F, C, T, G, H, P. Plants with smaller leaves and black fruits are considered **subsp. melanocarpa**. *C, T, C. Trailing plants with wider leaves **subsp. *periclymenum***. *C, T, G, P.

1. *Erica canariensis*
2. *Erica canariensis*
3. *Erica canariensis* (bark)
4. *Erica platycodon*
5. *Erica platycodon*
6. *Erica platycodon* (bark)
7. *Rubia occidens*
8. *Rubia fruticosa*
9. *Rubia fruticosa* (in flower)
10. *Rubia fruticosa* (in fruit)
11. *Rubia fruticosa* subsp. *melanocarpa*

Plocama

Herbs or shrubs, strong-smelling when crushed, with opposite leaves (sometimes seemingly whorled). Corolla funnel-shaped with (4)5–7 spreading lobes; stamens 4, inserted to projecting. Fruit a drupe or splitting into 1-seeded mericarps.

Plocama pendula BALO A strong-smelling, yellowish-green shrub to 2 m with *pendulous* branches. Flowers small, whitish, with 5 lobes; often crowded around the branch tips. Fruit a small, blackish berry. Bare and rocky slopes. Common in the lower zones of west-central islands; very rare in the east. *F, C, T, G, H, P.

Galium BEDSTRAW

Herbs with rounded stems, and leaves in whorls of 4–12. Flowers with a white to yellow, 4-lobed corolla. Fruit a pair of bristly nutlets with hooked bristles. Numerous species with fluctuating taxonomy; just a few are described here.

Galium scabrum RASPILLA DE SOMBRA A weak-stemmed perennial, 25–60 cm, with *markedly hairy* leaves; leaflets elliptic, *3-nerved*. Inflorescences sparse. Fruits ovoid with glandular hairs. Thickets and woods; common in the west. F, C, T, G, H, P.

Galium aparine RASPILLA CUAJALECHES, GOOSE-GRASS An annual with spreading stems to 1(3) m, often stouter and more hairy at the nodes, with strongly recurved prickles on the stems. Leaves 3–5 mm, narrowly oblong, in whorls of 6–8, rough-margined. *Corolla short-tubed, 1.5–2(3) mm across, whitish*. Nutlets 1.1–4.1 mm with hooked bristles. A common weed on disturbed ground. L, F, C, T, G, H, P. The following are similar. *G. spurium* has a greenish corolla just 0.8–1.2 mm. L, C, T, H. *G. verrucosum* is small and slender, to 25 cm, with weak, recurved bristles on the angles and leaves 4–15 mm. Flowers borne in cymes of 1–3, shorter than their subtending leaves. Fruits borne on recurved stalks, spherical, 4–6 mm, coarsely warty. C, T, G, H, P. *G. tricornutum* has very rough leaves and stems, and flowers borne in cymes of 2–5 that are shorter than, or equalling, their subtending leaves. Fruits minutely warted. L, F, C, T, G, P.

Galium intricatum (syn. *G. parisiense*) RASPILLA DE PARÍS A slender, intricately branched annual 35 cm. Stems slightly hairy above. Leaves 4–12 mm, rather broadly lanceolate, *scarcely* bristle-tipped. Flowers with rather broad, sharp-pointed to bristle-tipped, reddish-yellow corolla lobes, borne in ovoid inflorescences. Fruit <1 mm, with hooked hairs, hairless, or minutely warty. L, F, C, T, G, H, P.

Galium murale RASPILLA MENUDA A short, sprawling annual with stems to 25(30) cm, much-branched from the base and sparsely hairy. Leaves 1.7–6 mm long, narrowly elliptic with a *short spine-tip*, in whorls of 4–6. Flowers inconspicuous, yellowish, minute, just 0.4–0.65 mm, borne in lax, few-flowered inflorescences (appearing paired in leaf nodes); corolla lobes pointed, erect. Nutlets 1.1–1.7 mm, rather cylindrical with spreading lobes and unevenly hairy. Common on waste and fallow land. L, F, C, T, G, H, P.

Sherardia

Annuals with leaves in whorls of 4–6. Flowers lilac, borne in dense terminal and lateral clusters surrounded by 8–10 leafy bracts. Fruit a pair of nutlets with a persistent calyx.

Sherardia arvensis RASPILLA AZUL A small, slender, hairy or hairless annual with spreading stems to 40 cm or less. Leaves 5–18 mm, in whorls of 4–6, soon withering below, pale green. Flowers borne in clusters of 4–10; corolla pale lilac, 4–5 mm, with tube longer than the lobes. Common on fallow land and in grassy places. L, F, C, T, G, H, P.

Valantia

Small annuals with leaves in whorls of 4. Flowers with 3–4 lobes, whitish or yellowish, borne in clusters of 3 in the leaf axils. Fruit 2 hairless nutlets.

Valantia hispida RASPILLA ENANA A small annual to 20 cm, with stems bristly towards the apex. Leaves 60 mm–10 cm, borne in whorls of 4, narrowly obovate to oblanceolate, broadest above the middle. Flowers minute, 1.5–2 mm, cup-like. Fruits *with long, white bristles*; nutlets usually 2, brown or blackish and rough. Rocky crevices. L, F, C, T, G, H, P.

Phyllis

Leaves opposite or whorled. Flowers with a 5-toothed (male) or 2-toothed (cosexual) calyx; corolla 5-parted. Fruit dry, black.

Phyllis nobla CAPITANA A small, hairless to hairy subshrub. Leaves entire, lanceolate to ovate. Flowers small, whitish, borne in lax terminal and lateral panicles. Fruiting stalks downcurved. Laurel forest tracksides, gaps and cliffs; very locally common in the west. C, T, G, H, P. *P. viscosa* CAPITANA PEGAJOSA is similar with smaller, narrow, sticky leaves and short, dense inflorescences. Local on cliffs and slopes in the sclerophyllous woodlands zone. *T, G, H, P.

1. *Plocama pendula* (habit)
2. *Plocama pendula*
3. *Galium scabrum*
4. *Sherardia arvensis*
5. *Phyllis nobla*
6. *Phyllis nobla* (OLIVER WHITE)

Theligonum

Virtually hairless annuals with opposite to alternate leaves. Monoecious; male flowers borne in clusters of 2–3, each with 2–5 lobes; female flowers 1–3, each with 2–4 poorly defined lobes. Fruit nut-like.

Theligonum cynocrambe QUEBRADIZO A small, prostrate or ascending annual 80 mm–40(50) cm, unpleasant smelling. Leaves opposite below, alternate above, 10–20(40) mm, untoothed and slightly succulent. *Flowers inconspicuous*, 2–3 mm, solitary or in small lateral clusters of 2–3; stamens conspicuously exserted. Fruit 1.6–2.3 mm, slightly fleshy. Damp, shady habitats. C, T, G, P.

GENTIANACEAE | GENTIAN FAMILY

Hairless annuals or perennials with opposite, untoothed leaves. Flowers with 4–5(8) petals fused into a tube; calyx deeply lobed; as many stamens as petals; styles 1–2 (if 1, branched). Fruit a 2-parted, splitting capsule.

Centaurium CENTAURY

Hairless annuals or perennials, often small. Flowers pink (rarely white) with 4(5) calyx and corolla lobes; stamens 5; style 1, divided.

Centaurium erythraea CENTAURA, COMMON CENTAURY A short biennial to 50 cm with a solitary stem, branched above. Leaves 20–60 mm, elliptic to oval, 3–7-veined, the lowermost in a *distinct rosette*; stem leaves smaller. Flowers pink, purplish or sometimes white, virtually stalkless, borne in *flat-topped* clusters; corolla lobes 4.5–6 mm. Grassy habitats; local. C, T, H, P. *C. tenuiflorum* SLENDER CENTAURY is similar, with stems *erectly branched* above; flowers borne in *dense, narrow* clusters; corolla lobes 2–4 mm. Damp, marshy, coastal places. L, F, C, T, G, H, P.

Centaurium maritimum YELLOW CENTAURY A short annual to biennial to 30 cm with a solitary stem, branched above. Leaves 3–10 mm, fleshy, elliptic to oval, the lower 2 small, *not* forming a distinct rosette, the upper leaves much longer. *Flowers pale yellow*, borne in loosely branched clusters, sometimes solitary; petals lobes 5, elliptic, 4–7 mm. Sandy or grassy habitats. C.

Schenkia

Small annuals with straight, erect stems. Similar to, and still widely included in, the genus *Centaurium*.

Schenkia spicata (syn. *Centaurium spicatum*) SPIKED CENTAURY An erect annual to 50 cm, similar to *Centaurium* but easily distinguished by its *spike-like inflorescences 60 mm–25 cm with erect-pointed, pink flowers*; flowers 10–16 mm, short-stalked (1.5 mm); corolla lobes 3.5–5 mm. Damp, grassy habitats, especially coastal. G.

Blackstonia YELLOW-WORT

Herbs with pairs of stem leaves fused at the base. Flowers yellow, short-tubed, with 6–8–numerous spreading lobes; style 2-lobed.

Blackstonia perfoliata YELLOW-WORT An erect, hairless, bluish plant to 40(50) cm with stems branching from the base or from the middle, from a basal rosette of broadly oval leaves that soon wither; stem leaves opposite, oval–triangular, pointed and narrowed at the base and joined there, *almost encircling the stem*. Flowers yellow, with a short corolla tube and 6–8 spreading lobes 4.2– 10 mm; calyx divided into 12 *narrowly linear segments to 1 mm wide, free almost to the base*. Grassy habitats and scrub. C.

Ixanthus

Perennial herbs. Leaves opposite, entire, often stalkless. Flowers cosexual, 4–5-parted; calyx fused; ovary superior. Fruit a many-seeded capsule.

Ixanthus viscosus REINA DEL MONTE A tall, sticky subshrub to 180 cm with herbaceous stems. Leaves lanceolate to ovate with 3–5 parallel veins. Inflorescence branched; flowers yellow, large. Capsule oblong, rather fleshy; seeds many, black. Locally common in the laurel forests along tracksides. *C, T, G, H, P.

1. *Centaurium tenuiflorum*
2. *Ixanthus viscosus* (habit)
3. *Ixanthus viscosus*
4. *Vinca major*
5. *Calotropis procera* (flowers)
6. *Calotropis procera*
7. *Gomphocarpus fruticosus*

APOCYNACEAE | PERIWINKLE FAMILY

Trees, shrubs, climbers and herbs with opposite, untoothed leaves, often with a milky latex when cut. Corolla with a distinct tube and 5 lobes; stamens 5, inserted in the tube; style 1. Fruit a pair of follicles; seeds often with hairy tufts. A large family with several subfamilies, subject to many recent changes in taxonomy; now including species formerly classified in the family Asclepiadaceae.

Vinca PERIWINKLE

Perennials with trailing stems and pairs of leathery leaves. Flowers characteristically periwinkle-like; solitary in the upper leaf axils; corolla propeller-shaped. Fruit a pair of follicles.

Vinca major VINCA, GREATER PERIWINKLE A spreading evergreen shrub with long trailing stems to 1.5 m, often rooting down. Leaves *large*, 25–90 mm shiny dark green, oval with *minutely hairy margins*. Flowers *bluish-violet*; corolla 30–50 mm across (tube 12–15 mm; calyx lobes 7–17 mm), hairy on the margin. Follicles to 50 mm; often not produced. Cultivated and widely naturalised. C, T, G, H, P.

Calotropis

Small desert trees with large oblong–round leaves. Fruit a large, inflated, smooth follicle, spongy within.

Calotropis procera ÁRBOL DE LA SEDA A tall, erect shrub to 3(5) m with *corky, white* bark and milky sap. Leaves large, 50 mm–15(20) cm, oblong–round, stalkless and mealy-velvety, especially below. Flowers 25 mm across with 5 oval, spreading lobes, whitish, flushed with dark pink-mauve. Fruit a pair of *strongly inflated, soft*, ovoid to spherical follicles 65–95 mm (12 cm). Native to tropical and subtropical Africa to the Middle East; naturalised in dry, disturbed areas. F (south), C.

Gomphocarpus SILKWEED

Erect perennials. Flowers with a corona of 5 horns *without smaller projections*. Fruit 1(2) inflated, bristly, erect-spreading follicles.

Gomphocarpus fruticosus SEDERO, BRISTLE-FRUITED SILKWEED An erect shrub to 1.2(2) m with linear–lanceolate, hairless leaves 39 mm–12 cm long. Flowers *white*, borne in hairy, stalked umbels; corolla lobes 6.2–8.3 mm, hairy along the margins. Fruit conspicuous; follicles green, *inflated, pointed*, hairless between the prominent bristles, *25–45(60) mm across* with 120 cottony seeds. Naturalised along roadsides. L, F, C, T, G, H, P.

Cynanchum

Climbing herbaceous perennials or shrubs with heart-shaped leaves. Flowers with a corona of 5 long appendages. Fruit 1(2) pendent, spineless follicles.

Cynanchum acutum MATACÁN, STRANGLEWORT A hairless, twining, blue-green, woody climbing perennial to 3(4) m with slender stems. Leaves 22–78 mm (10 cm), *arrow-head-shaped*, deeply heart-shaped at the base and stalked. Flowers tubular, scented, borne in lateral or terminal umbels, with a corolla of 5 triangular, projecting lobes 4.9–7.2 mm. *Fruits long and spindle-like*, pointed, 17 mm–18 cm long. Introduced, naturalised. L, F.

Apocynaceae (subfamily Asclepiadoideae) species: A. *Calotropis procera*; **B.** *Gomphocarpus fruticosus*; **C.** *Apteranthes burchardii*; **D.** *Periploca laevigata*.

Periploca

Climbing perennials with oval–lanceolate leaves and greenish-brown flowers; corolla lobes hairy and fleshy; corona lobes without conspicuous veins, and with 5 slender appendages; filaments free. Fruit 1–2 follicles.

Periploca laevigata CORNICAL A *branched shrubby perennial* to 3 m with branches weakly climbing, often with few leaves. Flowers borne in few-flowered clusters, with *yellow-green* lobes 6–7 mm, a brown-purple centre, and whitish appendages; hairless except for woolly spots on the lobes. Fruit with follicles spreading horizontally. Locally very common in a range of lower zone habitats. L, F (south), C, T, G, H, P.

Apteranthes

Succulents, often still treated under the separate genus *Caralluma,* but with squarer, creeping stems and inflorescences pushed aside by the continuously growing stems into a lateral position.

Apteranthes burchardii (syn. *Caralluma burchardii*) CUERNÚA, COLMILLO DE PERRO A spreading succulent with 4-angled, greyish-green or purplish stems 10–20 cm. Flowers 5-lobed, hairy, starfish-like, borne in winter. Fruit a pair of long, horn-like follicles to 10 cm; seeds with a pappus of long, white hairs. Rocky hillsides and malpaís; rare and local. *L (north – rare), F (centre-north, especially Betancuria), T (very rare). **Subsp.** *burchardii* is the form of the Islands.

Ceropegia

Succulents with erect to sprawling, typically pencil-like stems. Corolla with a slender tube. Fruit erect.

Ceropegia dichotoma CARDONCILLO, SAYÓN An erect or sprawling succulent 60–75 cm with green to olive-grey stems; leaves soon-falling. Inflorescences few-flowered; flowers 20–40 mm, pale *yellow* with darker lobes. Fruits 10–12 cm, greenish-brown. Locally common in rocky habitats in the lower zones. Recognised forms include: **subsp.** *dichotoma* (*T, H, P) and **subsp.** *krainzii* (*T, G). Previously recognised forms now included within this species include *C. chrysantha* (with whitish corolla and yellow lobes: *T), and *C. ceratophora* (corolla lobes united in the apex: *G).

Ceropegia fusca (syn. *C. dichotoma* **subsp.** *fusca*) MATAPERROS is similar to *C. dichotoma* but with grey-brown stems and *reddish-brown* flowers. Locally common on rocky slopes (mainly south-facing). *C, T.

Cascabela

Shrubs native to Central and South America with trumpet-shaped, showy flowers.

Cascabela thevetia A shrub with willow-like, linear leaves to 16 cm. Flowers few, trumpet-shaped, usually yellow; tube 40–50 mm; lobes 10–20 mm. Fruit pear-shaped, reddish-black, single-seeded, fleshy, 20–40 mm. Commonly planted in gardens throughout.

---

1. *Periploca laevigata*
2. *Periploca laevigata* (flower)
3. *Periploca laevigata* (in fruit)
4. *Apteranthes burchardii*
5. *Apteranthes burchardii* (pale form)
6. *Apteranthes burchardii* (fruits)
7. *Ceropegia dichotoma* (habit)
8. *Ceropegia dichotoma*
9. *Ceropegia fusca* (habit)
10. *Ceropegia fusca*
11. *Ceropegia fusca*, Gran Canaria
12. *Cascabela thevetia*

BORAGINACEAE | BORAGE FAMILY

Annual or perennial herbs or shrubs, often with bristly stems and leaves; bristles often with swollen bases. Leaves simple and alternate. Flowers in spiralled clusters, short-stalked, blue in many species; stamens 5, borne on the corolla tube; ovary deeply 4-lobed; style 1 (sometimes split). Fruit 4 nutlets, often concealed in a persistent calyx.

Arnebia

Erect or prostrate, annual or perennial bristly herbs. Leaves stalkless, alternate. Flowers usually borne in terminal inflorescences; calyx 5-lobed to the base, exceeded by the corolla; corolla throat hairless. Nutlets (1–)4, rough and warty, *enveloped by stiff calyx bristles*.

Arnebia decumbens An erect, bristly annual to 17(30) cm, branching from the base, with long, yellowish hairs to 1.4 mm. Lower leaves linear–oblong, 11–40 mm, upper leaves linear to lanceolate, pointed. Flowers yellow; corolla tube 10–11 mm (1–2 x as long as the calyx); stigma 2–4-parted. Fruiting raceme *elongated and lax*. Nutlets irregularly warty, 2–2.8 mm long, beaked, grey-brown. Sandy places. L, F, C, T.

Buglossoides

Bristly annuals or perennials. Flowers often borne in branched panicles; stamens 5, included within the (typically) whitish or blue corolla. *Nutlets minutely warty*.

Buglossoides arvensis (syn. ***Lithospermum arvense***) An erect annual 10–35 cm with adpressed hairs. Leaves oblong–lanceolate. Flowers 5–9 mm with a cylindrical tube and without scales in the throat, *white* (rarely pale blue); calyx teeth linear–lanceolate with stiff white hairs, *long* (exceeding the corolla). Nutlets of fruit 3–4 mm, pale brown, without prominent swellings; fruit stalks not thickened. Open habitats. L, F, C, T.

Heliotropium

Small, bristly annuals or perennials. Flowers white or violet, borne on the upper side of outwardly-coiled branches; stamens included in the corolla (or virtually so); style terminal (not basal). Species rather similar.

(a) Low, often sprawling annuals to perennials to 50(70) cm.

Heliotropium europaeum HELIOTROPE A short, greyish to greenish, erect or spreading, softly hairy annual to 50(60) cm. Leaves 30–70 mm, oval to elliptic, stalked. Flowers white with a yellow throat, small, 2–4 mm across, borne in distinctly 1-sided, spiralled spikes; *calyx divided almost to the base*; style short or virtually absent. Fruit splitting into 4 nutlets. Bare, cultivated and waste ground. L, F, C, T, P. **H. supinum** is similar, with spreading, *almost prostrate* stems and leaves green above, and a *flask-like calyx with short lobes, divided >⅓ its length*, that encircles the fruit. L, F, C, T, P.

Heliotropium bacciferum CAMELLERA A perennial 20–50 cm, rather similar to *H. europaeum* (see above) but *woody* at the base and bristly hairy. Leaves ovoid–rhomboid, to 30 mm. Flowers white, fragrant, borne in dense, 2-sided inflorescences; corolla 3–5 mm long; calyx divided almost to the base; style to 0.7 mm (*not* absent); nutlets 2 or 4. Dunes and other open, dry habitats; common. L, F, C, T, G, P. In the Canary Islands, the plant has been commonly treated as **H. ramosissimum** but this species has flowers borne in lax, 1-sided inflorescences.

Heliotropium curassavicum A *greyish, fleshy perennial* with prostrate stems to 70 cm. Leaves narrow and *hairless*, blunt. Flowers 3 mm across, white with a greenish-yellow centre. Nutlets 4. Native to America; locally naturalised. F, C, T.

(b) Shrubs to 3 m (formerly *Ceballosia*).

Heliotropium messerschmidioides (syn. *Ceballosia fruticosa*) A shrub to 3 m with greyish-green stems. Leaves stalked, narrowly lanceolate to ovate–lanceolate, bristly. Flowers dull whitish-cream, rather inconspicuous; corolla starry with a long tube; calyx lobes acuminate, 2 mm. Fruit fleshy, wrinkled, black when ripe. Common on dry slopes at low elevation. *L, F, C, T, G, H, P.

Neatostema

Small, bristly annuals. Flowers *yellow*, borne in branched or spike-like inflorescences; stamens included in the corolla.

Neatostema apulum ALACRANILLO, YELLOW GROMWELL A short annual to 25(30) cm with erect, bristly stems, branched above. Leaves 45(70) mm long, narrowly oblong, and bristly along the margins, those on the stems *erect*. Flowers yellow, borne in dense inflorescences to 90 mm long; corolla 2–3.5 mm across. Nutlets pale brown and warted. Bare and stony habitats. L, F, C, G, T.

Mairetis

Small bristly annuals, similar to *Neatostema*. Flowers *blue*; corolla very small (<7 mm).

Mairetis microsperma MOCO GUIRRE A short, erect, bristly annual to 20 cm. Leaves obovate to narrowly oblong. Flowers very small, blue, borne in coiled, 1-sided cymes; calyx tubular, *conspicuously expanding* in fruit. Bare flats and sands; frequent in the east. L, F, C, T.

1. *Heliotropium bacciferum*
2. *Heliotropium messerschmidioides*
3. *Mairetis microsperma*
4. *Neatostema apulum*

Echium BUGLOSS

Bristly herbs or shrubs. Flowers with a zygomorphic, funnel-shaped corolla, borne in spiralled cymes often making up dense or lax panicles; stamens unequal and often at least some exserted, the fruits rough.

With the exception of *E. plantagineum* and *E. horridum*, the Canary Island species of *Echium* are all members of a single Macaronesian radiation. Reproductive barriers between species in this group are weak and where distributions overlap, hybridisation may occur and complicate identification.

(a) Annual or perennial herbs.

Echium auberianum TAGINASTE PICANTE Perennial with woody root stock. Stems erect, *leaves in a basal rosette* covered with long, yellowish bristles. Corolla blue, stamens included or the lower two slightly exserted. Restricted to the subalpine zone of Las Cañadas. *T.

Echium triste VIBORINA TRISTE Annual or perennial herb with spiny linear leaves. Flowers in a lax inflorescence, *corolla pinkish, small and narrow*, with slightly exserted stamens. Dry lower slopes. *C, T, G.

Echium bonnetii VIBORINA CANARIA Small annual herb with lanceolate, shortly petiolate, basal leaves, *small, blue corolla (up to 1.5 cm long) and exserted lower stamens*. Dry, rocky areas. *L, F, C, T. *E. lancerottense* LENGUA DE VACA, VIBORINA DE LANZAROTE differs from *E. bonnetii* in having ovate leaves and *stamens that are not exserted*. Restricted to the north of Lanzarote. *L. *E. plantagineum* PALOMINA, ZUAJA, LENGUA OVEJA, VIBORINA is also similar but is more robust, annual or short perennial, the leaves are sessile and it has *larger flowers (>2 cm)*. A common weed in the western-central islands. C, T, G, H, P. *E. horridum* is similar but covered in long, stiff, *broad-based, bristle-like hairs* and the corolla is also covered in hairs (in *E. plantagineum* hairs on the corolla are confined to the nerves and margin only). Recorded from the arid southeast of Gran Canaria, near Juan Grande. C.

(b) Unbranched shrubs that have a dense leaf rosette and a single, very large inflorescence.

Echium simplex PALOMINO, ARREBOL Leaves elliptic–lanceolate, covered in silvery hairs, arranged in a rosette on a short stem. Inflorescence cylindrical, up to 2 m in height with *white flowers* and exserted stamens. Restricted to the heavily disturbed lower margins of the laurel forest in the Anaga peninsula. *T.

Echium pininana PININANA Leaves elliptic–lanceolate, rough, arranged in a rosette on a long stem up to 2–3 m in length. Inflorescence cylindrical, up to 4 m in height with *blue flowers*. Restricted to the laurel forests in the northeast. *P.

Echium wildpretii TAGINASTE ROJO Leaves lanceolate, covered in white hairs, arranged in a rosette on a short stem. Inflorescence up to 2 m in height, with *red flowers*. Restricted to the subalpine zone of Las Cañadas. *T. The closely related *E. perezii* (syn. *E. wildpretii* subsp. *trichosiphon*) TAGINASTE ROSADO is similar but differs in the broader leaves and the corolla, which is *pink rather than red*. Restricted to the high mountains. *P.

1. *Echium auberianum*
2. *Echium auberianum* (RACHEL GRAHAM)
3. *Echium triste* (habit)
4. *Echium triste* (RACHEL GRAHAM)
5. *Echium triste*
6. *Echium bonnetii*
7. *Echium lancerottense*
8. *Echium plantagineum*
9. *Echium plantagineum* (RACHEL GRAHAM)

1. *Echium simplex* (habit) (RACHEL GRAHAM)
2. *Echium simplex* (RACHEL GRAHAM)
3. *Echium pininana* (RACHEL GRAHAM)
4. *Echium wildprettii*
5. *Echium perezii*
6. *Echium giganteum*
7. *Echium giganteum* (RACHEL GRAHAM)
8. *Echium aculeatum* (habit) (RACHEL GRAHAM)
9. *Echium aculeatum*
10. *Echium brevirame* (habit) (RACHEL GRAHAM)
11. *Echium brevirame* (RACHEL GRAHAM)
12. *Echium bethencourtii* (RACHEL GRAHAM)
13. *Echium bethencourtii* (habit) (RACHEL GRAHAM)
14. *Echium leucophaeum* (habit) (RACHEL GRAHAM)
15. *Echium leucophaeum* (RACHEL GRAHAM)
16. *Echium gentianoides* (habit) (RACHEL GRAHAM)
17. *Echium gentianoides* (RACHEL GRAHAM)

(c) Branched shrubs with several inflorescences and laterally compressed corollas.

Echium giganteum TAGINASTE GIGANTE   Leaves <2 cm wide, lanceolate to oblanceolate, hispid to hirsute. Inflorescence conical, corolla white with unequal lobes. North coast up to 700 m elevation. *T. *E. aculeatum* AJINAJO is similar but has *linear leaves* with densely spiny margins and midribs and calyx lobes that are as long as the corolla tube. Common in lower elevation habitats of the W and SW of Tenerife, La Gomera and El Hierro. *T, G, H. *E. brevirame* ARREBOL, TAGINASTE also has densely spiny margins and midribs to the leaves but the leaves are broader than in *E. aculeatum* and the *calyx lobes are shorter than the corolla tube*. A widespread species of lower-elevation habitats. *P.

Echium bethencourtii ARREBOL DE GUELGUÉN   Leaves lanceolate to oblanceolate, *>2 cm wide*, with a rough, silvery indumentum. Inflorescence conical; corolla white to bluish, the lobes subequal. North coast. *P.

Echium leucophaeum TAGINASTE DE ANAGA   Leaves linear–lanceolate, <1.5 cm wide, with stiff hairs. Inflorescence with a *flattish dome shape*, often branched near the base; corolla white with unequal lobes. Locally common on lower slopes of the Anaga peninsula. *T.

(d) Branched shrubs with several inflorescences, and corollas not laterally compressed.

Echium gentianoides TAGINASTE PALMERO DE CUMBRE   A shrub with distinctive lanceolate, bluish-green leaves, the upper surface of which is covered in *distinctive short, large-based hairs* and the lower surface is ± glabrous. Calyx tube spiny and often longer than the lobes. Corolla an intense blue-violet colour with the stamens more or less equal in length to the corolla or slightly exserted. Restricted to the high mountains above 1,800 m. *P.

Echium strictum TAGINASTE ROSADO   A variable shrub with *ovate–lanceolate to ovate leaves*, covered in short, stiff hairs. Inflorescence lax, comprising few, widely spaced lateral cymes. Corolla pink or blue, sometimes with deep blue lines. Locally frequent in lower zones. *C, T, G, H, P.

Echium decaisnei TAGINASTE BLANCO   A shrub with *lanceolate leaves and large, conical, broad-based inflorescences*. Corolla funnel-shaped, white with blue stripes and equal lobes. **Susbp. decaisnei** has large-based spines evenly spaced on the upper surface of the leaves; on the lower side they are restricted to the margins and midrib. It is locally common from 100 m to 1,000 m. *C. **Subsp. purpuriense** (syn. *E. famarae*) TAGINASTE BLANCO ORIENTAL has leaves that are almost smooth to the touch, with disk-like hairs, and inflorescences that are laxer. Restricted to the Famara and Jandía massifs. *L, F.

Echium virescens TAGINASTE AZUL DE TENERIFE   A shrub with lanceolate, acute leaves, covered in short, white hairs. Inflorescence *a long slender spike*. Flowers pink or bluish; corolla lobes equal; stamens protruding. Common in open places of the laurel or pine forests. *T. *E. sventenii* TAGINASTE DE ADEJE is similar but differs in its taller and more densely branched habit, its more slender inflorescence, and *corollas that appear 4-lobed* (the 2 posterior lobes more or less fused). A rare endemic found only in the Adeje massif. *T. *E. webbii* ARREBOL AZUL can be distinguished by its broader leaves and always blue corolla. Frequent in laurel and pine woods. *P.

Echium hierrense AJINAJO HERREÑO, TAJINASTE HERREÑO   A compact shrub with *short lanceolate to ovate leaves* (<8 cm long) covered in silvery hairs. Inflorescence short, cylindrical. Corolla pink or blue with protruding stamens. From 100 m to 1,200 m in the El Golfo region of El Hierro. *H.

1. *Echium strictum*   (RACHEL GRAHAM)
2. *Echium strictum*
3. *Echium decaisnei* subsp. *decaisnei*
4. *Echium decaisnei* subsp. *purpuriense*
5. *Echium virescens* (habit))
6. *Echium virescens*
7. *Echium virescens*   (RACHEL GRAHAM)
8. *Echium virescens*

1. *Echium webbii* (RACHEL GRAHAM)
2. *Echium hierrense* (habit) (LEOPOLDO MORO)
3. *Echium hierrense* (RACHEL GRAHAM)
4. *Echium acanthocarpum*
5. *Echium acanthocarpum* (RACHEL GRAHAM)
6. *Echium callithyrsum* (habit) (RACHEL GRAHAM)
7. *Echium callithyrsum* (RACHEL GRAHAM)
8. *Echium onosmifolium* (RACHEL GRAHAM)

Echium acanthocarpum A tall shrub with *large, rough, ovate to lanceolate leaves* (up to 10 cm wide), covered in stiff hairs. Inflorescence long and dense. Flowers blue. Laurel forest cliffs in the central zone. *G.

Echium callithyrsum TAGINASTE AZUL DE GRAN CANARIA A shrub with lanceolate to ovate leaves, the upper surface with a covering of large-based stiff hairs, and the lower surface with simple hairs and prominent, hairy veins. *Inflorescence short, ovate to oblong.* Corolla deep blue, reddish or white. A locally common plant of laurel forest, slopes and cliffs. *C.

Echium onosmifolium TAGINASTE NEGRO A shrub to 1 m tall, with linear to lanceolate leaves, the margins often rolled down. Upper surface of the leaves with large-based stiff hairs; the lower surface usually with simple hairs and bristly hairs restricted to the midrib. Inflorescence *narrow, cylindrical*; corolla whitish, pale pink (rarely blue) with a narrow tube. South and southwest, in the lower zone and montane scrub. *C. **Subsp. *spectabile*** is similar but with wider leaves (up to 4 cm wide) and longer inflorescences. Locally in the west. *C.

Echium handiense TAGINASTE DE JANDÍA A small shrub with more or less *elliptic leaves* that are covered in bristly hairs on both surfaces. *Inflorescence lax*, corolla blue. A very rare species restricted to the summit of the Jandía massif. *F.

Borago BORAGE

Bristly annuals or perennials. Flowers blue (rarely white), short-tubed with widely spreading lobes, and stamens equal, forward-pointing in a cone (included to exserted).

Borago officinalis BORRAJA, BORAGE A bristly annual with robust, generally branched stems to 60 cm. Basal leaves oval and light green, stalked; stem leaves smaller and unstalked. Flowers borne in loosely branched cymes, bright blue and star-like with a white centre, deflexed lobes 7–10 mm, and an exposed *cone of blackish anthers*; calyx 7–15 mm (20 mm in fruit). Cultivated and waste ground. L, F, C, T, G, H, P.

Anchusa ALKANET

Bristly annual or perennial herbs. Flowers with blue, purple, white, funnel-shaped corolla tube; stamens equal and not exserted.

Anchusa azurea LENGUAZA, ALICANEJA, LARGE BLUE ALKANET A robust species to 1.8 m with dense bristles, often with white, pimple-like bases. Leaves to 25(40) cm long and 50 mm wide; lanceolate. Flowers large, the corolla 8–15(20) mm across, deep blue or purple with a whitish centre (sometimes all white or cream); flowers borne in a large, loose and much-branched inflorescence; calyx divided almost to the base into linear lobes, erect in fruit. Frequent on bare ground. L, F, C, T, G, H, P.

Cynoglossum HOUND'S TONGUE

Hairy biennial to perennial herbs. Flowers borne in branched cymes without bracts; corolla with a short tube and 5 spreading petal lobes with scales closing the throat; stamens not protruding. Nutlets egg-shaped and with barbed spines.

Cynoglossum creticum LENGUAPERRO, BLUE HOUND'S TONGUE A short, robust, softly hairy biennial 30–70(90) cm with erect, angular stems branched above. Leaves to 30 cm, lanceolate, untoothed and densely hairy to felted on *both* surfaces, sometimes clasping the stem. Flowers purplish in bud, opening pale blue-violet with fine, *darker blue venation*, the corolla (5)7–9 mm across, borne in branched cymes that elongate in fruit. Nutlets without a distinct border, and with dense hooked spines, 6 mm. Disturbed ground. C, T, G, P.

Myosotis NOMEOLVIDES, FORGET-ME-NOT

Hairy annuals or perennials. Flowers with corolla tube short and the throat enclosed by 5 short scales; stamens equal and included.

(a) Perennials with large, pink or blue flowers to 10 mm across.

Myosotis latifolia A rhizomatous perennial with stems to 60 cm, often woody below. Leaves rather large, ovate. Corolla to 10 mm, blue or pink. Frequent in laurel forests. C, T, G, H, P.

(b) Annuals to biennials (not woody), with blue flowers ≤3 mm across.

Myosotis ramosissima A low, *bristly*, erect to spreading, short-lived annual to 25 cm. Flowers pale blue, just 0.8–2(3) mm across with corolla tube shorter than the calyx; calyx teeth spreading, flower stalks equalling calyx, and inflorescence longer than the leafy part of the stem when in fruit. Wet, grassy habitats. C, T, H, P.

Myosotis discolor An annual or biennial to 40 cm. Lower leaves spathulate; stem leaves oblong, all with soft, straight, spreading hairs. Flowers borne in long racemes, rather congested above; calyx tube with hooked or deflexed hairs; corolla 3 mm, bright blue with a yellow centre. Nutlets narrowly bordered. Damp habitats in otherwise dry areas. **Subsp.** *canariensis* is the form present. C, T, G, H, P.

CONVOLVULACEAE | CONVOLVULUS/BINDWEED FAMILY

Erect shrubs or subshrubs, or more typically climbing annuals or perennials with alternate leaves (leafless parasites in *Cuscuta*). Flowers 1–few in leaf axils or rarely in many-flowered panicles, large and showy in some genera; sepals and petals 4–5, fused at the base or to form a funnel-shaped corolla tube; stamens 4–5; style(s) 1–2. Fruit a capsule.

Convolvulus BINDWEED

Spiny or unarmed shrubs or herbs that are prostrate or erect with stems often twining or trailing. Leaves sessile or petiolate, entire or divided. Flowers 1–many; bracteoles not obscuring the calyx; style 1; stigmas 2, linear. Fruit unlobed.

(a) Spiny or unarmed shrubs.

Convolvulus caput-medusae CHAPARRO CANARIO Hummock-forming undershrub with white, *sericeous stems, spinescent when old*. Leaves oblanceolate to spathulate, small (1.5–2.5 cm x 0.4–1.5 cm), densely pubescent. Flowers more or less sessile, axillary, 1(–2) together. Restricted mainly to coastal sands and rocks. *F, C.

Convolvulus floridus GUAIDIL Erect shrub to 4 m in height. Leaves oblong, 2–14 cm x 0.5–2.6 cm, dark green. Flowers in terminal, many-flowered, *panicle-like inflorescences*. *L, F, C, T, G, H, P.

Convolvulus scoparius LEÑANOEL Erect shrub to 2 m in height. *Leaves small and filiform (0.5–4.5 cm x c. 0.1 cm), caducous*. Flowers in terminal and axillary cymes of (1–)5–6 flowers. *C, T, G.

1. *Myosotis latifolia*
2. *Convolvulus caput-medusae*
3. *Convolvulus floridus*
4. *Convolvulus scoparius*
5. *Convolvulus canariensis*
6. *Convolvulus lopezsocasi*
7. *Convolvulus fruticulosus*
8. *Convolvulus siculus*
9. *Convolvulus althaeoides*
10. *Convolvulus arvensis*
11. *Convolvulus farinosus*

(b) Trailing or twining plants with woody stems.

Convolvulus canariensis CORREGÜELÓN DE MONTE Liana or scrambling shrub to 10 m. Leaves petiolate, ovate to oblong ovate, 4–10 cm x 2–5 cm, *densely villous* with prominent veins on the lower surface. Flowers in groups of 3–7. Sepals densely hairy. Corolla blue with a white centre. Laurisilva forests. *C, T, G, H, P.

Convolvulus lopezsocasi CORREGÜELÓN DE FAMARA, ENREDADERA Scrambling shrub with *leaves glabrous or nearly so*. Leaves elliptic, mostly >4 x 1.5 cm, cymes 1–6-flowered. Sepals 9–10 mm long, corolla white with pink petaline bands. Restricted to the cliffs of the Famara Massif. *L. *C. volubilis* is similar but its leaves are lanceolate, the sepals are shorter (c. 5mm long) and the corolla white-bluish. *T, G.

Convolvulus fruticulosus Woody, scrambling plant with long, trailing stems. *Leaves shortly petiolate*, oblong, narrow (c. 0.5 cm wide and up to 6 x as long as wide), hirsute. Flowers solitary or in groups of up to 3. Sepals 6–9 mm long, outer sepals obtuse to subacute. Corolla pale blue. A species of *Euphorbia* communities. *T, H, P. *C. subauriculatus* is similar but has leaves that are sub-auriculate at the base. It is a rare cliff plant. *G. *C. glandulosus* (syn. *C. fruticulosus* subsp. *glandulosus*) differs in the leaves, which are broader (up to 1.5 cm and less than 6 x as long as wide) and typically only sparsely hairy, and the outer sepals, which are narrowly ovate or oblong–ovate and acuminate. *C. *C. perraudieri* is like *C. glandulosus* but densely hairy on both the leaves and the calyx. *C, T.

(c) Trailing or twining plants with herbaceous stems.

Convolvulus tricolor CORREGÜELA TRICOLOR, DWARF CONVOLVULUS A hairy, short, spreading annual to 40 cm. Leaves sessile, 13–40 mm long, oval to elliptic. Flowers solitary or paired on short stalks, 20–30 mm long, conspicuously *with 3 bands of colour*: yellow in the throat, then white, and bright blue on the perimeter. Ovary hairy. Local, on bare and sandy ground and waste places near the sea. C, T.

Convolvulus siculus CORREGÜELITA AZUL A twining, hairy, annual. Leaves petiolate, 12–60 mm long, lance-shaped to oval. Flowers solitary or paired on short stalks, *small (7–12 mm long)*, blue with a yellowish centre, distinctly 5-lobed. Ovary glabrous. Plants with bracteoles close to the flower are referrable to **subsp. *siculus***; plants with bracteoles that are remote from the calyx are referrable to **subsp. *elongatus***. Both occur in the archipelago. Dry, bare places. L, F, C, T, G, H, P.

Convolvulus althaeoides CORREGÜELA ROSADA A trailing or twining, often hairy, perennial to 2m long. Leaves petiolate, ovate–deltoid and typically dimorphic with the lower leaves irregularly crenate and the *upper leaves deeply lobed*. Flowers 1–2(–4) on stalks equal or longer than their leaves, corolla 1.8–4.5cm long, pink. Ovary glabrous. The Canary Island plants are **subsp. *althaeoides***. Wasteland, roadsides. L, F, C, T, G, H, P.

Convolvulus arvensis CORREGÜELA BLANCA, BINDWEED A creeping or twining, more or less hairless perennial to 2 m long. Leaves petiolate, 10–75 mm long, *arrow to oblong-shaped*. Flowers solitary or paired on stalks shorter than their leaves. Calyx <5mm long. Corolla 15–20(25) mm long, pale pink to white with paler stripes. Ovary glabrous. Fallow land. L, F, C, T, G, H, P.

Convolvulus farinosus A twining or prostrate perennial to 1 m long, often with a woody base. Vegetative parts hairy or farinose. Leaves petiolate, 3–9 cm long, ovate to triangular. Flowers solitary or in groups of up to 6 on stalks shorter than their leaves. Calyx 6–8 mm long. Corolla *small (10–15 mm long)*, white or pink. Ovary glabrous. Introduced species of tropical African origin, recorded from Los Realejos and Puerto de la Cruz. T.

Calystegia BINDWEED

Perennial herbs with prostrate stems, with petiolate, entire leaves. Flowers typically large and solitary, with large bracteoles immediately subtending the calyx; style 1, stigmas 2, globose. Fruit unlobed.

Calystegia soldanella CORREGÜELA DE PLAYA, SEA BINDWEED Prostrate, trailing herb to 50 cm, stems glabrous. Leaves kidney-shaped, deep green and veined, rather fleshy and long-stalked (30–90 mm). *Flowers pale pink* with white stripes, 40–45 mm; solitary. Coastal dunes. L.

Ipomoea

Climbing or scrambling annuals or tuberous perennials with petiolate, oval to heart-shaped, entire or lobed leaves. Flowers often with showy corollas; style 1, stigmas with 2–3 sub-globose lobes. Fruit unlobed.

Ipomoea indica BATATILLA DE INDIAS Twining perennial herb, stems pubescent. Leaves ovate or shallowly 3-lobed, both forms sometimes on the same plant. Sepals subequal, narrowly ovate, acuminate. Corolla *deep blue with violet mid-petaline bands*, 5–6 cm long. Ovary glabrous. Stigma typically 3-lobed. F, C, T, G, H, P.

Ipomoea purpurea CAMPANILLA MORADA Twining annual herb, stems pilose. Leaves ovate (rarely 3-lobed to half way). Sepals subequal, lanceolate to oblong–lanceolate. Corolla tube white, limb usually pink, sometimes cream or bluish, 4–5 cm long. Ovary glabrous. Stigma typically 3-lobed. Native to tropical America, naturalised. C, T.

Ipomoea nil (syn. *I. hederacea*) Trailing or twining herb, stems roughly pilose. *Leaves 3-lobed*. Sepals lanceolate, tapering into a long, linear point. Corolla tube white, limb blue, 3.5–4.5 cm. Ovary glabrous. Stigma typically 3-lobed. Planted in developed areas and occasionally naturalised. L.

Ipomoea cairica CAMPANILLA PALMEADA Twining perennial herb to 3 m, stems glabrous. Leaves digitately divided into *5–7 leaflets*. Sepals slightly unequal, outer oblong–ovate and acute; inner broadly ovate–elliptic, obtuse. Corolla pink, 4.5–7 cm long. Ovary glabrous. Stigma 2-lobed. Planted and naturalised near gardens. L, F, C, T, G, H, P.

Ipomoea pes-caprae BATATILLA DE PLAYA Trailing perennial herb to several metres long; stems stout, glabrous, rooting at the nodes. Leaves *leathery somewhat succulent, ovate to reniform or suborbicular*, the apex emarginate to shallowly bilobed. Sepals slightly unequal. Corolla 4–5 cm, pink. Ovary glabrous. Stigma 2-lobed. Plant of coastal sandy areas. L, F.

Ipomoea batatas BATATERA Creeping (rarely climbing) perennial herb to several metres long, rooting from the stem, the stem glabrous to coarsely pilose. Leaves very variable in form, ovate or shallowly to deeply 3–5-lobed, cordate, shortly acuminate. Sepals unequal, each with a fine point; outer sepals slightly shorter, oblong–elliptic to oblong–oblanceolate; inner sepals broadly elliptic, rounded and mucronate. Corolla pink, often with a dark centre, 4–4.5 cm long. *Ovary pubescent*. Stigma 2-lobed. The large, reddish tubers are cultivated as a food source. Cultivated. L, F, C, T, G, H, P.

Ipomoea corymbosa (syn. *Turbina corymbosa*) Climbing liana, stems becoming woody, usually glabrous. Leaves cordate with rounded auricles. Flowers in lax, compound cymes. Sepals slightly unequal. Corolla *cream with dark centre* and yellow midpetaline bands, 2.5–3 cm long. Ovary glabrous. Stigma 3-lobed. Introduced species native to tropical America. T, C.

1. *Ipomoea indica*
2. *Ipomoea cairica*
3. *Cuscuta planiflora* (habit)
4. *Cuscuta planiflora*
5. *Cuscuta approximata* (habit)
6. *Cuscuta approximata*
7. *Cressa cretica*
8. *Nicandra physalodes*

Ipomoea imperati (syn. *Pharbitis preauxii*) Perennial herb; stems trailing, rooting at the nodes, glabrous. Leaves *slightly succulent, linear, lanceolate or shortly oblong or 3–5-lobed* with a longer terminal lobe. Sepals unequal. Corolla white with a yellowish tube, 3.5–4 cm long. Ovary glabrous. Pantropical species of sandy beaches, recorded only from the Maspalomas area and considered probably native, although extinct in the area today. C.

Cuscuta GREÑA, DODDER

Leafless, parasitic twining herbs virtually without chlorophyll. Flowers small with 4–5-lobed corolla; styles 2. Fruit a 2–4-valved capsule.

Cuscuta planiflora A twining annual with very *slender* yellowish stems (reddish in places and in juveniles) 0.5 mm wide. Flowers in *compact, stalkless clusters* of 5–20(30), (4)5-parted, *whitish*, 2.5–3.5 mm; *calyx lobes with crystalline margins*. Parasitic on various herbs and subshrubs. L, F, C, T, G, H, P. *C. approximata* is similar but with calyx lobes with smooth (not crystalline) edges and often bright golden yellow, fleshy and shiny tips. Parasitic mainly on *Launaea arborescens*; common in the east. L, F, C, T.

Cressa

Herbs, often woody at the base. Leaves sessile, entire. Flowers small, in bracteate clusters at tips of branchlets. Sepals exserted. Styles 2, exserted, stigmas capitate. Capsule 2–4-valved.

Cressa cretica Branched herb with pilose stems. Leaves lanceolate to ovate. Corolla 5–6 mm long, whitish-pink. Probably introduced. Wet saline soils. C.

Dichondra

Prostrate perennial herbs. Leaves petiolate, cordate-reniform. Flowers small, solitary in leaf-axils. Corolla somewhat shorter than calyx. Styles 2, stigmas capitate. Capsule 2-lobed.

Dichondra micrantha CÉSPED DE RIÑÓN Plant with thin stems (<1 mm in diameter), leaves sparsely pubescent, the corolla about as long as the calyx at anthesis, and the calyx-lobes twice as long as they are broad, or less, shorter than the fruits. F, C, T, P.

SOLANACEAE | POTATO FAMILY

Herbs or shrubs with simple, entire or pinnately divided, alternate leaves. Flowers with a star- or bell-shaped corolla, the 5 petals fused below; stamens 5, attached to the corolla tube; style 1; ovary superior with 2 compartments. Fruit a berry or 2–4-valved capsule.

Nicandra

Herbaceous, hairless annuals with toothed to lobed leaves. Corolla broadly bell-shaped and shallowly lobed; calyx much-inflated. Fruit a dry berry.

Nicandra physalodes TOMATERA DE CULEBRA, APPLE OF PERU A hairless, unpleasant-smelling, vigorous, much-branched annual to 80 cm with stalked, oval and toothed to lobed, large leaves. Flowers borne singly in the leaf axils; corolla 12–20 mm, blue to violet with a white centre, soon closing; calyx much enlarging in fruit (25–35 mm). Disturbed habitats. L, C, T, G, P.

Lycium

Woody, almost hairless shrubs with simple, deciduous leaves. Flowers dull purple, rather deeply lobed, often with stamens protruding. Fruit a berry.

Lycium europaeum ESPINO BLANCO, TEA TREE A deciduous, robustly spiny shrub to 3(4) m. Leaves elliptic, broadest above the middle, 20–73 mm long. Flowers borne in clusters of 2–5, with a pink or white corolla, narrowly funnel-shaped, 10–17 mm long (less slender than *L. schweinfurthii*); lobes 3–5 mm; stamens usually protruding; calyx 2–3.5 mm. Fruit a spherical, reddish berry, 5–6 mm. Thickets. T, G, P.

Lycium intricatum ESPINO A slightly *succulent* shrub with arched, grey, spiny stems. Leaves 2–26 mm long. Corolla tube 11–18 mm long; lilac, pink or white with stamens *not* protruding; calyx 1.7–3 mm. Fruit an orange-red or blackish berry 3–7(9) mm. Malpaís; common, abundant in the east. L, F, C, T, G, H, P.

Hyoscyamus HENBANE

Glandular-hairy annuals to biennials with simple, toothed to lobed leaves. Flowers borne numerously in rows along outwardly-coiled, leafy stems; corolla funnel-shaped. Fruit a splitting capsule.

Hyoscyamus albus BELEÑO, WHITE HENBANE A sticky annual or short-lived perennial with long, sparsely branched stems to 80 cm, often woody below. Leaves 6–20 cm long, broadly oblong and wedge-shaped or heart-shaped at the base, with wide teeth along the margin. *Flowers stalkless*, at least above, borne in long, dense, 1-sided spikes; calyx densely hairy and swollen below, ending in short, triangular teeth; corolla 20–30 mm, greenish or *yellowish-white* with a green or purple throat; *stamens not, or scarcely protruding*. Capsule 10 mm. Occasional on waste ground. L, F, C, T, G, H, P.

Withania

Shrubs with opposite leaves and bell-shaped flowers. Fruit a berry *surrounded by an inflated calyx*.

Withania aristata OROBAL A shrub to 2 m with virtually hairless stems. Leaves to 14 cm, ovate, often unequal at the base. Flowers yellowish-green; calyx with 5 long teeth, inflating to surround the orange-red berry in fruit. Frequent in barranco beds in the west. *C, T, G, H, P.

Withania somnifera A shrub to 1.5 m with sparingly branched stems, somewhat hairy or *hairless* when young. Leaves 40 mm–10 cm, wedge-shaped at the base. Flowers borne 4–16 in crowded clusters arising from the axils; calyx 5 mm, densely downy; corolla small (7–17 mm), yellowish-green with spreading lobes. Fruit a red berry 5–8 mm. Dry habitats; uncommon. C, T, G, P. *W. frutescens* is similar but with *small* leaves and flowers; corolla just 4.5–7.5 mm with young branches *white-hairy*. Rocky habitats; rare. F, C.

Solanum NIGHTSHADE

Herbs or shrubs with simple leaves. Flowers borne in 1-sided cymes or umbels; leaf-opposed; corolla star-shaped with spreading petal lobes; stamens protruding, forming a cone around the stigma. Fruit a berry.

(a) Annuals.

Solanum nigrum YERBAMORA, BLACK NIGHTSHADE A variable, hairless to hairy *annual* to 70 cm with stems spreading and blackish. Leaves oval–lanceolate, toothed, lobed or entire; stalked.

1. *Lycium intricatum* (habit)
2. *Lycium intricatum*
3. *Lycium intricatum* (fruits)
4. *Hyoscyamus albus*
5. *Withania aristata*
6. *Solanum nigrum*
7. *Solanum vespertilio* subsp. *vespertilio*
8. *Solanum mauritianum*

Flowers with white petals, star-like with a yellow cone of anthers borne in clusters of 5–10; corolla 5–6 mm. Berry 6–10 mm, round and green ripening matt-black, borne on erect (to spreading) stalks. A common weed on cultivated land. L, F, C, T, G, H, P. *S. nigrum* has a number of subspecies and is one of a complex group of taxa. **S. villosum** is *densely long-hairy* with more deeply lobed leaves, clusters of 3–6 flowers, and fruits *reddish or yellowish* (not black), 6–10 mm, often longer than wide. Common in disturbed or bare places. Throughout. L, F, C, T, G, H, P.

(b) Shrubs and woody-based climbers.

Solanum vespertilio REJALGADERA An erect shrub to 1.5 m with densely spiny stems and leaf stalks. Leaves ovate to ovate–rhomboid, 5–15 cm, yellowish, felted beneath. Flowers with 4 petals, borne in clusters of 5–10, blue-purple. Fruits 20 mm, red. Ravines, slopes and open places 100–700 m; rare and local. *C (**subsp.** *doramae*), *T (Anaga: **subsp.** *vespertilio*). *S. lidii* PIMENTERA is similar but procumbent with remotely spiny stems; leaves narrower, flowers with 5 petals and fruits to just 10 mm, orange. Mountain slopes; rare. *C (southeast).

Solanum laxum (syn. *S. jasminoides*) ENREDADERA DE PAPA, POTATO VINE An ornamental *evergreen, climbing perennial* to 5 m. Leaves 25–75 mm, bright green, leathery and shiny; pointed and lanceolate with a heart-shaped base; stalked. Flowers with a *white* corolla 15–18 mm across, star-like, with yellow cones of anthers, borne in lax, many-flowered, showy cymes. Fruit 4–5 mm. Widely planted in developed areas. C, T. **S. mauritianum** is an ornamental shrub native to South America; 2–4 m; leaves simple, elliptic–lanceolate, felted, entire, large, to 40 cm; stems and leaf stalks *woolly*. Flowers purple with yellow anthers. Berries yellow-orange when ripe, to 20 mm. Planted; locally naturalised. T.

Datura (including *Dutra*)

Erect annuals with simple leaves. Flowers regular, showy, trumpet-shaped, upward-pointing; calyx with 5 teeth. Fruit a large, spiny capsule.

Datura stramonium ESTRAMONIO, HIERBA DEL DIABLO A stout and vigorous, unpleasant-smelling, normally *hairless* annual to 1.5 m with stout, spreading stems. Leaves oval to elliptic, lobed with jagged teeth, 50 mm–18 cm. Flowers with a white corolla, sometimes flushed with purple, trumpet-like, 50 mm–10 cm long, borne solitary in the leaf axils of the upper leaves; calyx large, to ½ the length of the corolla (30–50 mm), sharply angled. Fruit an erect, large, spiny capsule 35–70 mm long. Frequent in damp, disturbed habitats. L, F, C, T, G, H, P.

Datura innoxia BURLADORA A perennial (now classified by some authors under the separate genus *Dutra*), *softly hairy,* to 50 cm (2 m), with *large flowers* that have a corolla 14–16.5 cm long, hairless outside, and capsules 30–50 mm, *nodding* when mature. Disturbed habitats, roadsides; casual. L, F, C, T, G, H, P. **D. metel** is similar, *not* softly hairy and with *warted* fruits. L, F, T.

Nicotiana TOBACCO

Shrubs or perennials, sticky, with simple, entire leaves. Flowers with an elongated, funnel-shaped corolla, borne in branched, leafless clusters. Fruit a 2-valved capsule. Native to South America, widely planted and naturalised.

Nicotiana glauca TABACO MORO, BOBO, SHRUB TOBACCO A hairless, lax shrub *or small tree to 6 m* with long grey branches and sparse, *grey-green*, stalked, elliptic–lanceolate, pointed leaves 21 mm–12 cm. Flowers greenish-yellow, borne in lax panicles; corolla tubular, 27–45 mm long. Fruit an egg-shaped capsule 8.5–15 mm, exceeding the persistent, papery calyx. Native to South America; abundant across all lower zones. L, F, C, T, G, H, P. **N. paniculata** is a similar but smaller, sticky subshrub. Cliffs near the coast. T (naturalised in the north), H (rare).

1. *Datura stramonium*
2. *Datura innoxia*
3. *Datura innoxia* (fruits)
4. *Datura metel* (fruits)
5. *Nicotiana glauca*
6. *Nicotiana tabacum*
7. *Nicotiana glutinosa*
8. *Solandra maxima*
9. *Salpichroa origanifolia*

Nicotiana tabacum TABACO An annual to 1 m, sticky throughout, sparingly branched. Leaves large, to 30 cm, ovate to elliptic, pointed and *clasping* the stem at the base. Flowers scented; corolla with a long tube, *white* flushed pink or purple, 30–55 mm. Fruit a capsule to 20 mm. Naturalised in rural areas. L, F, C, T, G, H, P. *N. glutinosa* is similar but smaller with leaves *stalked*; corolla tube short. Rare, in stone walls. T.

Solandra

Vigorous shrubby vines native to Central and South America with shiny foliage. Flowers very large, showy. Fruit a large capsule.

Solandra maxima TROMPETERO, HAWAIIAN LILY A shrubby climber with oval, dark green leaves and *large, showy, trumpet-shaped flowers to 20 cm long*; yellow-cream with purple veins and prominent stamens. Fruit a berry. A striking exotic ornamental widely planted in towns, gardens and resorts in the region. Planted.

Salpichroa

Fleshy, hairy, spineless perennial herbs native to South America with simple leaves. Flowers solitary, with regular, tubular corolla. Fruit a berry.

Salpichroa origanifolia HUEVITO DE GALLO A much-branched, hairy perennial to 1 m with woody stems below, usually scrambling among surrounding vegetation. Leaves oval, short-stalked and entire, 50 x 37 mm. Flowers with corolla white and bell-shaped, 6.5–11 mm long. Fruit a small, cream-white berry, 10–15 mm. Naturalised locally on roadsides, waste places and crops. F, C, T.

OLEACEAE | OLIVE FAMILY

Trees and shrubs, usually with opposite, simple leaves. Flowers in cymes or panicles; calyx with 4 small teeth; corolla with 4(–6) free or fused petals; stamens usually 2; style 1. Fruit a 2-valved capsule, 2–4-seeded berry or winged nut or achene.

Jasminum JASMINE

Woody climbers and shrubs with alternate compound leaves (rarely simply and opposite). Flowers yellow or white with 4–6 petals united into a tube. Fruit (usually) a 2-lobed black berry.

Jasminum odoratissimum JAZMÍN CANARIO, LEÑA BLANCA A shrub to 4 m. Leaves alternate with 3–5 leaflets, oblong to obovate. Flowers yellow with corolla lobes shorter than their tube. Rocky slopes and lower forest zones. F, C, T, G, H, P.

Olea OLIVE

Small evergreen trees with opposite, entire leaves and flowers with 4 corolla lobes and sepals borne in axillary clusters. Fruit a berry (the olive).

Olea europaea OLIVO, OLIVE A much-branched silvery tree to 15 m with a grey trunk. Leaves 7–60 mm, opposite; grey-green, silvery beneath, minutely scaly, lanceolate, untoothed and short-stalked. Flowers small, 6–8.5 mm, whitish, borne in erect clusters. Fruit an olive, 6–15(20) mm. Abandoned pastures and extensively cultivated. L, F, C, T, G, H, P.

OLEACEAE | 293

Olea cerasiformis ACEBUCHE   A shrub or small tree to 6 m, similar to the familiar olive tree, with narrower leaves, glossy dark green above, white-scaly beneath; corolla white. Fruit fleshy, becoming brown or black. Rocky habitats, barrancos and cliffs. Rather patchy and local. *L, F (centre), C, T, G, H, P.

Phillyrea

Evergreen shrubs with simple leaves. Flowers greenish or yellowish with 4 corolla lobes and projecting stamens. Fruit a berry.

Phillyrea angustifolia OLIVILLO   An olive tree-like evergreen shrub to 2(4) m with upright branches and grey bark. Leaves 31–78 mm x *3–12 mm,* opposite, linear–lanceolate, entirely glabrous, untoothed to finely toothed, with 4–6 pairs of *obscure* veins. Flowers small, 2.7–4.2 mm, greenish, 4-lobed and borne in lateral clusters to 10 mm across; *fragrant*. Fruit small, 4–6.6 mm, fleshy with a point, blue-black when ripe. Rocky habitats, cliffs; local. L, C.

Picconia

Shrubs and trees with whitish bark. Leaves opposite, simple. Flowers cosexual; corolla and calyx 4-parted. Fruit a dry drupe.

Picconia excelsa PALO BLANCO   A shrub or tree to 10 m. Leaves lanceolate to obovate. Corolla white, deeply 4-lobed. Fruit ovoid, black. Frequent in laurel forests and on cliffs from the centre-west. F (very rare), C, T, G, H, P.

1. *Jasminum odoratissimum*
2. *Olea europaea*
3. *Olea cerasiformis*
4. *Phillyrea angustifolia*
5. *Phillyrea angustifolia* (in fruit)

PLANTAGINACEAE | PLANTAIN FAMILY

Annual or perennial herbs or shrubs with opposite or whorled, simple or compound leaves. Flowers variable, but usually zygomorphic and 2-lipped; sepals and petals 2–4; stamens 4; style 1. Fruit a capsule or 1-seeded nut. The family includes numerous genera traditionally in the Scrophulariaceae (though the revised classification and taxonomy are not universally accepted).

Plantago PLANTAIN

Small annual or perennial herbs with a basal rosette of leaves, opposite or alternate along the stem. Flowers small and inconspicuous, borne in dense heads or spikes; 4-parted; corolla papery; stamens protruding. Fruit a splitting capsule.

(a) Annuals. *Stems leafy and branched* (spikes borne in axils opposite the leaves).

Plantago afra ZARAGATONA A markedly *sticky* annual to 30(50) cm with much-branched stems. Leaves 40–80 mm, *linear* to linear–lanceolate, opposite or whorled, not fleshy and normally untoothed. Flowers brownish-white, to 4 mm, borne in round or conical spikes 5–15 mm, on spreading stalks; anthers pale yellow. Waste ground; common. L, F, C, T, G, H, P. *Divergently branched* annuals identified as *P. phaeostoma* in the east may in fact be a form of *P. afra* in the authors' opinion. Further work is needed. L, F.

(b) Annuals to perennials. *Leaves borne in a rosette*, linear or narrowly lanceolate, stems not ribbed; spikes borne on leafless stems.

Plantago coronopus (incl. *P. aschersonii*) RABO CORDERO, BUCK'S HORN PLANTAIN A very variable, low annual or perennial to 20 cm with solitary or clustered leaf rosettes. Leaves 20 mm–20 cm, linear–lanceolate, usually *pinnately lobed*, though sometimes unlobed, *not* particularly fleshy; hairless or finely hairy. Flowers yellowish-brown, to 3 mm, borne in spikes 40–70 mm long, terminating from smooth (not grooved), curved stems exceeding the leaves; anthers pale yellow. Disturbed habitats; common. L, F, C, T, G, H, P.

Plantago bellardii A low, *densely hairy* annual to 80 mm (16 cm) high with 1 or more leaf rosettes with linear–lanceolate leaves 15–60 mm, scarcely toothed or entire, 3-veined and white-hairy. Flowers brownish, borne in spreading spikes 8–20(48) mm, rather large relative to the leaves, *borne on stalks exceeded by the leaves* (to 13 cm); bracts hairy; corolla lobes oval–lanceolate. Tracksides. T, H. *P. ovata* PELOTILLA has flower stalks with *adpressed* (not spreading) hairs and flat, sub-equal sepals. Bare habitats. L, F, C, T, G, P.

Plantago albicans LLANTÉN BLANCO, SILVERY PLANTAIN A short, tufted, *silver-woolly* perennial to 28 cm with a woody stock. Leaves 30 mm–15 cm, linear and often slightly twisted, 3-veined (obscured by hairs), and untoothed. Flowers greenish, borne in small, oblong spikes to 5–11 cm on, long, spreading or erect stems; stamens not markedly protruding. *Seeds 4–5 mm*. Bare ground. L, F, C, T. *P. amplexicaulis* is similar, less hairy, and *less silvery* with elliptic leaves 20–50 mm broadest above the middle, tapered at the base, and faintly veined. Spikes 10–20(30) mm. *Seeds just 2.5 mm*. L, F, C, T, G, H, P.

1. *Plantago phaeostoma*
2. *Plantago coronopus*
3. *Plantago ovata*
4. *Plantago amplexicaulis*
5. *Plantago lagopus*
6. *Plantago major*
7. *Plantago major* (inflorescence and leaf)
8. *Plantago arborescens* (habit)
9. *Plantago arborescens* (flowers)
10. *Plantago arborescens* (in fruit)
11. *Plantago webbii*
12. *Plantago famarae*

(c) Annuals to perennials. Leaves borne in a rosette, linear or narrowly lanceolate, *stems grooved or ribbed*; spikes borne on leafless stems.

Plantago lanceolata RIBWORT PLANTAIN A variable hairy or hairless perennial to 50 cm with 1–several leaf rosettes. Leaves 15 mm–20 cm, linear–lanceolate or lanceolate, toothed or untoothed, 3–5-veined, *strongly ribbed* and stalked. *Bracts hairless*. Flowers brown, borne in short, blackish spikes 40(80) mm long on grooved stalks that markedly exceed the leaves; anthers pale yellow. A weed on fallow land. C, T, H. *P. lagopus* LENGUA DE OVEJA is similar to *P. lanceolata* but annual, smaller to 15(47) cm and more *white-hairy*, *especially the bracts*. Spikes 10–30 mm. Common in various habitats. L, F, C, T, G, H, P.

(d) Annuals to perennials. Leaves borne in a rosette, *broadly* oval or elliptic; spikes borne on leafless stems.

Plantago major LLANTÉN, GREATER PLANTAIN A short, hairy or hairless perennial with broadly oval to elliptic leaves, 50 mm–37 cm, in a *single* basal rosette, 3–9-veined, narrowing abruptly into a broad stalk at the base; stalk equalling the blade. Spikes *long, dense and slender, 30 mm–32 cm* borne on smooth (not furrowed), hairy stalks, *shorter than the leaves*; corolla whitish, anthers yellowish. Very common in cultivated and grassy places. L, F, C, T, G, H, P.

(d) Small, branched *shrubs*.

Plantago arborescens PINILLO A small shrub to 60 cm with ascending branches. Leaves densely crowded at the stem tips, *thread-like*, ascending, finely hairy with ciliate margins, not conspicuously succulent. Spikes few-flowered, ovoid, with stalks 30–40 mm. Common in low, rocky habitats from the centre west. C, T, G, H, P. *P. webbii* is similar but erect with greyish, silky-hairy stems with upswept leaves, and a short corolla tube. Very locally common in high altitude rocky habitats. C, T, P.

Plantago famarae PINILLO DE FAMARA A compact shrub to 40 cm, superficially similar to *P. arborescens*; stems clothed in persistent dead leaves below; leaves *linear, flat, rather succulent*, congested around the stem tips. Spikes ovoid, dense, terminal; stamens exserted; anthers cream. Very rare and local on a few remote sea cliffs or nearby in Famara. *L.

Antirrhinum SNAPDRAGON

Dwarf shrubs or woody-based herbs with entire leaves. Flowers zygomorphic and 2-lipped; stamens 4. Capsule with 3 apical pores.

Antirrhinum majus CONEJITO MAYOR, BOCA DE DRAGÓN, SNAPDRAGON A variable (many forms described, accepted by some as separate species), bushy perennial, to 65 cm (1 m) much-branched below, with stems woody at the base. Leaves lanceolate to linear and wedge-shaped at the base, opposite or alternate. Flowers with corolla 33–45 mm, bright pink-purple (pale yellow in cultivated forms); calyx 6–10 mm. Fruit capsule 12–15 mm. Widely naturalised. L, C, T, G, H, P.

Misopates

Annuals similar in form to *Antirrhinum* with distinctly unequal, rather long, linear calyx lobes; stamens 4.

Misopates orontium CALABACILLA, LESSER SNAPDRAGON A short, sparingly branched, more or less hairless annual to 30(70) cm. Leaves 10–55 mm, linear to elliptic, untoothed, opposite below and alternate above. Flowers with corolla 10–17 mm, pale *pink*, snapdragon-like; *calyx 12–20 mm with long lobes*. Fruit capsule 5–10 mm, glandular-hairy. Common in a range of habitats. L, F, C, T, G, H, P. The following are similar: **M. calycinum** is usually hairless with a raceme that elongates in fruit, and *larger, whitish* flowers with corolla 18–22 mm, *exceeding the calyx lobes*; calyx 14–20 mm. Fields, tracksides. L, C, P. **M. salvagense** has much smaller, whitish flowers, more slender (subcylindric) capsules, and seeds with branched furrows (invisible without magnification). Malpaís, bare ground; local in the east. L, F, T, H.

Linaria LINARIA

Herbs with simple unstalked leaves, opposite or whorled, alternate above. Flowers in spikes or racemes, snapdragon-like but small; calyx unequally 5-lobed and short; stamens 4. Fruit a capsule opening by slits.

Linaria micrantha A leafy, slightly blue-grey annual to 45(55) cm, hairless throughout except on the inflorescence. Leaves 5–40 mm, linear–lanceolate. Racemes dense in fruit; flowers *very small*; corolla just 2.5–5 mm, *lilac*; calyx 2.5–5 mm; *spur minute*, 0.5–1 mm, straight or curved; borne in rather inconspicuous inflorescences; flowers borne in small terminal clusters of 8–25. Capsule 3.5–6 mm. Cultivated and waste places. L, F, C, T. ***L. simplex*** is similar but with pale yellow flowers; corolla 5–9 mm; spur 2–3.5 mm. C, T.

Linaria spartea A more or less hairless, erect annual to 40(55) cm. Leaves *small*, 6–29 mm, linear, mostly alternate but whorled or opposite below; *distant*. Racemes short, *sparsely hairy and lax*; even more so in fruit; flower stalks erect-spreading, 2–11 mm long, and greatly *exceeding* their bracts (4–22 mm in fruit); corolla bright *yellow*, 12–24 mm long with a yellow spur 4–12 mm; stigma distinctly 2-parted. Capsule 2.5–5 mm. T, H.

Kickxia PICO PÁJARO

Annuals with oval to elliptic or arrow-shaped leaves, entire to sparsely toothed, with pinnate veins. Corolla strongly zygomorphic; stamens 4. Capsule opening by 2 oblique lids.

(a) Flowerstalks *hairless*. Flowers whitish or yellowish, with at least some purple markings.

Kickxia commutata A spreading, glandular-hairy perennial with slender stems to 80 cm, sometimes rooting at the nodes. Leaves 10–45 mm, oval to arrow-shaped, long-hairy. Corolla 7–17 mm, whitish or yellowish with a blue or violet upper lip and spotted palate; spur strongly curved; flowerstalks 5–25 mm, *hairless*. Capsule leathery, 2–4 mm. Cultivated and waste ground. C, T, P.

(b) Flowerstalks *hairy*. Flowers whitish, yellowish, with at least some violet markings.

Kickxia elatine An annual to 60 cm, similar to *K. commutata* with leaves 3–10 mm. Corolla *yellow* with a purple upper lip, 7–15 mm long; *spur straight*; flowerstalks 5–20 mm, *sparsely hairy* and *longer* than their bracts. Capsule 3–5 mm, hairy above. C, T, P.

(c) Flowerstalks *hairless*. Flowers typically bright yellow throughout.

Kickxia heterophylla A delicate *prostrate or clambering* perennial, much-branched. Leaves rather fleshy, narrow, pointed, with 2 divergent lobes at the base. Flowers solitary, bright yellow, sometimes with darker spots on the palate, with a long and conspicuous, *forward-pointing spur*; flowerstalks hairless. Malpaís, cliffs, sandy areas, thickets and roadsides; very common in the east. L, F, C, T. ***K. scoparia*** is similar but *erect and shrubby* in habit. Leaves linear, short-stalked. Flowers with more or less erect spurs. C, T, G, P.

Cymbalaria IVY-LEAVED TOADFLAX

Trailing herbs with very slender stems and palmately veined leaves. Flowers with zygomorphic, 2-lipped corolla with spur at the base. Species all similar.

Cymbalaria muralis PALOMILLA, IVY-LEAVED TOADFLAX A trailing, purplish, tufted, hairless perennial to 60 cm. Leaves alternate, circular with 5–9 lobes; long-stalked. Flowers small; corolla 9–15 mm, lilac with a yellowish palate, borne on long, slender stalks; spur 1.5–3 mm long (about equalling the calyx). Capsule hairless. Damp, shady places, walls; common. L, F, C, T, G, H, P.

Digitalis FOXGLOVE

Tall biennial to perennial herbs with alternate leaves. Flowers showy with long, tubular-bell-shaped, 2-lipped corolla; stamens 4. Capsule opening by 2 valves.

Digitalis canariensis (syn. *Isoplexis canariensis*) CRESTA DE GALLO A shrub to 1.5 m. Leaves lanceolate to ovate, glossy, sparsely hairy beneath, with toothed margins. Flowers bright orange-red, 30 mm, borne in rather dense, erect, spike-like inflorescences held on upward-arching stems. Capsule exceeding the calyx. Laurel forests; locally frequent. *T, G, P. ***D. chalcantha*** has *narrowly* lanceolate leaves densely hairy beneath; flowers <20 mm, borne in laxer inflorescences; flower stalks equalling the calyx. Laurel forests in a few valleys; very rare. *C (north). ***D. isabelliana*** is similar but with leaves more or less *hairless* beneath and smaller, darker reddish flowers with shorter stalks. Pine forest areas; rare. *C (centre).

Veronica VERÓNICA, SPEEDWELL

Annual or perennial herbs (sometimes shrubs) with opposite leaves. Flowers often blue, short-tubed and with 4 unequal lobes longer than their tube; stamens 2. Capsule opening by 2 valves.

(a) Flowers in terminal clusters, not borne in the leaf axils.

Veronica serpyllifolia THYME-LEAVED SPEEDWELL A more or less hairless *perennial* herb with creeping, rooting stems to 30 cm as well as *erect or ascending*, flowering stems. Leaves oval, 8–20 mm. Flowers 5–10 mm across, pale blue to white, borne in lax terminal spikes; flower stalks longer than the calyx. Capsule wider than long, with an equal style. T. ***V. arvensis*** WALL SPEEDWELL is an *erect annual* to 30 cm with oval leaves 2–35 mm, many hairs non-glandular. Flowers 2–3 mm across, blue. *Fruits hairy* and as long as they are *broad, and heart-shaped*. L, F, C, T, G, H, P.

(b) Flowers in clusters borne in the lower leaf axils, with a leafy, non-flowering shoot at the tip of the plant.

Veronica anagallis-aquatica MAJAPELO, BLUE WATER-SPEEDWELL A hairless or slightly hairy, *erect perennial* to 50 cm with branched or unbranched stems. Leaves opposite, oval–lanceolate and scarcely toothed; stalked below, unstalked and semi-clasping the stem above. Flowers 4–9 mm across, pale blue with darker veins, borne in slender, paired racemes. Capsules 2.5–4 mm, rounded or elliptic; hairless. Aquatic habitats. C, T, G, P.

---

1. *Misopates orontium*
2. *Misopates calycinum*
3. *Misopates salvagense*
4. *Kickxia commutata*
5. *Kickxia heterophylla* (habit)
6. *Kickxia heterophylla*
7. *Kickxia scoparia*
8. *Cymbalaria muralis*
9. *Digitalis canariensis* (habit)
10. *Digitalis canariensis*

(c) Flowers solitary in the leaf axils.

Veronica persica COMMON FIELD-SPEEDWELL A prostrate, hairy annual with stems to 50 cm and triangular–oval leaves 15–17 mm, coarsely toothed (>7 teeth), hairy below. Flowers 8–12 mm across, *bright blue with a paler or white lower lip*, borne solitary in the leaf axils; calyx lobes with rounded bases. Capsule with spreading hairs. Bare and cultivated land. C, T, G, P. *V. polita* GREY FIELD-SPEEDWELL is a similar, hairy annual with *dull green* leaves 6–17 mm and *flowers bright blue throughout,* 4–8 mm across. Capsule with many short, arched hairs (some glandular). C, T, G. *V. hederifolia* IVY-LEAVED SPEEDWELL has *kidney-shaped leaves* 8–28 mm, *with 3–7 large, shallow teeth near the base.* Flowers 4–9 mm across, pale lilac, the corolla shorter than the calyx; calyx lobes heart-shaped at the base. Capsule hairless. C, T.

Globularia MOSQUERA

Perennial herbs or small shrubs with alternate, undivided leaves, and flowers in dense rounded heads with an involucre of bracts; stamens 4. Fruit a 1-seeded nut.

Globularia salicina A shrub to 1.5 m. Leaves narrowly to broadly lanceolate, entire, erect or spreading. Inflorescences lateral and terminal, often crowded near the stem tips; flowerheads *small,* <1.5 mm, whitish with a darker blue centre. Locally frequent on rocky slopes. C, T, G, H, P.

Globularia sarcophylla A small, pendulous cliff shrub with *small*, obovate, fleshy leaves to 3 cm. Flowerheads solitary, lateral, on *long stalks* (5–10 cm), whitish, blue-centred. High basalt cliffs in mountains, very rare. *C (centre). *G. ascanii* is similar but with *larger* leaves 5–10 cm and *short flower stalks* (10–20 mm). Pine forest cliffs in the Tamadaba region; very rare. *C.

SCROPHULARIACEAE | FIGWORT FAMILY

Herbs (rarely shrubs or trees) with opposite or alternate leaves. Flowers zygomorphic, usually in spikes or racemes; calyx 4–5-lobed or 2-lipped; corolla 5-lobed, often 2-lipped; stamens 2 or 4. Fruit usually a 2-parted capsule. Many genera previously classified in the family now transferred to the Plantaginaceae.

Scrophularia FISTULERA, FIGWORT

Annuals or perennials, often with square stems and opposite leaves. Flowers yellow or greenish flowers; corolla with 5, small, spreading lobes; calyx with 5 lobes; fertile stamens 4 (and 1 sterile).

(a) Annuals.

Scrophularia arguta ORTIGUILLA MANSA A small annual to 50 cm. Leaves ovate with double-toothed margins. Corolla lobes reddish-brown. Capsule beaked. Dry rocky habitats; widespread but rather local. L, F, C, T, G, H, P.

(b) Robust perennials to shrubs. Leaves simple.

Scrophularia glabrata A dense, much-branched shrub. Leaves virtually hairless, with double-toothed margins. Flowers small, corolla with *dark reddish-purple lobes*; sepals rounded. Capsules ovoid. Rocky slopes and pine forests, but can grow at lower altitudes; common locally (Cañadas del Teide). *T, P.

1. *Globularia salicina*
2. *Scrophularia glabrata*
3. *Scrophularia smithii* subsp. *smithii*
4. *Scrophularia calliantha*
5. *Verbascum thapsus* (habit)
6. *Verbascum thapsus*
7. *Verbascum virgatum*
8. *Myoporum laetum*
9. *Myoporum tenuifolium*

Scrophularia scorodonia BALM-LEAVED FIGWORT A more or less *downy* perennial to 1.75 m with slightly wrinkled, oval, pointed leaves 55 mm–14 cm; *with toothed margins*. Flowers purple, 6.5–10 mm long, borne in erect, sparse inflorescences; calyx lobes with *broad membranous margins*. Local in woods and in damp habitats on higher ground. T.

Scrophularia smithii A woody-based perennial herb with woolly hairy stems. Leaves hairy on both sides with 1–2 toothed margins. Flowers lax, borne in mostly unbranched inflorescences; *corolla green to whitish*. Several regional forms, of which **subsp.** *smithii* (greenish corolla) and **subsp.** *langeana* (red corolla) are the most common. Laurel forests. *C, T, G, H, P.

(c) Robust perennials. Leaves with 3 leaflets.

Scrophularia calliantha An ascending to erect perennial to 1.5 m. Leaves *with 3 leaflets*, the terminal largest. Corolla large, >10 mm with a yellow throat and orange to red lobes. Laurel forests; rare. *C.

Verbascum GORDOLOBO

Herbs with large basal rosettes, and often grey-hairy leaves. Flowers usually with yellow (sometimes white or purple) 5-lobed corolla; calyx with 5 equal lobes; stamens 5; style 1.

Verbascum sinuatum A stout, erect, *grey or yellow-woolly* biennial to 1.5 m. Basal leaves 15–45 cm, dense, *forming distinctive rosettes*; *conspicuously wavy-pinnately lobed*. Flowers yellow, 13–25 mm across, borne in clusters on a *widely branching inflorescence*; stamens with violet-hairy filaments; bracts 1.5–4 mm. Scrub. C.

Verbascum thapsus GREAT MULLEIN A tall, white-woolly biennial to 2 m. Basal leaves 10–50 cm, elliptic, rather wavy-margined and blunt, toothed or not, with a narrow, winged stalk. Stem leaves with *winged bases running down the stem (decurrent) almost to the leaf beneath*. Flowers yellow, 18–23 mm across, borne in an often solitary, woolly, *dense* terminal spike-like inflorescence; petals with 5 more or less equal lobes; stamens 5, the upper *filaments 3 with yellow-white-hairy filaments, the lower 2 almost hairless*. Rocky slopes and scree. T.

Verbascum virgatum TWIGGY MULLEIN An erect biennial to 1.5 m, glandular hairy and with flowers 28–35 mm *borne on short stalks* 2–4 mm (shorter than the calyx) in clusters of *2–5 per bract* (not solitary); bracts 6–10 mm. Roadsides and disturbed habitats. C, T, G, P.

Myoporum

A genus of trees and shrubs native to Australasia, formally classified in the Myoporaceae. Flowers with 5 corolla lobes; stamens 4; ovary 2–4-parted. Fruit a berry.

Myoporum laetum TRANSPARENTE An evergreen shrub or tree to 9(13) m with sticky shoots. Leaves 4–10(17) cm, alternate, narrowly lanceolate, with inconspicuous *forward-pointing teeth* towards the tips, hairless and dotted with oil glands. Flowers 10–15 mm across, borne in lateral clusters of 5–6(10), white with purple markings; stamens 4. Fruit a purple berry 7–10 mm. Frequently planted; naturalised locally. L, F, C, T, G, H. **M. tenuifolium** SIEMPREVERDE is a similar shrub or round-crowned tree to 8 m with leaves 45 mm–10 cm, lanceolate, *thin and without teeth*. Flowers borne in *dense cymes* of 5–9 (occasionally solitary). Fruit 7–9 mm, very dark purple when mature. Naturalised in urban areas. F, C.

Camptoloma

Small shrublets with more or less opposite lower leaves. Flowers borne in short racemes; corolla soon falling, 5-lobed; stamens 4, included. Capsule dry, 2-valved.

Camptoloma canariense SALADILLO   A dwarf shrub. Lower leaves opposite or bunched, broadly spathulate to rounded, densely glandular-hairy with coarse-toothed margins. Flowers long-stalked, whitish with *conspicuous dark pink-purple veins*; corolla soon-falling. Moist vertical rock faces and overhanging cliffs; infrequent. *C.

Campylanthus

Shrubs with alternate, linear and succulent leaves. Calyx and corolla 5-parted; corolla tube long; stamens 2 with short filaments. Capsule dry, 2-valved.

Campylanthus salsoloides ROMERO MARINO, PALILLO   A shrub to 2 m with green, fleshy, linear leaves. Flowers pink with an orange centre (rarely white) with a hairy tube, borne in terminal, arched racemes. Barrancos; locally frequent but patchy. *L, F (centre), C, T, G, P.

ACANTHACEAE

Herbaceous perennials with simple, often lobed leaves and erect, (usually) unbranched stems; bracts conspicuous and spiny. Flowers borne in dense spikes; calyx 4-lobed and 2-lipped; corolla zygomorphic, 1–2-lipped, the lower lip 3-lobed; stamens 4, not protruding; style 1. Fruit a capsule.

Acanthus ACANTO, BEAR'S BREECH

Leafy perennials with flowers borne in long, dense, terminal spikes. Easily distinguished by the robust habit, large pinnately lobed leaves, spiny bracts and zygomorphic flowers.

Acanthus mollis   A robust, bushy perennial 75 cm (1 m). Leaves large, 20 cm–1 m, shiny dark green, pinnately lobed, and soft; hairless or nearly so, and *long-stalked*; stem leaves small and few. Bracts purple-tinged and spiny. Flowers white, 35–50 mm long, borne in dense spikes; corolla 3-lobed; calyx purple and hairless. Various habitats, often in woods or in damp, shady places; also a garden escape. C, T, G, P.

Justicia

Herbs or shrubs with entire, opposite leaves. Corolla 2-lipped; stamens 2. A large genus, mainly tropical.

Justicia hyssopifolia MATAPRIETA   A small shrub to 1 m. Leaves rather dark, dull green, opposite or in whorls of 4–6, entire, ovate–lanceolate, to 40 mm. Flowers solitary, whitish to cream with faint brownish marbling; calyx 5-lobed. Fruit 15–20 mm, club-shaped. Locally frequent in lower, rocky habitats. *T, G.

1. *Camptoloma canariense* (habit)
2. *Camptoloma canariense*
3. *Campylanthus salsoloides*
4. *Justicia hyssopifolia* (habit)
5. *Justicia hyssopifolia*
6. *Spathodea campanulata*
7. *Tecoma stans*
8. *Pyrostegia venusta*

BIGNONIACEAE

A large, tropical and subtropical family of trees and climbers with opposite, sometimes pinnately divided leaves. Flowers with *a large, tubular* corolla with *4 stamens in 2 pairs*. Fruit a pod-like capsule splitting lengthways into 2, containing numerous, often winged seeds.

Spathodea

Evergreen trees. Leaves opposite, pinnate with 9–17 leaflets. Flowers trumpet-shaped. Fruit a capsule.

Spathodea campanulata TULIPERO AFRICANO A tree to 10 m. Leaves pinnate, to 45 cm, stalkless or short-stalked. Flowers with a spathe-like calyx 10–12 cm; corolla *bright orange-red,* lobes crimped; stamens slightly exserted. Fruit capsule 15–20 cm oblong; seeds winged. Native to Africa; widely planted.

Tecoma

Small trees, shrubs or climbers. Leaves simple or pinnately divided with a terminal leaflet. Flowers typically with a tubular corolla; stamens 4. Fruit a hairless, linear capsule. Seeds winged.

Tecoma stans TECOMA AMARILLA is an evergreen, scrambling (to climbing) shrub to 3(7) m with pale, eventually furrowed bark. Leaves glossy green, opposite, pinnately lobed, to 13(15) cm long with 3–9 leaflets, and broadly tubular-bell-shaped *yellow* flowers 30–43 mm long with reddish veins. Widely planted in hot, landscaped areas; rarely naturalised. F, T.

Pyrostegia

Climbers with tendrils and angled branches. Leaves opposite with 2–3 leaflets, with or without a terminal leaflet. Flowers borne in terminal panicles; calyx bell-shaped; corolla tubular. Fruit a linear capsule.

Pyrostegia venusta BIGNONIA DE FUEGO A climbing, woody perennial. Leaflets to 12 cm, oval to lanceolate, leathery and rather scaly above; hairless to finely hairy beneath. Flowers *bright orange*, borne in clusters; calyx 4–6 mm with 5–10 veins; corolla tubular, 35–70 mm, curved and hairless, with lobes 9–14 mm. Fruit smooth and leathery, 24–30 mm. C, T.

VERBENACEAE

Herbaceous annuals, perennials or shrubs with opposite leaves and square stems. Flowers borne in clusters or heads; corolla a slender tube and flat limb, often 2-lipped; stamens 4, not protruding; style 1. Fruit berry-like or 4 1-seeded nutlets.

Lantana camara LANTANA A small, prickly shrub to 1.5(4) m with square and prickly branches that are hairy when young. Leaves 40 mm–13 cm, opposite, oval, toothed and short-stalked. Flowers 10–11 mm, bright yellow or orange, ageing red, congested in tight, often paired heads carried on long stalks; corolla slightly 2-lipped. Fruit a small black berry 4–7 mm. A common garden escape (invasive). L, F, C, T, G, H, P.

VERBENACEAE–LAMIACEAE

Verbena

Herbaceous annuals or perennials. Flowers borne in elongated or flat-topped clusters; calyx and corolla 5-lobed. Fruit 4 nutlets.

Verbena officinalis VERBENA, VERVAIN A rough-hairy perennial to 1.8 m with slender, stiffly erect and square stems; superficially mint-like. Leaves 40–80 mm, opposite, lanceolate to diamond-shaped in outline but deeply pinnately lobed; stalked below, stalkless and often unlobed above. Flowers stalkless, pink, 4.5–6.5 mm, slightly 2-lipped, borne in long, slender, leafless spikes to 30(55) cm long. Fruit separating into 4 nutlets. Bare waste ground. L, F, C, T, G, H, P.

LAMIACEAE | MINT FAMILY

A large and important family of herbs and shrubs; often glandular and aromatic. Leaves opposite and simple, often toothed or lobed. Flowers zygomorphic, often borne in whorls around the stem; calyx 5-lobed, often a tube with teeth; corolla 2-lipped with 3–5 lobes, the upper lip sometimes reduced; stamens 4, 2 long and 2 short; style 1. Fruit comprising 4 1-seeded nutlets concealed within the persistent calyx.

Ajuga BUGLE

Annual or perennial herbs with entire to deeply divided leaves. Flowers with a 5-lobed calyx; corolla pink, white, blue or yellow with a very short upper lip and a conspicuous 3-lobed lower lip, and with a ring of hairs within; stamens 4, shorter than the lower lip.

Ajuga iva HIERBA CLÍN, SOUTHERN BUGLE   A tufted or sprawling, short, softly hairy perennial to 15 cm. Leaves 15–40 mm, broadly linear, normally with short lobes, sometimes unlobed. Bracts similar to the leaves and exceeding the flowers. Flowers yellow, often with small red spots; corolla 13–24 mm long, the *upper lip entire and highly reduced*. Common on bare and stony ground. Var. *pseudoiva* is the common form of the islands. L, F, C, T, G, H, P.

Teucrium GERMANDER

Herbs or shrubs with toothed to lobed leaves. Flowers with a 2-lipped calyx and corolla with a *single, 5-lobed lower lip*; stamens 4, shorter than the lower lip.

Teucrium heterophyllum JOCAMA   A shrub to 2 m with lanceolate to ovate, densely hairy leaves with crimped to toothed margins. Flowers lateral, borne in clusters of 1–4; corolla pink to red, 2-lipped with a very short, split upper lip, the lower lip 3-lobed; stamens and styles long-exserted (2 x the corolla). Dry rocky slopes; locally common except in the east. Regional forms include: subsp. *hierrense* (*C, H) and subsp. *brevipilosum*. *C, T, G, P.

Teucrium spinosum   A low, hairy dwarf shrub with woody-based, erect stems, *thorny* from the base, with spine-tipped branches. Leaves stalked, soon-dropping, *lobed*; the uppermost stalkless and toothed. Flowers borne in distant whorls of 2–4 with stiff, *spine-like* bracts exceeding the calyces; calyx with a concave, oval upper lip and 4-lobed lower lip; corolla *white*. Dry fields. C, T.

Prasium

Small shrubs with flowers that have a 2-lipped corolla, the upper lip arched over the stamens, the lower 3-lobed; stamens 4.

Prasium majus PRASIUM   A subshrub to 1 m, hairless or slightly hairy. Leaves 16–43 mm, *shiny* dark green; oval, pointed and toothed, heart-shaped at the base; all stalked; bracts similar. Flowers with a white corolla forming a tube 10 mm long, upper lip 6–8 mm, lower lip 8–11 mm, borne in terminal racemes; calyx 2-lipped with bristle-tipped teeth; corolla with the middle lobe the largest. Nutlets shiny black when ripe. T (introduced).

Marrubium HOREHOUND

Perennial herbs with toothed leaves. Calyx with 10 teeth, spreading when in fruit; corolla white (cream or purple) with a flat upper lip; stamens 4, none protruding.

Marrubium vulgare MARRUBIO, WHITE HOREHOUND   An erect, white-downy, aromatic, rather nettle-like perennial to 85 cm with erect, square, *branched, white-cottony stems*. Leaves 17–65 mm, oval, heart-shaped at the base and wrinkled on the surface, with stalks shorter than their blades. Flowers

---

1. *Ajuga iva*
2. *Teucrium heterophyllum* subsp. *brevipilosum*
3. *Marrubium vulgare*
4. *Sideritis macrostachyos* (habit)
5. *Sideritis macrostachyos*

small and rather inconspicuous, borne in dense, *many-flowered* whorls in the leaf axils up the stem; corolla 2-lipped and white, the tube 3.5–5 mm, upper lip 2–3.5 mm, lower lip 1.8–3.5 mm; *calyx with 10 or more equally short, hooked teeth*. Common on rocky hill slopes. L, F, C, T, G, H, P.

Sideritis CHAHORRA

Erect, aromatic herbs, perennials and shrubs. Flowers with a bell-shaped calyx, 5-toothed and 10-veined, corolla usually yellow and 2-lipped with a flat upper lip; stamens 4, not protruding.

(a) Inflorescence erect or ascending, *compact* without remote false whorls of flowers (internodes absent).

Sideritis macrostachyos A shrub to 1 m. Leaves large, ovate, with heart-shaped bases; lower surface densely white-felted, grey above. Inflorescence a *very dense*, erect spike; flowers small, rather inconspicuous, corolla whitish with brown lips; calyx white-woolly. Damp cliffs and laurel forests; rare and local. *T.

(b) Inflorescence erect or ascending, with at least a few remote false whorls of flowers.

Sideritis lotsyi A shrub to 80 cm. Leaves ovate to elliptic with heart-shaped bases; lower surface densely white or yellow-felted; margins crimped. Inflorescences branched at the base; corolla white or pale yellow with reddish-brown lips; false whorls compact above, distant at the base. Cliffs. *G.

(c) Inflorescence erect or ascending, *lax*, with marked internodes between false whorls of flowers (at least below when mature).

Sideritis dasygnaphala A tall, erect to spreading shrub to 80 cm. Leaves lanceolate to ovate, densely white-woolly on both surfaces. Flowers yellow; calyx teeth with apical spines. Mountain scrub. *C.

Sideritis infernalis A small shrub with thin, wrinkled green leaves. Flowers few and lax in distant false whorls; corolla whitish with reddish-brown lips. Damp cliffs; rare. *T. *S. cystosiphon* is similar but much-branched with narrower, densely hairy leaves. Flowers pale yellow or white, expanded at the apex with minute, brownish lips. Dry rocks; very local. *T.

Sideritis dendrochahorra A tall shrub. Leaves lanceolate to narrowly ovate with a dense, white-felted lower surface; margins entire. Inflorescences short, dense, branched from below; false whorls remote; corolla yellow. Locally abundant in hill scrub, sclerophyllous woodlands and open zones in lower laurel forests (Anaga). *T.

Sideritis nervosa A small shrub to 50 cm. Leaves narrowly ovate with heart-shaped bases and conspicuous veins beneath, with yellowish hairs. Flowers yellow, borne in short inflorescences, branched from beneath. Scrub; rare. *T (Punto de Teno). *S. kuegleriana* is similar but taller, with rather thin leaves that have short hairs and an inflorescence branched more or less throughout, bending in fruit. Basalt rocks in forests; rare. *T (north).

Sideritis canariensis A tall shrub. Leaves ovate, with heart-shaped bases; margins minutely toothed. Inflorescences lax and erect, many-flowered; corolla yellowish with darker lips. Common in laurel forests. *T, H, P. *S. discolor* is similar but with leaves green above, cream-felted below. Corolla white with a long lower lip. Laurel forests in Los Tilos de Moya; virtually extinct. *C.

Sideritis oroteneriffae A subshrub with ovate to ovate–oblong leaves. Inflorescences lax; flowers yellow with brown lobes. High-elevation pine forests. *T.

Sideritis pumila A subshrub with ovate to ovate–oblong leaves with slightly heart-shaped bases; branches, leaf undersides and margins white-felted. Flowers borne in dense, ovoid spikes; corolla dull greenish-cream; calyx 5–12 mm. Rocky cliffs and crevices; rare. *L (Famara), F.

1. *Sideritis dendrochahorra*
2. *Sideritis dendrochahorra*
3. *Sideritis kuegleriana*
4. *Sideritis canariensis* (habit)
5. *Sideritis canariensis*
6. *Sideritis canariensis*
7. *Sideritis oroteneriffae* (habit)
8. *Sideritis oroteneriffae*

1. *Sideritis discolor*
2. *Sideritis pumila*
3. *Sideritis soluta* subsp. *gueimaris*
4. *Sideritis soluta* subsp. *soluta*
5. *Sideritis soluta* subsp. *gueimaris* (left) and *S. soluta* subsp. *soluta* (right)
6. *Sideritis gomeraea* (habit)
7. *Sideritis gomeraea*
8. *Nepeta teydea*

Sideritis soluta A *densely white-felted* shrub. Leaves ovate–lanceolate, felted on both sides, with regularly crimped-toothed margins. Flowers borne on elongated, erect, branched, rather leggy stalks with *long internodes*; inflorescence to 25 cm; calyx 5–7 mm, woolly; corolla yellowish-white with brownish lips. Pine woods, rocky slopes. Two co-occurring forms described: **subsp. *gueimaris*** (restricted to the Ladera de Güímar) and **subsp. *soluta*** (slopes and barranco beds). *T.

(d) Inflorescence *pendulous*.

Sideritis nutans A cliff plant with a woody stock. Leaves lanceolate with heart-shaped bases and crimped edges, densely glandular-hairy. Inflorescence unbranched and *downward-facing* (pendulous); corolla white with brownish lips. Basalt cliffs; rare. *G (west). *S. gomeraea* is similar but *densely white-woolly all over* and with longer hanging inflorescences. *G (east).

Nepeta CATMINT

Perennial herbs with toothed leaves, plants often male-sterile. Flowers with 5-toothed calyx; corolla white, blue or purple, with flat to slightly hooded upper lip; stamens 4, shorter than upper lip of corolla.

Nepeta teydea TONÁTICA A tall, white-hairy perennial to 1 m. Leaves opposite and regularly spaced, lanceolate–oblong with coarsely toothed margins. Inflorescence superficially lavender-like, branched at the base; corolla blue-purple (rarely white); middle corolla lobe largest. Rocky slopes at high altitude. *T, P.

Melissa BALM

Perennials with toothed leaves. Flowers borne in distant whorls in the leaf axils, with a 2-lipped calyx; calyx usually pale yellow; stamens 4.

Melissa officinalis TORONJIL A *lemon-scented*, rhizomatous perennial to 90 cm (1.5 m) with few branches. Leaves stalked, 30–70 mm, oval, with toothed margins. Flowers borne in few, distant whorls of 4–10; calyx bell-shaped, 13-veined; corolla cream to yellow (or pinkish), (6)8–12(15) mm. Sometimes escaped from gardens. C, T, G, P.

Prunella SELF-HEAL

Perennial herbs with entire to divided leaves. Flowers with 2-lipped calyx, the upper broad with 3 short teeth, the lower with 2 narrow lobes; corolla yellow, blue, pink or white with strongly hooded upper lip; stamens 4, shorter than the upper corolla lip.

Prunella vulgaris MAZOQUERA, COMMON SELF-HEAL A *hairy*, tufted perennial with stems ascending from a creeping base, 50 mm–60 cm. Leaves 17–96 mm, *broadly lanceolate*, toothed or not, stalked. Flowers blue-purple (rarely white), 11–12 mm long, emerging from very dense, small, cylindrical heads of conspicuous purple, spiky calyces, with *both bracts and a pair of leaves immediately below*. Damp and wooded places. C, T, P.

Lamium DEAD-NETTLE

Annual or perennial herbs with crowded whorls of flowers. Flowers with a white, pink or purple corolla with a hooded upper lip and tubular or bell-shaped calyx with 5 fine-pointed lobes; stamens 4, exceeded by upper corolla lip. Small annuals 10–25 cm, with flowers <15 mm long.

Lamium amplexicaule ORTIGUILLA MANSA, HENBIT DEAD-NETTLE A short, scarcely branched, hairy annual 10–40 cm. Leaves 9–20 mm, circular or oval, blunt-toothed and stalked below; bracts kidney-shaped and distinctly *clasping* the stem. Flowers *upward-spreading*; pink-purple, (13)15–20 mm long, with a slender, straight tube; calyx hairy with teeth about equalling the tube. A common weed of cultivated ground. L, F, C, T, G, H, P.

Lamium purpureum RED DEAD-NETTLE is similar but with stalked, not clasping upper leaves and bracts, and densely leafy inflorescences, flushed purple above. Flowers 8–12 mm; corolla somewhat exceeding the calyx. Disturbed ground; uncommon. C, T, G.

Ballota

Perennials with sterile leaf rosettes. Flowers with a 2-lipped corolla that has a concave upper lip; calyx funnel-shaped, 10-veined and usually 5-lobed; stamens 4, exceeded by the upper corolla lip.

Ballota nigra MARRUBIO NEGRO A moderately branched, weak-stemmed perennial to 1.5 m with short, deflexed hairs. Leaves oval and toothed. Flowers borne in distant whorls of 10–20; calyx 7–10 mm, scarcely expanded into 5 *long*, equal, triangular lobes ending in *teeth*; corolla 9–15 mm, pink to reddish-purple with a hooded upper lip. Widespread and variable with many forms described. Woods. H.

Stachys

Annual or perennial herbs, often with toothed leaves. Flowers borne in dense, spike-like inflorescences; calyx tubular or bell-shaped with 5 equal teeth; corolla yellow, pink or purple, 2-lipped with a flat or hooded upper lip and a 3-lobed lower lip; stamens 4, exceeded by upper corolla lip.

Stachys ocymastrum ALFABEGA An erect, pale green, hairy annual to 70 cm (1.1 m). Leaves 16–65 mm long, oblong and pointed, slightly heart-shaped at the base, toothed and wavy along the margin. Flowers borne in dense whorls of 4–6 along the stem, congested above, *laxer below*; calyx densely hairy with 2 upper teeth as long as the tube and 3 shorter lower teeth; *corolla white*, 10–16 mm (shorter than the calyx), the upper lip entire. Dry hillsides. L, F, C, T, G, H, P.

Stachys germanica MATAGALLOS, DOWNY WOUNDWORT A *densely white-felted*, erect perennial to 1.3 m. Leaves 58 mm–17 cm, *oblong and heart-shaped at the base*, grey-woolly below and grey-green and less hairy above; calyx with unequal teeth, the upper 2 less than half as long as the tube, but longer than the lower 3. Flowers with corolla to 20 mm; cream-white or bright pink-purple, borne in congested terminal inflorescences with equal or longer bracts. **Subsp. *cordigera*** is the form in the region. Dry grassy places. T.

Stachys arvensis HIERBA GATO, FIELD WOUNDWORT A small, erect, hairy annual to 20(45) cm with sparse, spreading hairs. Leaves 15–50 mm, oval and heart-shaped at the base, hairy, toothed along the margin and wavy-edged. Flowers borne in whorls of (2)4–6, crowded above, distant below; calyx with teeth as long as the tube, purple-tinged; corolla 6–8 mm, white or pale pink and scarcely exceeding the calyx; upper lip entire. Cultivated land; very common. L, F, C, T, G, H, P.

1. *Lamium amplexicaule*
2. *Stachys ocymastrum*
3. *Stachys germanica* subsp. *cordigera*
4. *Stachys arvensis*
5. *Micromeria teneriffae*
6. *Micromeria teneriffae*

Clinopodium

Perennial or annual herbs with entire or toothed leaves. Flowers borne in stalked axillary clusters (sometimes reduced to solitary flowers); calyx 5-lobed, tubular with (11)13(15) veins; corolla tube 2-lipped and straight to curved; stamens 4, not protruding. The genus now includes species traditionally classified under *Acinos* and *Calamintha*.

Clinopodium nepeta (syn. *Calamintha nepeta*) NAUTA, LESSER CALAMINT A mint-scented, greyish, hairy perennial 20–75 cm with creeping rhizomes; stems erect and branched. Leaves 17–70 mm, oval, virtually *untoothed*; stalked. Flowers with corolla 6–17 mm, pale pink-purple with darker markings, borne in leafy whorls; calyx ribbed and purplish, with white *hairs protruding from the mouth*. Fallow land, track margins in laurel forests. C, T, G, H, P.

Micromeria TOMILLO SALVAJE

Small, often thyme-like subshrubs. Flowers with a calyx that has 13–15 veins and 5 pointed lobes. A very complex genus on the islands; many closely related species, difficult to identify in the field.

(a) Leaves *ovate* to lanceolate, hairless (or virtually so), with margins straight (or scarcely revolute).

Micromeria teneriffae A small shrublet with opposite to bunched, sub-circular, *hairless* leaves, with glands beneath. Flowers white or pink, ‹5 mm; calyx hairy. Lower cliffs; local. *T.

LAMIACEAE | 315

Micromeria helianthemifolia A subshrub with long, lax branches. Leaves rather large (20 mm), lanceolate, hairy. Flowers borne prolifically, in dense, terminal clusters; calyx tube long with slender teeth; corolla large, bright pink. Slopes and cliffs; rare. *C (south). **M. densiflora** is similar but with very narrow, glabrous above, erica-like leaves <2 mm across. Rare. *T.

Micromeria glomerata A subshrub with opposite, regularly arranged leaves, ovate to rounded with tapering points. Flowers large in terminal heads, bright pink-violet, style included. *T. **M. rivas-martinezii** is similar but with oblong–lanceolate, non-tapering leaves and whitish flowers with the style exserted in lax whorls. *T.

(b) Leaves narrow (*linear* to lanceolate), variably *hairy*, with *revolute* margins.

Micromeria pineolens A greenish shrub to 75 cm. Leaves to 20 mm, ovate, densely long-hairy with revolute margins; arranged regularly along the stems. Inflorescence terminal; calyx with long hairs and blunt teeth; corolla large, pink. Pine forests in Tamadaba – common. *C.

Micromeria lanata A densely branched, low, *white*-felted shrublet. Leaves opposite, narrowly elliptic, white-woolly with revolute margins. Flowers to 4 mm in terminal heads; calyx ovoid, woolly <3mm; corolla small and white; *upper lip reduced*. *C. **M. benthamii** is similar but with numerous slender, erect branches. Flowers rather few, pink or white, borne in interrupted false whorls forming long spikes; calyx tube longer >3 mm with slightly longer lower lip. *C. **M. tenuis** is a *grey-hairy* plant, with a more lax habit. Inflorescences pedunculate; flowers white (5–6 mm long); with a conspicuous lower calyx lip. *C. **Subsp. *linkii*** is similar but has sericeous leaves, inflorescences with shorter peduncles and longer corollas (6–7 mm). *C.

Micromeria herpyllomorpha A spreading, *thyme-like* shrublet with bunched, small, linear–lanceolate green leaves, hairless above, hairy beneath. Inflorescences lax; flowers pinkish-white to deep pink; calyx with long teeth. Local in woods and high mountains. *P.

Micromeria ericifolia A variably hairy, low, thyme-like shrublet with bunched, small, linear–lanceolate greyish leaves. Flowers white; calyx with acute teeth. Common on southern and western slopes. *T. The following are similar: **M. tragothymus** with glabrescent leaves, green sometimes suffused with red; flowers pink. Common on north-facing slopes. *T (Recently, populations of the latter species from Teno Massif have been proposed as a separate species, **M. tenensis**). **M. pedro-luisii** is a spreading shrublet to 50 cm (often shorter when exposed). Leaves often slightly velvety-hairy and red-tipped, with revolute margins. Flowers white. Cliff ledges and slopes. *G.

Micromeria mahanensis A small subshrub to 20(40) cm. Leaves 3–7(10) mm, lanceolate, minutely hairy on both surfaces. Flowers (including calyx) rather *dark pinkish-purple*, *3–4 mm* with purplish anthers, scarcely exserted. Scattered in the east. *L, F.

Micromeria lachnophylla A grey-green shrublet with numerous erect to ascending branches. Leaves covered in *short, adpressed bristles* (strigose); woolly beneath. Calyx tubular; corolla white with lower lip upward-pointed. *T. **M. lasiophylla** is similar but with leaves hairy, without adpressed bristles; calyx bell-shaped; corolla pinkish with lower lip backward-pointed. *T.

1. *Micromeria glomerata*
2. *Micromeria rivas-martinezii*
3. *Micromeria pineolens*
4. *Micromeria pineolens*
5. *Micromeria lanata* (habit)
6. *Micromeria lanata*
7. *Micromeria benthamii* (habit)
8. *Micromeria benthamii*
9. *Micromeria tenuis* subsp. *linkii*
10. *Micromeria ericifolia*

Thymus THYME

Dwarf shrubs, woody at the base, and characteristically aromatic, with entire leaves. Flowers borne in heads; calyx 2-lipped, the upper lip with 3 short teeth, the lower with 2 long teeth; corolla 2-lipped; stamens 4, protruding (in cosexual flowers).

Thymus origanoides TAJOSE A small, spreading, subshrub. Leaves small, ovate to lanceolate, glandular. Flowers borne few, in heads, rather interrupted below; corolla pink. Rocks and cliffs; rare. *L (Famara).

Thymus vulgaris TOMILLO A variable, short, dark grey-green, densely branched subshrub to 40 cm, strongly aromatic, with spreading, woody stems. Leaves 3.5–6.5 mm, narrowly elliptic, with leaf clusters in the axils, *hairy and with down-turned margins* (without long hairs at the base). Flowers whitish, pink or purple, borne in rounded heads; calyx 3.5–5.5 mm, 10–13-veined and stiffly hairy. Introduced, sometimes escaping from gardens. T, P.

Origanum MARJORAM

Woody-based perennials and small shrubs. Flowers clustered in terminal heads; calyx bell-shaped and white-hairy within, with 13 veins and 5 lobes; corolla purplish (or white); stamens 4, protruding in cosexual flowers.

Origanum vulgare ORÉGANO, MARJORAM A rather thyme-like, lax, aromatic perennial with erect, purplish stems to 70 cm (1.3 m). Leaves 15–42 mm, oval and scarcely toothed or untoothed; short stalked or unstalked and with leaf tufts in the axils. Flowers with corolla 4.5–10 mm, pink to reddish purple, darker in bud, borne in *broad, branched, panicle-like clusters*; *calyx bell-shaped with 13 veins* and *5 almost equal teeth*. Grassy places. C, T, G, H, P.

Mentha MINT

Aromatic perennial herbs with creeping rhizomes. Flowers cosexual or female in dense whorls; calyx regular or weakly 2-lipped, corolla weakly 2-lipped with 4 subequal lobes; stamens 4, protruding (in cosexual flowers). Frequently hybridising.

Mentha pulegium MENTA POLEO, PENNYROYAL A strong-smelling, spreading or ascending perennial 12–78 cm with erect flowering stems; leaves roughly oval, 8.5–30 mm and finely downy and blunt-toothed. Flowers mauve, borne in *rounded, well-separated* heads, *without* a clear terminal head; corolla 4–5 mm; calyx 2.5–3.5 mm. Pond-sides. C, T, G, H, P.

Mentha spicata HIERBA HUERTO, SPEARMINT A variable, green, mint sauce-smelling perennial 43–84 cm with lanceolate, toothed, *scarcely hairy leaves* 17–88 mm. Flowers pink or white, borne in long, cylindrical, interrupted, *lax spikes*; corolla 2.5–4 mm; calyx 1.5–2.5 mm, the *tube hairless* (teeth sometimes hairy). C, T, G, H, P. **M. longifolia** MASTUERZO, HORSE MINT is similar, though greyish and distinctly *white-downy below*, and leaves 15 mm–12 cm, *grey and densely-silkily hairy*. Flowers white or lilac, borne in long, terminal, *dense spikes* which separate as they mature; calyx 2–3 mm; corolla 3–3.5 mm. Wet habitats. C, T, G, H, P.

---

1. *Micromeria tragothymus*
2. *Micromeria pedro-luisii* (habit)
3. *Micromeria pedro-luisii*
4. *Micromeria pedro-luisii*
5. *Micromeria mahanensis*
6. *Micromeria lachnophylla*
7. *Micromeria lasiophylla*
8. *Micromeria lasiophylla*
9. *Thymus origanoides*

Mentha suaveolens APPLE MINT A rather small perennial to 40(87) cm with a sickly-sweet scent and stems variably hairy. Leaves 18–52 mm, *bright green*, stalkless or virtually so, circular to oblong and broadest near the base, toothed, hairy above, and grey-hairy beneath. Flowers borne in many congested whorls forming long, dense inflorescences; often branched; corolla 3–3.8 mm, pale pink or white; calyx 1.2–2.5 mm, hairy with subequal teeth. Damp habitats. F, C, T.

Lavandula LAVENDER

Aromatic shrubs with narrow leaves and distinct bracts. Flowers borne in crowded, long-stalked terminal spikes; calyx with 5 small teeth, 13-veined; corolla purple, 2-lipped, weakly zygomorphic; stamens 4, not protruding.

(a) Leaves linear, entire.

Lavandula stoechas ROMANILLO, LAVANDA, FRENCH LAVENDER A greyish shrub to 1.5 m with erect, much-branched stems. Leaves 6–37 mm, *linear and untoothed*. Flowers with a deep mauve corolla 4.5–5 mm borne in short, dense spines, topped by *conspicuous purple flower-like bracts* 8–36 mm long; inflorescences longer than their spikes. Scrub. C, T.

(b) Leaves divided; stems hairless.

Lavandula canariensis A woody shrub to 1.5 m with hairless stems. Leaves green, ovate in outline, *singly or double-lobed* with rounded, flat segments. Flowers borne in spikes to 10 cm; *calyx green*, flushed violet-pink in the upper half. Common on rocky slopes. Very similar regional forms include **subsp.** *canariae* (*C), **subsp.** *canariensis* (*T), **subsp.** *fuerteventurae* (*F), **subsp.** *gomerensis* (*G), **subsp.** *hierrensis* (*H), **subsp.** *lancerottensis* (*L) and **subsp.** *palmensis* (*P). L, F, C, T, G, H, P. *L. bramwellii*, a woody shrub to 50 cm, is similar. Leaves green, ovate, 25–35 mm, regularly dissected into narrowly ovate lobes. Calyx *violet-blue throughout*. Macizo de Güigüí; rare. *C (southwest).

(c) Leaves divided; stems minutely hairy to woolly.

Lavandula minutolii A woody shrub with *woolly* stems and leaves (hairs highly branched under magnification). Leaves narrowly ovate in outline, singly lobed with oblanceolate segments. Inflorescence bracts oval, equalling or exceeding the calyx. Barrancos; very local. *C, T. **L. pinnata** is similar but with leaves pinnate (divided to the midrib) with broad, flat lobes and hairs short and dense. Calyx much exceeded by the bracts. Locally abundant on sea cliffs and mountains in the north. L. **L. buchii** has *ashy,* grey leaves with inflorescence bracts shorter than or equal to the calyx. *T.

Salvia SAGE

Herbs and shrubs with distinct whorls of (often purple) flowers forming a lax inflorescence. Flowers with both calyx and corolla 2-lipped, the upper corolla lip hooded, the lower 3-lobed; stamens 2, hinged in the middle, joined beneath the corolla hood (shorter than upper corolla lip).

(a) Leaves linear.

Salvia herbanica (syn. *Pleudia herbanica*) CONSERVILLA MAJORERA A much-branched, white-hairy, procumbent shrub. Leaves linear, bunched, superficially rosemary-like, 20 mm long. Flowers pale lilac with a white, spotted palate, borne in 6–8 whorls; calyx flushed purple, long-white-hairy. Rocky crevices in high cliffs; very rare (only 10 populations). *F (south-central).

---

1. *Lavandula canariensis* (habit) (RACHEL GRAHAM)
2. *Lavandula canariensis*
3. *Lavandula canariensis* (leaf and inflorescence)
4. *Lavandula canariensis* subsp. *canariae*
5. *Lavandula minutolii*
6. *Lavandula pinnata* (habit)
7. *Lavandula pinnata*
8. *Lavandula buchii* (habit)
9. *Lavandula buchii*

Salvia rosmarinus (syn. *Rosmarinus officinalis*) ROMERO, ROSEMARY A familiar evergreen shrub to 1.8 m, characteristically *aromatic*; branches brown and woody, erect to spreading. Leaves needle-like, 10–41 mm, linear and leathery, mid to dark green, sharply pointed, with down-turned margins. Flowers white, flushed pale purple, 8.5–13.5 mm long, borne in small lateral clusters; corolla with 2 protruding stamens; calyx bell-shaped. Widely planted and naturalised. C, T, G, H, P.

(b) Leaves narrow, with divergent lobes at the base.

Salvia canariensis SALVIA CANARIA A shrub to 2 m. Leaves lanceolate, 50 mm–15 cm, white-woolly beneath, *with two divergent lobes at the base* (hastate). Flower bracts conspicuous, ovate, papery, purple, and exceeding the calyx. Corolla purple-pink, rarely white. Common on lower rocky slopes in the west. *C, T, G, P (introduced elsewhere).

Salvia fruticosa (syn. *S. triloba*) THREE-LOBED SAGE An aromatic subshrub to 1.5 m with felted stems. Leaves grey-green and wrinkled above, grey-felted below, and stalked; narrowly ovate usually with 3(5) *lobes at the base*. Inflorescence spike-like with 2–6-flowered false whorls; corolla blue-purple or pink, 15–30 mm long; calyx 5–11 mm, bell-shaped and indistinctly 2-lipped, with 5 triangular teeth 1.5–3.5 mm long. Planted; rarely naturalised. T, H.

(c) Leaves broadly ovate, heart-shaped at the base (cordate).

Salvia broussonetii OREJA DE BURRO A shrub to 75 cm. Leaves *broadly ovate*, 10–20 cm, with *coarsely crimped* margins, heart-shaped at the base. Flowers white to pinkish. Cliffs; scattered, very locally common. *T.

Salvia officinalis SALVIA, SAGE A strongly aromatic, greyish shrub to 60 cm with erect branches that are woody beneath. Leaves broadly elliptic, greenish above and white-felted below, with a finely toothed margin; slightly heart-shaped at the base. Flowers pale violet, blue, pink or white; corolla 15–25(35) mm long; calyx 10–14 mm, flushed purple, the upper lip 3-toothed; bracts oval and hairy. Dry, stony pastures. T.

(d) Leaves narrowly ovate to linear–elliptic, crimped to toothed.

Salvia aegyptiaca (syn. *Pleudia aegyptiaca*) CONSERVILLA, BROTONA A woody-based, tufted dwarf shrub with leafy, erect to ascending stems to 25 cm with short to long hairs. Leaves narrowly ovate to linear–elliptic, 12–25 mm with crimped to toothed margins. Flowers borne in simple racemes (sometimes branched) with distant whorls of 2–6; calyx 5 mm (larger in fruit); corolla violet-blue or white with darker markings in the lip, 6–8 mm with a straight to deflexed upper lip, exceeded by the lower. Nutlets smooth and black. Arid and semi-arid habitats. L, F, C, T, G.

(e) Leaves pinnately lobed.

Salvia verbenaca GALLOCRESTA, WILD CLARY A perennial to 80 cm with erect stems, glandular above. Basal leaves forming a rosette, wrinkled, variably deeply *pinnately lobed*, long-stalked below, and short-stalked or stalkless above. Flowers pale blue or violet, 6–16 mm long, borne in lax whorls forming a spike; calyx 5–12 mm, bell-shaped and distinctly veined; enlarged in fruit. Fallow land, grassy places, roadsides; common. L, F, C, T, G, H, P.

1. *Salvia herbanica*
2. *Salvia canariensis* (white form)
3. *Salvia canariensis* (purple form)
4. *Salvia canariensis*
5. *Salvia broussonetii*
6. *Salvia aegyptiaca*
7. *Salvia verbenaca*
8. *Bystropogon canariensis*
9. *Bystropogon canariensis*

LAMIACEAE

Bystropogon POLEO DE MONTE

Strongly aromatic shrubs. Leaves entire, often crimped. Flowers small, white or pink, in dense inflorescences. Corolla 2-lipped, the upper lobe divided or with 2 small teeth; the lower lip 3-lobed; stamens included; styles exserted.

Bystropogon canariensis A shrub to 2.5 m. Leaves green, ovate–lanceolate, green, variably hairy (especially beneath); margins coarsely crimped. Flowers very small, white, borne in *small, spherical clusters*; calyx teeth narrow. Fairly common in laurel forest scrub. *C, T, G, H, P. *B. plumosus* has *silvery*, ovate–rhomboid leaves, usually *without* crimped edges, and corolla exceeding the calyx. Rocky slopes. *T. *B. origanifolius* has lanceolate leaves and very long, slender calyx teeth exceeding the corolla. Dry rocky slopes and woods. *C, T, G, H, P.

Cedronella

Herbaceous perennials, sometimes woody-based. *Leaves trifoliate*. Inflorescence a dense spike or terminal head; corolla 2-lipped; stamens included or slightly exserted.

Cedronella canariensis ALGARITOFE An aromatic perennial to 1.5 m. Leaves green, trifoliate with lanceolate leaflets, hairy beneath, with toothed margins. Flowers borne in rather congested terminal spikes; corolla pink. Common in laurel forests. C, T, G, H, P.

OROBANCHACEAE | BROOMRAPE FAMILY

A distinctive family of root parasitic (hemiparasitic or holoparasitic) herbs, attaching to the roots of host plants as seedlings. Flowers borne in spikes or racemes; corolla zygomorphic, more or less 2-lipped; stamens 4; style 1. Fruit a 2-parted capsule. Hemiparasitic species were previously classified with the Scrophulariaceae.

Parentucellia ALGARABA

Hemiparasitic annuals with opposite leaves. Flowers with tubular to bell-shaped, regularly 4-lobed calyx; corolla long-tubed with open mouth. Capsule with numerous seeds.

Parentucellia viscosa YELLOW BARTSIA   A short, glandular-hairy annual with erect, normally unbranched stems to 35(60) cm. Leaves 17–35 mm, opposite, lanceolate, deeply toothed and sessile; bracts similar, and decreasing in size along the stem. *Flowers yellow* (rarely white); corolla 18–23 mm, 2-lipped, the upper lip hooded, the lower lip 3-lobed; calyx teeth linear–lanceolate and long, 4.5–6.5 mm. Damp grassy places; local. C, T, G, H, P.

Parentucellia latifolia SOUTHERN RED BARTSIA   A short annual to 20(40) cm with *triangular–lanceolate and deeply lobed leaves* 6–15 mm. Flowers small, with a corolla 8–10(15) mm long, pale *reddish-purple* (rarely white); calyx teeth 1.4–1.6 mm long. Grassy habitats; rare and local. T.

Bartsia

Hemiparasitic perennials with opposite, toothed leaves. Flowers with a tubular to bell-shaped, regularly 4-lobed calyx and long-tubed corolla with an open mouth. Capsules with few, winged seeds.

Bartsia trixago (syn. *Bellardia trixago*) GALLOCRESTA   A short, glandular-hairy annual with erect, simple stems to 60(70) cm. Leaves 20–40 mm, opposite, linear–lanceolate, toothed and stalkless; bracts similar, and decreasing in size along the stem. Flowers with corolla white, normally flushed with pink, or bright yellow, 17–24 mm, borne in a dense, *4-sided spike*. Grassy places; local. C, T, G, H, P.

Cistanche

Robust, obligate parasites with no true leaves or green pigment. Flowers borne numerously in dense spikes; calyx tubular with 5 teeth; corolla with 5 almost equal teeth. Fruit an ovoid capsule. Distributions unclear due to confusion between species.

Cistanche phelypaea RABO CORDERO   A robust, bright yellow herb 10–30 cm with hairless, unbranched, thick stems with oval bracts to 20 mm. Flowers *bright lemon-yellow* (or tinted dull purple); calyx 12–18 mm and 5-lobed; bracts much shorter than the flowers; corolla 30–50 mm with 5 more or less equal, oval recurved lobes, broadly bell-shaped with a sharp bend. Parasitic on shrubby Amaranthaceae on coastal dunes, flowering after winter rain. Eastern only and rather local. L (north), F (south).

---

1. *Bystropogon origanifolius*
2. *Cedronella canariensis*
3. *Parentucellia viscosa*
4. *Parentucellia latifolia*
5. *Cistanche phelypaea*

Orobanchaceae species: A. *Cistanche phelypaea*; **B.** *Orobanche minor*; **C.** *Phelipanche lavandulacea*; **D.** *Phelipanche gratiosa*; **E.** *Orobanche cernua*.

Phelipanche JOPO, BROOMRAPE

Branched or unbranched holoparasites, often bluish. Flowers with *bracteoles* below the calyx; corolla zygomorphic, 2-lipped. Capsule with numerous minute seeds. Host identification useful but difficult in mixed vegetation. Closely related to *Orobanche*. Reliable identification for closely related species not always possible in the field, and some species may be recorded in error.

Phelipanche ramosa BRANCHED BROOMRAPE A small, blue-purple (sometimes yellow-white), glandular-pubescent annual often with *branched stems*, *short*, 50 mm–20(30) cm. *Flowers small*, 10–15(22) mm long, pale blue, violet or cream with a white patch at the base; stigmas white or pale blue; calyx teeth 5–15 mm (shorter than the corolla), lower corolla teeth *blunt*. Parasitic on various hosts, usually *cultivated* (e.g. Solanaceae and Leguminosae). Disturbed habitats. L, F, C, T, G, H, P. The following species are similar and their distributions unclear due to confusion: *P. mutelii* has more erect flowers with a distinctly *open* corolla mouth and *divergent*, rounded, broadly elliptic lower teeth; anthers usually hairless (not always). On various *uncultivated* hosts, especially Asteraceae, less often Leguminosae. L, F, C, T, G, H. *P. nana* is usually *unbranched*, with *pointed* lower corolla lobes, and calyx teeth almost as long as the corolla. Various wild hosts. C, T, G, H, P.

Phelipanche gratiosa A fairly robust, branched or unbranched annual with yellowish-brown stems. Flowers erect-spreading; calyx 7–11 mm with narrowly triangular teeth, longer than their tube; corolla pale blue with spreading upper lobes and spathulate–obovate, rather pointed lower lobes. Dunes, hillsides and roadsides, parasitic on *Launaea*; common in the east. *L, F, C. *P. purpurea* has been recorded in error for this species.

Phelipanche lavandulacea A robust annual to 45(59) cm with *numerous, dense* flowers and with a *dark purple or blackish appearance*. Corolla 19–22 mm with *lower lip obtuse*; stigma white, bluish or yellowish; filaments slightly hairy; calyx teeth *not exceeding* the corolla tube. Lower portion of the stem with stalked flowers. Various hosts, particularly *Bituminaria bituminosa*. Local, on cliffs. C, T.

Orobanche JOPO, BROOMRAPE

Unbranched holoparasites, often dull reddish or yellowish, similar to *Phelipanche* but *lacking 2 bracteoles* below the calyx; corolla zygomorphic, 2-lipped. Capsule with numerous minute seeds. Host identification useful but difficult in mixed vegetation; reliable identification for closely related species not always possible in the field.

(a) Corolla patent, narrow, tubular, and small, 10–25 mm; unscented. Species very similar.

Orobanche minor COMMON BROOMRAPE A variable annual with lax to dense spikes to 60 cm. Flowers *small* with an *evenly arched* corolla 10–18 mm, dull cream, tinged brownish-violet; lower lip evenly lobed without a hairy margin, stigma pink; filaments *slightly hairy*, usually inserted close to the corolla base; bracts equalling or exceeding the corolla; calyx lobes variably toothed. Parasitic on various hosts across many families, in disturbed habitats; fairly common. L, F, C, T, G, H, P.

Orobanche castellana affin. Like *O. minor* but yellowish. Corolla 15–20(25) mm, strongly downward-curved (cernuous), pale yellow tinged red-brown; stigma reddish-brown. Parasitic on *Andryala* on volcanic slopes. L. This confused taxon has been recorded as *O. castellana*, *O. amethystea* and *O. calendulae* and needs further examination.

Orobanche hederae IVY BROOMRAPE A short-lived biennial like the previous species in form; corolla 10–22 mm, cream-tinted maroon, *inflated* at the base, *narrow* and *straight-backed*, with *notched lobes*, the middle lobe square; stigma lobes always *yellow*; filaments scarcely hairy. Parasitic on *Hedera* (rarely other species e.g. *Geranium*). Woods in the west. T, G, P.

(b) Corolla patent, tubular-bell-shaped, large, 20–30 mm, *fragrant*.

Orobanche crenata BEAN BROOMRAPE An annual to 1 m with densely flowered spikes. Stems purplish, flowers large; corolla (15)20–30 mm, broadly campanulate and *white with violet veins*; *fragrant*; stigma variable in colour. Parasitic on various legumes, particularly on cultivated peas and beans, sometimes in large numbers. Cultivated and fallow ground; local. L, C, T, H, P.

(c) Corolla downward-curved.

Orobanche cernua A lax, erect, bluish or brownish plant to 23 cm with flowers more or less *from the base* to the apex of the spike. Corolla 12–18 mm, white to yellow with a *blackish margin*; *downward-pointing*; stigma lobes yellow-white. Parasitic on Asteraceae, often coastal; widespread but uncommon. L, F, C, T, G, H, P.

AQUIFOLIACEAE | HOLLY FAMILY

Evergreen trees and shrubs with alternate, often spiny leaves. Flowers dioecious, borne in cymes in the leaf axils, regular, with 4 free sepals and 4 white petals fused at the base; stamens 4; stigma 4-lobed. Fruit a 2(4)-seeded drupe.

Ilex canariensis ACEBIÑO, CANARY ISLAND HOLLY An evergreen shrub or small tree to 7(10) m with grey-brown bark. Leaves glossy, dark green, ovate, 50–70 mm, leathery, entire or remotely spiny; often with *fungal dark spots* beneath. Flowers small and white, borne in lateral or sub-terminal clusters. Fruit 10 mm, *fleshy, red*, borne on small stalks (3–8 mm). Very frequent in laurel forests and heaths in the west. C, T, G, H, P. *I. perado* NARANJERO SALVAJE is similar, but with large, ovate to sub-circular leaves, smooth and bright green with *conspicuous forward-directed teeth*. Regional forms include **subsp. *lopezlilloi*** (*G) and **subsp. *platyphylla***. Locally frequent in more wet habitats. *T, G, P.

CAMPANULACEAE | BELLFLOWER FAMILY

Annual or perennial herbs, exuding a latex when cut, with alternate leaves. Flowers often large and showy, clustered or solitary; corolla bell-shaped, lobed, often blue or purple; stamens 5, fused or free; style 1. Fruit a capsule.

Campanula BELLFLOWER

Annuals or perennials. Flowers with a 5-lobed calyx fused to the ovary, and funnel- or bell-shaped corolla. Fruit a capsule or berry.

1. *Phelipanche nana*
2. *Phelipanche gratiosa* (habit)
3. *Phelipanche gratiosa*
4. *Phelipanche gratiosa* (fruiting capsules)
5. *Phelipanche gratiosa* (cross section)
6. *Orobanche castellana* affin. (habit)
7. *Orobanche castellana* affin.
8. *Orobanche hederae*
9. *Orobanche crenata* (habit)
10. *Orobanche crenata*
11. *Orobanche cernua* (habit)
12. *Orobanche cernua*

Campanula erinus CAMPANILLA, ANNUAL BELLFLOWER A slender, rough-haired, spreading, branched annual to 35 cm with oval- to wedge-shaped leaves 10–20 mm, and tiny, pale blue (or reddish or white) flowers with an *inconspicuous* corolla just 2–5 mm, borne in the axils of the branches; flowers very short-stalked. Capsule 2.5–3.6 mm. L, F, C, T, G, H, P.

Campanula occidentalis CAMPANILLA CANARIA An annual to 35 cm, *bristly*. Leaves ovate–oblong with wavy-toothed margins. Flowers typically borne in clusters of 1–2; corolla tubular, 15 mm, with a whitish tube and pale blue-violet lobes. Capsule opening laterally. Bare ground; locally common. *L, F, T, G.

Solenopsis

Delicate annual or perennial herbs with simple, thin leaves. Flowers bell-shaped, zygomorphic, *2-lipped*, blue or lilac. Capsule 2-parted.

Solenopsis laurentia A small, *slender*, more or less *hairless annual* 60 mm–20 cm. Leaves 12–20 mm, oblong to spatula-like, entire, often forming a basal rosette; small *stem leaves* present. Flowers solitary; small, 3–5(6) mm borne on stems with 1 or 2 bracteoles; calyx 5-parted; corolla blue, lilac or white, 3.5–6 mm. Capsule 2–3.5 mm. Widespread but uncommon; damp habitats. C, T.

Legousia

Annual herbs. Flowers borne in panicles or racemes; calyx 5-lobed; corolla flattish, 5-lobed, fused at the base; stamens 5; style 1, stigmas 3. Fruit a cylindrical, 3-valved capsule.

Legousia hybrida CAMPANILLA BASTARDA, VENUS'S LOOKING GLASS An annual to 30 cm similar to the above species but with *markedly wavy leaves* to 30 mm, and few, pink, violet or white flowers borne in terminal clusters with petal lobes just 2.5 mm, about ½ the length of the calyx lobes (5–6 mm); *calyx teeth erect in fruit*. Capsule 15–25 mm, constricted above. Fallow land and olive groves. L, T, C.

Legousia falcata CAMPANITA DE HOZ, SPICATE VENUS'S LOOKING GLASS An erect to spreading annual, 15–50 cm, with hairless or bristly stems. Leaves 20–40 mm, oval, slightly wavy-margined and usually short-stalked below. Flowers violet to blue-purple, solitary or paired with lobes 7 mm, borne in a *spike-like inflorescence*; calyx teeth linear and curved, 4–8 mm (equal or ½ length of the tube), spreading, often exceeding the petals. Capsule 10–22 mm, narrowed at the apex. Stony pastures. C, T, G, H, P.

Canarina

A small genus of hairless, scrambling perennials native to Macaronesia and Africa. Leaves opposite. Flowers *large, bell-shaped*. Fruit a pendulous berry.

Canarina canariensis BICÁCARO A tuberous, clambering or pendulous perennial with hollow stems with a latex. Leaves triangular–hastate with toothed margins. Flowers lateral, solitary; corolla bell-shaped, 30–60 mm, orange to brownish-red, borne in winter and spring. Fruit a large, ribbed, fleshy berry, ovoid, 30–40 mm, yellowish-orange; pendulous. Very locally frequent in rocky thickets, laurel forests and forest margins. *C, T, G, H, P.

---

1. *Ilex canariensis* (flowers)
2. *Ilex canariensis* (fruits)
3. *Ilex canariensis* (note fungal dark spots)
4. *Ilex perado* subsp. *platyphylla* (fruits)
5. *Ilex perado* subsp. *platyphylla*
6. *Campanula erinus*
7. *Campanula occidentalis*
8. *Legousia falcata*
9. *Legousia falcata*
10. *Canarina canariensis* (fruit)
11. *Canarina canariensis* (habit)
12. *Canarina canariensis*

Wahlenbergia

Annual or perennial herbs with erect stems, circular in section, leafy below. Flowers solitary or in cymes; sepals, petals and stamens 5. Fruit a many-seeded capsule.

Wahlenbergia lobelioides ALMIRÓN, SANGUINARIA An annual with fleshy, *reddish stems* and *linear, toothed leaves*. Flowers to 10 mm long, borne nodding in bud, becoming erect in flower and fruit; corolla *pale* blue, pink or white with petals fused below and pointed at the tips. Fruit a capsule opening by valves at the top. Malpaís and disturbed habitats; frequent. L, F, C, T, G, H, P.

ASTERACEAE

The largest family of flowering plants (see also Orchidaceae). Herbs, perennials or shrubs with alternate, opposite or rosette leaves, simple or compound. Flowerheads typically with an involucre of closely overlapping bracts around the base; flowers (*florets*) borne in congested flowerheads (*capitula*); those in the centre (*disk florets*) tubular and often distinct from the edge (*ray florets*); stamens 5, fused around the style; ovary inferior. Fruit an achene, often (but not exclusively) with an appendage of bristles, hairs or scales (a *pappus*).

SUBFAMILY ASTEROIDEAE

Plants *without* white latex, and stem leaves spirally alternate (or absent; rarely opposite). Flowerheads usually *yellow* and *radiate* (with a central disk of tubular flowers, surrounded by ray flowers) or *discoid* (with tubular – often orange – flowers only).

Filago YESQUERILLA, CUDWEED

Downy or woolly annuals with alternate, untoothed leaves. Flowerheads inconspicuous, often numerous (8–40) per cluster. Pappus of simple hairs or absent.

(a) Plant erect, with flowerheads (15)20–40 in rounded clusters *not exceeded by their leaves*.

Filago germanica (syn. *F. vulgaris*) COMMON CUDWEED An erect, white-woolly annual to 30(40) cm, branched or unbranched below, always branched above in 2–3 forks. Leaves erect, lanceolate, to 20 mm long, wavy-edged and untoothed; widest in the basal ½. Flowerheads (15)20–40, borne in dense *rounded* clusters 10–12(20) mm across, *not* exceeded by the leaves immediately below; flower bracts 4–4.5 mm with a long, transparent bristle-tip. Common on bare, disturbed, sandy ground. L, F, C, T, G, H, P.

(b) Plant erect, with flowerheads (5)10–20(25) per cluster, *exceeded by their leaves*.

Filago desertorum YESQUERILLA RASTRERA A small, sparsely grey-downy annual with a short primary stem surrounded by a ring of longer, prostrate to ascending branches. Most leaves congested below the flowerheads, oblong–spathulate. Flowerheads 3.5–5 mm long, borne in congested, yellowish clusters of 6–12, exceeded by the leaves. Bare and arid habitats, flowering after winter rain in the far southeast. L, F, C, T.

Filago pyramidata BROAD-LEAVED CUDWEED An erect, white-woolly annual always branched and spreading *from the base*, often almost prostrate. Leaves oval and bristle-tipped. Flowerheads 5-angled, 10–20 mm, borne in dense clusters of (5)10–20(25) in the branch axils and terminally, *exceeded* by 2–4(5) leaves immediately below; outer flower bracts *with yellow points*, curving inwards in fruit. Open fallow and sandy waste places. L, F, C, T, G, H, P.

ASTERACEAE | 331

Filago lutescens YESQUERILLA AMARILLA, RED-TIPPED CUDWEED An erect, often yellowish-woolly annual with irregularly branched *stems with yellow (not white) hairs*, and broader, not wavy, bristle-tipped leaves. Flowerheads borne in clusters of (5)10–20(25), overtopped by (0)1–2 leaves; *flower bracts yellowish*, with prominent, erect, *red-brown or purplish* bristle-tips. Disturbed habitats. L, F, C, T, H.

(c) Plant often short (to virtually stalkless), with leaves and flowerheads congested, rosette-like (many species previously described under *Evax*).

Filago pygmaea (syn. *Evax pygmaea*) A very small, grey-felted annual to 40 mm, branched at the base. Leaves oblong, *narrow* and blunt, to 15 mm long and 5 mm wide, produced in a rather flat rosette; upper leaves 2–3 x longer than flowerheads; all leaves without distinct stalks. Flowerheads borne in very compact clusters to 35 mm across; brownish-yellow. Fairly common on dry, bare, and stony places. L, C, T, H.

Logfia CUDWEED

Small, woolly annuals. Previously treated either as distinct or under *Filago*, but recently resurrected based on DNA analysis; characterised by flowerheads that are solitary or in small clusters and *achenes that are always hairless*.

Logfia minima (syn. *Filago minima*) SMALL CUDWEED A short, grey silky-hairy annual to 25 cm. Stems very slender, branched above the middle. Leaves linear–lanceolate, 4–10 mm long. *Flowerheads in clusters of 3–7, with 5 marked angles*, 2–5 mm, ovoid to pyramidal, *not* overtopped by the leaves immediately below; outer flower bracts *woolly at the base*, but yellow and *hairless at the tip*. Fallow and sandy waste places. T, H.

Logfia gallica (syn. *Filago gallica*) YESQUERILLA DE LA FIEBRE, NARROW-LEAVED CUDWEED An annual similar to *L. minima*, to 33 cm with linear, *thread-like leaves (4)12–20(26) mm long*, the most apical *overtopping* the flowerheads; flowerheads leafy, borne in clusters of 3–14, 3.5–5 mm; bracts woolly and yellowish at the tip. Coastal scrub, pastures and tracksides. L, F, C, T, G, H, P.

Ifloga

Typically small, desert-dwelling annuals. Female florets with divided styles. *Pappus feathery* (plumose) above.

Ifloga spicata A variable annual 10(50) mm–10(12) cm, branched from the base. Leaves crowded, with adpressed, silky hairs on the upper surface, often with sand grains adhered, *linear*, 6–20 mm long with rather down-turned margins. Flowerheads yellowish or reddish, made up of 26–38 female florets and 19–26 cosexual florets; heads 3–4 mm long, borne terminally and congested in leafy racemes, much exceeded by the leaves; bracts translucent and whitish to golden. Dry, bare places. L, F, C, T, G, H, P.

Gnaphalium BORRIZA, CUDWEED

Inconspicuous annuals and perennials, often white woolly-hairy. Flowerheads small, yellow-brown and bell-shaped with papery bracts, borne in clusters of (3)5–40. Pappus of simple hairs.

Gnaphalium antillanum (syn. *Gamochaeta antillana*) A grey, sparsely downy annual herb to 40 cm. Leaves to 40 mm. Flowerheads small and congested, 2–3 mm; disk florets purplish-orange; ray florets absent. Roadsides and thickets; native to North America, naturalised. T, H, P.

1. *Filago desertorum*
2. *Filago lutescens*
3. *Logfia gallica*
4. *Ifloga spicata*
5. *Gnaphalium antillanum*
6. *Pseudognaphalium luteoalbum*
7. *Helichrysum monogynum*
8. *Helichrysum gossypinum* (habit)
9. *Helichrysum gossypinum*

Pseudognaphalium BORRIZA

Perennial and annual herbs with alternate leaves, *woolly*. Achenes hairless.

Pseudognaphalium luteoalbum (syn. *Gnaphalium luteoalbum, Laphangium luteoalbum*) JERSEY CUDWEED A *slender annual* (more similar to the group above, in which it is often still included), to 50 cm, *white-woolly*, unbranched or with branches spreading from the base, then erect. Leaves broadly lanceolate, blunt, and running down the stem, *woolly on both sides*. Flowerheads terminal and densely ovoid, and *not overtopped* by their surrounding leaves (at least when mature); bracts elliptic, *shiny, straw-yellow*, only the outermost woolly below, and not bristle-tipped; florets pale yellow, with red stigmas. Disturbed habitats. L, F, C, T, G, H, P.

Helichrysum CURRY PLANT

Dwarf, greyish aromatic shrubs with alternate, untoothed leaves. Flowers borne in dense clusters with papery bracts. Pappus absent.

Helichrysum orientale A tufted, white-woolly, woody-based perennial dwarf shrub with almost erect stems to 30 cm. Basal leaves rather *broad*, egg-shaped and more or less blunt; those above smaller and linear-oblong, all leaves *whitish on both sides*. Flowerheads almost spherical, 6–9 mm, borne in dense heads 12–25 mm; bracts pale lemon yellow, shiny and translucent. Cultivated and perhaps locally escaped. L, C.

Helichrysum monogynum YESQUERA ROJA A white-woolly cliff plant with woody stems to 30 cm. Leaves lanceolate to spathulate, entire, blunt. Flowerheads yellow, ageing brown, borne in dense corymbs; heads to 6 mm with large, papery bracts. Sea cliffs; locally common. *L.

Helichrysum gossypinum YESQUERA AMARILLA, ALGODONERA A white-woolly, loosely mounded shrub, similar in form to the above species but larger and with *red* flowerheads. North-facing cliffs and locally in malpaís; locally common. *L.

Phagnalon YESCA

Grey dwarf shrubs with alternate leaves and flowerheads borne solitary at the tips of the branches; flowerheads solitary, with densely overlapping bracts. Pappus of bristles.

Phagnalon rupestre A small shrub 15–40 cm with erect to ascending, white-felted stems. Leaves *narrowly oval and more or less toothed*, small, 5–35 mm x 1.5–4.5 mm, green or whitish above and white-felted below, with down-turned margins. Flowerheads *solitary*, long-stalked (20–90 mm); yellowish; flower bracts brownish, membranous, somewhat hairy and closely overlapping. L, F, C, T, G, H, P.

Phagnalon saxatile A shrub to 50 cm with *linear* leaves 14–50 mm x 1.2–4 mm, sometimes broadest above the middle, *green* (not normally white-felted above) and white-felted below. Flowerheads *very long-stalked* (25 mm–13 cm); bracts with somewhat wavy margins, the tips often slightly recurved. Rocky hillslopes; frequent. L, F, C, T, G, H, P.

Phagnalon purpurascens A shrub to 40 cm with white-woolly stems. Leaves 20–30 mm, *linear with strongly down-turned margins* (revolute), densely woolly beneath. Flowerheads borne on long slender stalks, solitary; bracts slender, the innermost deep purple. Common on interior, dry slopes. L, F, C, T, G, H, P.

Phagnalon umbelliforme An erect shrub to 75 cm with yellowish-woolly stems. Leaves to 50 mm, linear to narrowly lanceolate, without revolute margins. Flowerheads borne in *dense umbels* of up to 40. Dry, often north-facing slopes, locally common. *C, T, G, H, P.

Pulicaria FLEABANE

Annual or perennial herbs with simple, alternate leaves. Flowerheads yellow and daisy-like, borne terminally. Achenes with scales around the pappus.

Pulicaria arabica* subsp. *hispanica HIERBA PULGUERA A rhizomatous, hairy perennial herb with downy or woolly, erect, branched stems to 1 m. Leaves narrow, *linear*, rigid. Flowerheads yellow. Achenes with erect to spreading hairs. C, T, G. *P. odora* is similar but lacks creeping stolons and has basal leaves *not withered* at flowering time. C.

Pulicaria burchardii DAMA A prostrate, silvery-floury shrub. Leaves entire, linear–lanceolate, blunt. Flowerheads small, 50–80 mm across, borne solitary at the ends of the lateral branches; ray florets short, pale yellow; bracts linear. Rare; coastal. F (Jandía).

Pulicaria canariensis PULICARIA PURPURARIA A small shrub. Leaves broadly oblanceolate, densely woolly with entire to remotely wavy-toothed margins. Flowerheads large, to 30 mm across; outer bracts hairy; ray florets to 7 mm, golden yellow. Mainly on coastal cliffs; very locally common. **Subsp. *lanata*** is a subspecies endemic to the sea cliffs of Lanzarote. *L, F.

Dittrichia

Annual or perennial herbs or small shrubs with sticky stems and simple, alternate leaves; very similar to *Inula*. Flowerheads borne in branched inflorescences. Achenes abruptly contracted below the pappus.

***Dittrichia viscosa* (syn. *Inula viscosa*)** ALTABACA A densely glandular, sticky perennial 40 cm–1.3 m with stems woody at the base. Lower leaves bright green, linear, scarcely and sparsely toothed; upper leaves stalkless and semi-clasping the stem. Flowerheads 10–15(20) mm, 5–6 mm across, bright yellow, *with ray florets to 8–12 mm long*, much exceeding the flower bracts; bracts adpressed. Common to abundant in disturbed habitats. L, F, C, T, G, H, P.

Erigeron FLEABANE

Annuals or perennials with linear to oblong, entire to toothed, unstalked to shortly stalked leaves. Flowerheads with numerous ray florets in several rows, narrow and strap-like; bracts numerous and overlapping. Pappus 1 row of hairs (or with additional shorter hairs).

(a) Ray florets white to lilac, erect and *longer than the disk florets*.

Erigeron calderae POMPÓN DE LAS CAÑADAS A sprawling, rock-dwelling annual with hairless, linear–lanceolate, blunt-tipped leaves. Flowerheads borne rather few on long, erect to ascending stalks; ray florets whitish-violet; disk florets yellow; bracts linear, reflexed at the tips. Rare. *T (Las Cañadas del Teide).

---

1. *Phagnalon saxatile*
2. *Phagnalon purpurascens*
3. *Phagnalon umbelliforme*
4. *Pulicaria burchardii*
5. *Pulicaria canariensis* (habit)
6. *Pulicaria canariensis*
7. *Pulicaria canariensis* subsp. *lanata* (habit)
8. *Pulicaria canariensis* subsp. *lanata*
9. *Dittrichia viscosa*
10. *Erigeron calderae*

Erigeron karvinskianus TERESITA A sprawling ruderal plant with copious pink and white, daisy-like flowerheads. Native to Central America; naturalised. Disturbed habitats and towns; fairly common on most islands. L, C, T, H, P.

(b) Flowerheads with very *inconspicuous or absent* ray florets. Previously described as *Conyza*, and still described under that genus by many floras.

Erigeron canadensis (syn. *Conyza canadensis*) SIMONILLO FLORIDO, CANADIAN FLEABANE A tall, sparsely hairy annual to 1.5 m. Leaves alternate, narrowly oblong, stalked and yellow-green, often withered below and deciduous before flowering. Flowerheads very small, 3–5 mm, whitish, borne abundantly in a *cylindrical inflorescence*; outer florets exceeding the involucre; disk florets with 4-lobed corolla. Pappus cream. A weed native to North America, very commonly naturalised on waste ground in towns. L, F, C, T, G, H, P. *E. bonariensis* (syn. *Conyza bonariensis*) SIMONILLO is a similar annual to 60 cm, more densely brown to reddish-grey hairy, with larger flowerheads 6–10 mm borne in an *inflorescence with long lateral branches*, overtopping the main axis; *terminal axis with ≤30 flowerheads*; outer florets *minute*; disk flowers with 5-lobed corolla. Pappus dirty-white to reddish. Native to tropical America, commonly naturalised. L, F, C, T, G, H, P. *E. sumatrensis* (syn. *Conyza sumatrensis*) is similar to *C. bonariensis*, with flowerheads 5–7 mm borne in a *pyramidal* inflorescence; *terminal axis with ›50 flowerheads*; disk flowers with 5-lobed corolla. Pappus cream to grey. Native to the Americas, naturalised. L, C, T, G, H, P.

Bellis PASCUETA, DAISY

Annuals or perennials with basal leaf rosettes; leaves simple. Flowerheads solitary, on long stalks; ray florets white to pink; disk flowers yellow. Pappus absent.

Bellis perennis DAISY A perennial to 12(20) cm with a *dense basal rosette*, leaves *abruptly* tapered at the base into a petiole, toothed. Flowerheads 12–25 mm across, borne on *unbranched, leafless stems*. Grassy habitats at high elevation. C, T.

Bellis annua ANNUAL DAISY A low, normally *soft-bristly-hairy annual* to 10 cm with short, erect, *leafy stems* (at least below). Leaves spathulate, toothed or not, stalked below, *not borne in distinct rosettes*. Flowerheads white with a yellow disk, to 15 mm across, the rays sometimes tinged purple. Grassy habitats at high elevation. T.

Artemisia WORMWOOD

Annual or perennial herbs, or low shrubs, often strongly aromatic. Leaves alternate, entire to divided. Flowerheads small, often nodding in a lax inflorescence. Achenes lacking (or virtually lacking) a pappus.

Artemisia thuscula INCIENSO A grey, much-branched shrub to 1 m. Leaves 30–70 mm, aromatic with *flat*, linear to oblanceolate, blunt lobes. Inflorescences dense and elongated; flowerheads spherical, golden, 4 mm across. Frequent west of Gran Canaria on low-elevation hills and scrub. *C, T, G, H, P. *A. ramosa* INCIENSO MORISCO has *ovoid* flowerheads and leaves with linear, *fleshy* (not flat) lobes. Dry slopes and barrancos; local. *C, T. *A. reptans* AMULEY is a *small* perennial (12–30 cm) with leaf lobes to just 5 mm. Flowerheads brownish and *pendulous*. Coastal hills and barrancos; very locally common. F, C, T.

1. *Erigeron karvinskianus*	4. *Erigeron bonariensis*	7. *Artemisia thuscula*
2. *Erigeron canadensis*	5. *Erigeron sumatrensis*	8. *Artemisia ramosa*
3. *Erigeron bonariensis* (habit)	6. *Artemisia thuscula*	9. *Artemisia ramosa*

Achillea

Perennial herbs with alternate, simple and wavy-edged to shallowly or deeply dissected leaves. Flowerheads congested into umbel-like flat-topped clusters; disk and ray florets white, pink or yellow. Pappus absent.

Achillea maritima (syn. *Otanthus maritimus*) COTTONWEED A short, *densely white-woolly*, spreading perennial to 30 cm with robust, ascending stems. Leaves oblong–lanceolate, untoothed or blunt-toothed, fleshy and unstalked. Flowerheads few, 6–9 mm across, *yellow, rayless* and button-like, borne in lax, flat-topped clusters; flower bracts white-woolly. Fixed dunes. Apparently extinct in the Islands. L, C.

Cladanthus

A genus closely related to *Anthemis* and *Chamaemelum* (many species previously included in the latter genus).

Cladanthus mixtus (syn. *Chamaemelum mixtum*) MANZANILLA ESTRELLADA A hairy, chamomile-like annual to 60 cm, normally with much-branched stems. Leaves oval in outline, 1–2-pinnately divided below, deeply toothed or 1-pinnately divided and *unstalked above*; leaf lobes linear–lanceolate and toothed or not. Flowerheads daisy-like with an involucre to 45 mm and spreading white rays and a yellow disk; rays 3-toothed at the tip; flower bracts with a wide, pale brown membranous margin. Achenes 1.2–1.6 mm, ovoid and weakly ridged. Fallow ground. C, T, G, H, P.

Anthemis MANZANILLA, CHAMOMILE

Slightly hairy, often aromatic herbs or dwarf shrubs with *leaves cut into linear segments*. Flowerheads usually with white or yellow rays. Achenes oval to obconical (not compressed), circular to square in cross-section, usually with about 10 smooth or rough *distinct* ribs. Pappus absent. Numerous species in the area; only a subset are included here.

Anthemis arvensis CORN CHAMOMILE A variable, hairy, aromatic annual to biennial with spreading or ascending branched, downy stems to 50 cm. Leaves to 50 mm long, oval in outline, 1–2-pinnately lobed with narrow, pointed segments, *woolly below*, especially when young. Flowerheads with white ray florets and a yellow disk to 40 mm across; bracts with brown, papery margins. Receptacle cylindrical–cone-shaped in fruit; achenes with 10 ridges. Disturbed, cultivated and fallow land. L, F, C, T, G.

Anthemis cotula STINKING MAYWEED An erect annual, *strong and unpleasant-smelling*, with *almost hairless leaves*, slightly fleshy and irregularly 2–3-pinnately lobed into linear segments. Flowerheads with ray florets deflexed when mature. Achenes with rough ridges. Various habitats. L, F, C, T, G, H, P.

Anacyclus PAJITO

Anthemis-like annual or perennial herbs with alternate, 1–2-pinnately divided leaves. Outer achenes 2-*winged*, inner achenes unwinged; pappus absent.

Anacyclus clavatus A short, widely branched, hairy annual 20–50 cm. Leaves alternate, 2–3-pinnately lobed with linear segments. Flowerheads daisy-like with *white recurved rays and a yellow disk*; flowers borne solitary on stalks *distinctly swollen* below the fruiting head; flower bracts with a narrow whitish or purplish margin. Achenes winged. Sandy ground. L, F, T, P.

Anacyclus radiatus A soft-hairy annual 40–60 cm with erect, rigid stems branched above. Leaves oval–oblong in outline with linear lobes. Flowerheads yellow, 25–30 mm across; bracts woolly, with white, transparent appendages. Wings of outer achenes broader than the achene body; inner achenes wingless. Two forms recognised: **subsp.** *radiatus* (all islands except C); **subsp.** *coronatus* PAJITO COLORADO (absent from the far west). L, F, C, T.

Glebionis (*Chrysanthemum*) PAJITO

Annuals with simple leaves. Flowerheads with yellow, cream or white ray florets; receptacle without scales. Achene without a pappus.

Glebionis segetum (syn. *Chrysanthemum segetum*) CORN MARIGOLD A short to tall, green, hairless, slightly fleshy annual to 60 cm with erect to ascending, branched or unbranched stems. Leaves blue-grey, alternate, narrowly oval in outline, slightly to *deeply toothed*, at least below. Flowerheads 30–70 mm across, bright yellow and daisy-like, with a flat disk. Achenes 2.5–3 mm, deeply ridged and unwinged. Common on disturbed fallow and cultivated land, roadsides and coastal waste places. F, C, T, G, P.

Glebionis coronaria (syn. *Chrysanthemum coronarium*) OJO DE BUEY, PAJITO, CROWN DAISY A slightly hairy annual to 80 cm with leaves *2-pinnately lobed*. Flowerheads 40–80 mm across, yellow and daisy-like, with rays cream-white in the upper ½ (or cream or yellow). Achenes 3–3.5 mm, deeply ridged and winged. Common to abundant on disturbed land, roadsides and coastal waste places. L, F, C, T, G, H, P.

Coleostephus

Similar to *Glebionis* but with leaves *regularly finely-toothed* (not deeply lobed or divided).

Coleostephus myconis GIRALDA A sparingly branched, virtually hairless annual. Leaves oval, *regularly fine-toothed* (*not* lobed), the lower leaves tapered gradually into a stalk; the upper leaves unstalked and semi-clasping the stem. Flowerheads yellow and daisy-like, to 22 mm across with yellow or paler rays and a yellow disk. Cultivated, fallow and damp ground. L, F, C, T.

Bethencourtia

An endemic genus similar to *Senecio* but with obtuse style branches.

Bethencourtia palmensis TURGAYTE A small shrub with fleshy, lanceolate leaves with toothed margins. Flowerheads golden yellow, slender, borne in dense corymbs. Cliffs at high elevation; rare. *T (Las Cañadas del Teide), P.

Senecio

A large and diverse genus of annual or perennial herbs and shrubs with alternate, pinnately veined leaves. Flowerheads often numerous, borne in flat-topped clusters, usually yellow. Achene with a white or greyish, hairy pappus.

(a) Annuals. Ray florets absent.

Senecio vulgaris CASAMELOS, GROUNDSEL A short, more or less hairy, rather succulent annual to 30 cm (usually less) with weak, sparingly and irregularly branched stems. Leaves coarsely lobed and toothed, oval in outline and short-stalked below, semi-clasping the stem at the base.

ASTERACEAE | 341

Flowerheads *small, without (or with few) rays*, to 5 mm across and yellow; involucre cylindrical with black-tipped bracts. Achenes <2.5 mm with much longer pappus hairs. A very common weed in a range of disturbed natural and urban habitats. L, F, C, T, G, H, P.

Senecio massaicus is similar to *S. glaucus* (below) but has flowerheads *without* ray florets. Leaves deeply divided. Towns and disturbed habitats. L, F, C, T.

(b) Annuals. Ray florets usually present.

Senecio leucanthemifolius MOQUEGUIRRE HEDIONDO An ascending, branched annual to 25(30) cm. Leaves *fleshy*, hairless or slightly white hairy; sometimes strongly purple-flushed. Flowerheads 7–8 mm with *black-tipped flower bracts*; ray florets conspicuous and spreading. Achenes 2–2.2 mm, hairless. Coastal habitats; absent from the west and north. Variable: **var. *falcifolius*** is one of several forms described. L, F, C, T, G, H, P.

Senecio glaucus SANGUINARIA An erect to spreading *hairless* (to sparsely hairy), greyish-blue, sometimes fleshy annual to 40(60) cm; similar to *S. leucanthemifolius*. Leaves oblong or lanceolate in outline, 10–50 mm (13 cm) long, the upper usually stalkless and clasping; all deeply lobed with toothed, linear segments. Flowerheads yellow, borne in lax cymes or corymbs; ray florets conspicuous; bracts *without* conspicuous black tips. Achenes 1.8–2.2 mm long. Coastal sands and bare ground. **Subsp. *coronopifolius*** is the form in the region. L, F, C, T, G, H.

Senecio flavus SANGUINARIA CENICIENTA A hairless, erect annual 50 mm–40 cm. Basal leaves glaucous, 20–40 mm long, *ovate*, coarsely toothed; upper leaves stalkless with semi-clasping bases. Flowerheads yellow with disk and ray florets, borne in lax corymbs, 4–6 mm across; bracts black-tipped. Pappus 4–5 mm, white. Dry, stony habitats; frequent. L, F, C, T.

Senecio incrassatus MOQUEGUIRRE DE COSTA A small, hairless annual to 30 cm, branched from the base. Leaves bright green, rather shiny, to 50 mm, with irregular teeth and downturned margins; somewhat succulent. Flowerheads large, golden yellow. Stony plains in the west. H, P.

(c) Creeping, shrubby perennials.

Senecio angulatus A *creeping*, slightly *succulent* perennial with stems to 2 m long and *bluntly lobed, shiny green leaves*. Flowers yellow, typically *Senecio*-like and scented. A native of South Africa, commonly planted in the region. Throughout. F, C, T, H, P.

Kleinia

A predominantly tropical genus generally recognised as distinct from *Senecio*. Plants with a woody, succulent habit. Pappus hairy.

Kleinia neriifolia VERODE A very succulent shrub with swollen grey stems becoming bottle-like with age. Leaves glaucous, linear–lanceolate to oblanceolate, pointed; deciduous in summer. Flowerheads borne in umbels at the ends of the branches in late summer. Abundant throughout, in a range of habitats at low–medium elevation. *L, F, C, T, G, H, P.

---

1. *Anacyclus radiatus* subsp. *coronatus*
2. *Glebionis coronaria* (habit)
3. *Glebionis coronaria*
4. *Coleostephus myconis*
5. *Bethencourtia palmensis*
6. *Senecio massaicus*
7. *Senecio leucanthemifolius* (habit)
8. *Senecio leucanthemifolius* var. *falcifolius*
9. *Senecio leucanthemifolius* (inland form)
10. *Senecio leucanthemifolius* var. *falcifolius*
11. *Senecio glaucus* subsp. *coronopifolius*

Delairea

Twining perennials with *ivy-like* leaves.

Delairea odorata YEDRA ALEMANA   A succulent climber to 4 m. Leaves ivy-like, 30 mm–10 cm with up to 10 triangular lobes. Flowerheads yellow, fragrant, borne in dense corymbs. Native to South Africa; naturalised. C, T, P.

Calendula   MARAVILLA, ALPOAERA, MARIGOLD

Annual or perennial, often aromatic herbs with alternate, undivided leaves. Flowerheads daisy-like with yellow or orange ray florets; bracts in 1–2 rows. *Achenes strongly curved*, without a pappus.

Calendula arvensis FIELD MARIGOLD   A short, slender, thinly hairy, ascending or spreading annual to 30 cm. Leaves oblong and finely toothed. Flowerheads 10–25(27) mm across; yellow-orange throughout or with a brownish disk. Fruiting head with an outer row of beaked and strongly incurved achenes. Common on fallow ground. L, F, C, T, G, H, P. *C. officinalis* is similar but with larger flowers 40–70 mm across, yellow-orange with an orange or brownish disk. L, C, T, G, H, P.

Calendula sventenii   A short, slender, glandular-hairy annual to 15 cm. Leaves linear–lanceolate, entire or toothed. Flowerheads solitary, yellow, *small*, 7–10(12) mm across; involucre with 2 rows of almost equal glandular-hairy bracts. Fruiting head with markedly differing achenes, the outermost beakless and with *3 prominent wings*; the intermediate boat-shaped, the innermost transversely marked with stripes of small warts. Arid habitats, soon-flowering after winter rain. L, F, C, T.

Calendula suffruticosa   A short to medium perennial, often with a woody stock. Leaves lanceolate, somewhat fleshy, pointed, with few, distant teeth; glandular. Flowerheads large, to 40 mm across, bright yellow, the rays sometimes red-tipped, to 20 mm long. Fruiting head with a *conspicuous outer row of spreading (weakly curved) achenes* 22 mm. Seacliffs; local. T.

Pallenis

Annuals to subshrubs. Flowerheads yellow with bracts in 2–3 rows, leafy, the outermost spineless or spine-tipped. Pappus of scales. A genus now established to include species traditionally placed in *Asteriscus*.

Pallenis spinosa ESTRELLADA ESPINOSA   A slender, softly hairy annual to 60 cm with rigid stems, woody at the base with branches overtopping the main stem. Leaves elliptic, stalked below, unstalked and semi-clasping the stem above. Flowerheads daisy-like and bright yellow with a large disk to 20 mm across. Flower *bracts spine-tipped and 2 x the length of the ray florets*; inner flower bracts papery and not spine-tipped. Achene 2–2.5 mm. Forests and thickets. T.

Pallenis hierichuntica (syn. *Asteriscus pygmaeus*) RESURRECTION PLANT   A small, *virtually stemless* (main stem 0–7 mm) annual, similar to *Asteriscus aquaticus* in general appearance (see below), but to just 15 cm with small lateral branches. Flowerheads generally exceeded by their outermost involucral bracts. Achene 1.2–1.9 mm. Bare, dry slopes. F.

---

1. *Senecio flavus*
2. *Senecio flavus*
3. *Senecio incrassatus* (LEOPOLDO MORO)
4. *Kleinia neriifolia* (habit)
5. *Kleinia neriifolia*
6. *Kleinia neriifolia* (in fruit)
7. *Delairea odorata*
8. *Calendula arvensis*
9. *Calendula arvensis* (fruit)
10. *Calendula sventenii*
11. *Pallenis spinosa*
12. *Pallenis hierichuntica*

Asteriscus

A genus very similar to *Pallenis* to which some species have recently been transferred.

Asteriscus aquaticus A similar species to the *Pallenis* group but an *aromatic annual* to 50 cm with erect to spreading stems, and flower bracts *greatly exceeding* the ray florets. Damp and sandy places, waste places and roadsides. Similar to *Pallenis spinosa* but lacking spine-tipped flower bracts. Achene 1.5–2 mm. Various habitats. C, T, G, H, P.

Asteriscus graveolens BOTONERA A small, densely branched shrublet to 40 cm with greyish stems. Leaves densely clothing the stems, oblanceolate, silky- to short-hairy. Flowerheads 20 mm across with lemon to golden yellow ray florets. Two forms recognised: **subsp. *stenophyllus*** is a compact shrublet densely silky-grey-hairy with linear leaves. *C; **subsp. *odorus*** is short-hairy, laxer, with broader leaves and flowerheads to 15 mm. Both forms found on dry slopes. C.

Asteriscus sericeus JORADO A tough, aromatic shrub with grey to blackish stems with leaves crowded near the tips, 30–50 mm, densely silver-hairy, broadly oblanceolate. Flowerheads 30–35 mm; ray florets with 3 teeth at the tips. Locally common in rocky areas, especially at high elevation. *F. ***A. intermedius*** TOJIA is similar but with silvery, not blackish stems, leaves less congested and narrower flowerheads. Frequent on the top of Famara. *L.

Asteriscus schultzii TOJIA BLANCA A compact dwarf shrub with succulent, spathulate, sparsely hairy leaves with toothed margins. Flowerheads 20–25 mm across; ray florets *pale yellow to cream*. Interior sand flats and coastal habitats; rare. L (common in coastal rocks at Famara), F (especially Lajares).

Xanthium COCKLEBURS

Cosmopolitan annual weeds, native to the Americas. Leaves alternate. Flowerheads discoid. Fruiting head *conspicuously spiny*. Pappus absent.

Xanthium spinosum CACHURRERA, SPINY COCKLEBUR A stiffly branched annual to 1.2 m with *prominent beige spines* projecting from the leaf bases. Leaves alternate and heart-shaped at the base, shallowly lobed and long-stalked. Flowerheads greenish, with male and female flowers borne separately in lateral clusters. *Fruiting heads covered in hooked spines*. Disturbed habitats. L, F, C, T, P.

Bidens BUR-MARIGOLD

Annuals with opposite leaves. Flowerheads rounded, often solitary, often without ray florets; bracts in 2 rows, the innermost papery. Pappus comprising 2–5 strong, barbed bristles. Native to Central and South America.

Bidens aurea TÉ CANARIO, BIDENS A virtually hairless, medium to tall, slender perennial to 1 m. Leaves linear–lanceolate or 2-pinnately divided with linear lobes; toothed. Flowerheads held erect, daisy-like, yellow, white or mottled both these colours, 30–35 mm across with only *5–6 large*, broad, grooved ray florets; flower bracts all similar and much shorter than the ray florets. Achenes with 3–4 bristles. Naturalised in damp places. F, C, T, G, H, P.

1. *Asteriscus graveolens* subsp. *stephophyllus*
2. *Asteriscus sericeus* (habit)
3. *Asteriscus sericeus*
4. *Asteriscus intermedius* (habit)
5. *Asteriscus intermedius*
6. *Asteriscus schultzii* (habit)
7. *Asteriscus schultzii*
8. *Xanthium spinosum*
9. *Bidens aurea*
10. *Bidens pilosa*

Bidens pilosa AMORSECO A sparse, erect annual to 1 m. Leaves pinnate with leaflet teeth wider than long. Flowerheads with *small, white* ray florets (when present) and yellow disk florets. Achenes with 2–3 bristles. Frequent in thickets and along roadsides, especially in the west or at high elevation. L, F, C, T, G, H, P.

Allagopappus

Shrubs with sticky leaves. Flowerheads <5 mm across, in terminal corymbs, *without* ray florets.

Allagopappus canariensis MADAMA DE RISCO A shrub to 1 m with rusty-brown stems. Leaves 20–30 mm, narrowly lanceolate, toothed, rather crowded and upward-swept. Flowerheads borne in flat corymbs, yellow. Rocky slopes and barrancos. *C, T, G. *A. viscosissimus* MADAMA PEGAJOSA is similar but with narrower, *linear* leaves with entire margins. Rocky slopes and barrancos of the south and west. *C.

Vieria

Like *Allagopappus* but succulent, with larger flowerheads *with* ray florets.

Vieria laevigata AMARGOSA A leafy shrub to 1 m with greyish stems. Leaves light green to glaucous, to 50 mm, fleshy, ovate to lanceolate and strongly toothed towards the tips. Flowerheads *large*, 20 mm across, borne 5–10 in sparse clusters, with bright yellow ray florets and a golden yellow disk. Basalt cliffs; rare. *T.

Argyranthemum MAGARZA

Shrubs with dissected leaves. Heads with ray and disk florets, ligules usually white but sometimes yellow. Outer and inner fruits within a fruiting head morphologically distinct. Pappus a marginal ridge, crown-like or absent.

A genus endemic to the Macaronesian region. Reproductive barriers between species in this group are weak and where species distributions overlap, hybridisation may occur and complicate identification. Hybridisation has also resulted in the evolution of two recognised species in Tenerife (*A. sundingii* and *A. lemsii*).

Argyranthemum adauctum Shrub to 90 cm branching throughout. Leaves 2–3 times divided, the lobes very variable in size (0.5–15 mm) and extending more or less to the base. Flowerheads small, 4–12 mm wide. Ligules white, up to 15 x 3.5 mm. Outer fruits of heads lacking wings and often in clusters. A widespread and very variable species, within which several subspecies have been recognised. It is found above 350 m in a range of habitats including laurel and pine forests and montane scrub. *C, T, H, P.

Argyranthemum frutescens MAGARZA, MARGARITA Trailing to erect shrub up to 80 cm tall. Leaves petiolate, once or twice (or sometimes three times) divided into lobes very variable in size. Flowerheads ranging from 6–22 mm wide. Ligules white, up to 15 x 5 mm. Outer fruits of heads with 2 large and one small wing. A very variable species, with numerous subspecies recognised. Widespread on the western islands in coastal and lowland areas, although in Gran Canaria it is confined to the north and west, and rare in La Palma. *C, T, G, H, P. Numerous subspecies have been recognised including **subsp. *succulentum*** with glaucous, succulent leaves (T: north coast), **subsp. *canariae*** with acute to cuspidate leaf lobes (GC: north coast) and **subsp. *pumilum*** with acute leaf lobes (GC: northwest). The Tenerife endemics *A. lemsii* MAGARZA DE LEMS (*T) and *A. sundingii* MAGARZA DE SUNDING (*T) are species of

1. *Allagopappus canariensis*
2. *Allagopappus viscosissimus*
3. *Vieria laevigata*
4. *Argyranthemum frutescens* subsp. *succulentum*
5. *Argyranthemum frutescens* subsp. *canariae*
6. *Argyranthemum frutescens* subsp. *pumilum*
7. *Argyranthemum lemsii*
8. *Argyranthemum sundingii*
9. *Argyranthemum broussonetii*

hybrid origin that are morphologically intermediate between the parental species, *A. frutescens* and *A. broussonetii*. Both are restricted to intermediate, mid-elevation habitats in the Anaga peninsula. *A. lemsii* is found on the eastern slopes and *A. sundingii* on the southern slopes.

Argyranthemum broussonetii MARGARITA DE MONTE A robust shrub up to 120 cm tall. Leaves large (up to 16 cm long), glabrous or hairy only on the central nerve, twice-divided into broad lobes (primary lobes up to 2 cm wide) and lacking a distinct petiole with the lobes extending more or less to the base. *Flowerheads large, up to 2.2 cm wide. Ligules white, large, 20–40 mm long.* Outer fruits of heads 3-winged or often wingless. Occurring in clearings in the laurel forest in Tenerife. *T.

Argyranthemum coronopifolium MAGARZA AZULADA Shrub up to 1.2 m tall. Leaves glabrous and fleshy, once or twice divided into broad lobes (primary lobes up to 1.5 cm wide), with a distinct basal petiole. *Flowerheads large, up to 2.2 cm wide. Ligules white, 11–24 mm long.* Outer fruits of heads winged. North-facing coasts in Teno and in Anaga (where it is extremely rare). *T.

Argyranthemum tenerifae MARGARITA DEL TEIDE *A compact, and often dome-shaped shrub to 50 cm high*, branching from the base. Leaves once or twice divided, with a distinct petiole and covered in very small bristly hairs. Flowerheads 7–15 mm across. Ligules white. Outer fruits of heads usually 3-winged. Restricted to the subalpine zone of Las Cañadas where it is locally common. *T.

Argyranthemum foeniculaceum MAGARZA DE SANTIAGO A spreading or ascending shrub to 1 m. *Leaves densely crowded around the base of the flower stalk*, 2–3 times divided; the lobes very narrow (<3mm), distinctly petiolate and covered in a grey-white bloom. Flowerheads large, up to 2 cm across. Ligules white, 15–22 mm long. Outer fruits of heads 3-winged. Locally common, often on dry cliffs in the south and southwest of Tenerife, at 200–1,100 m. *T. (**A. vicentii** is a taxon not officially described; similar to *A. foeniculaceum* but taller (up to 2 m), the leaves less densely crowded and bluish-green to green in colour. Open places in the pine forest of Tenerife at 800–1,800 m. *T.)

Argyranthemum gracile MAGARZA GRÁCIL Erect shrub to 1.2 m. Leaves arranged along the slender stems, usually trifid with very narrow lobes (<3 mm), petiolate, glabrous. Flowerheads small, up to 12 mm across. Ligules white, 10–14 mm long. Outer fruits of heads 3-winged. Frequent on southwest Tenerife from Adeje in the south, northwards to the Teno massif, typically between 100 m and 550 m. *T.

Argyranthemum callichrysum Shrub to 1 m tall, branching throughout. Leaves twice divided, with distinct petioles (i.e. with no lobes at the base) and glabrous. Flowerheads up to 1.5 cm wide. Ligules white (or yellow). Outer fruits of heads with narrow wings and sometimes grouped together. Two subspecies are recognised: in **subsp.** *callichrysum* MARGARITA GOMERA AMARILLA the primary leaf lobes are linear–lanceolate, divided to the midrib and narrow (2–7.5 mm wide); in **subsp.** *gomerensis* MARGARITA GOMERA DE MONTE (syn. *A. broussonetii* subsp. *gomerensis*), the primary leaf lobes are obovate, less deeply divided and wider (7.5–15 mm wide). Both subspecies are restricted to La Gomera. Subspecies *gomerensis* is found along roadsides and in clearings in laurel forests in the north of the island at 550–1,000 m. Subspecies *callichrysum* is found from the SW coast to the central mountains where it grows in xerophytic scrub on south-facing slopes. *G.

---

1. *Argyranthemum coronopifolium*
2. *Argyranthemum tenerifae* (RACHEL GRAHAM)
3. *Argyranthemum gracile*
4. *Argyranthemum callichrysum* (RACHEL GRAHAM)
5. *Argyranthemum sventenii* (LEOPOLDO MORO)
6. *Argyranthemum hierrense* (LEOPOLDO MORO)
7. *Argyranthemum winteri*
8. *Argyranthemum maderense* (habit)
9. *Argyranthemum maderense*

Argyranthemum filifolium MAGARZA FINA Slender, wiry shrub to 0.8 m, branching at the base. Leaves glabrous, once or twice divided, with few, narrow (<4 mm) lobes. Flowerheads, often numerous, small, less than 12 mm across. Ligules white, short, 6–12 mm long and narrow (<2.5 mm wide). Outer fruits in a head with very small (or no) wings and often grouped together. Dry, scrub vegetation in the south of Gran Canaria, up to 500 m. *C. *A. escarrei* MAGARZA DE LA ALDEA is similar but the plants are smaller (50 cm high), the leaf lobes broader and the ligules larger (up to 14 mm and 4–6.5 mm wide). Locally abundant in xerophytic communities in mountain regions of the south and southwest of Gran Canaria. *C.

Argyranthemum lidii MAGARZA DE LID Shrub to 60 cm tall, branching from the base. Leaves twice divided, with broad, obtuse lobes (primary lobes up to 15 mm), the lobes extending almost to the base, glabrous or slightly rough. Flowerheads 8–16 mm across. Ligules white, up to 15 mm long and 4 mm wide. Outer fruits of heads with distinct lateral wings. Local and rare endemic, known only from rocky slopes on the west coast of Gran Canaria. *C.

Argyranthemum haouarytheum BAINENA Shrub to 1 m tall, branching throughout. Leaves glabrous, twice deeply divided to the midrib into narrow lobes (primary leaf lobes <4 mm wide) and with a distinct petiole. Flowerheads 11–15 mm across. Ligules white, large, up to 20 x 5 mm. Outer fruits of heads with 2 large and one small wing. Locally common on the central, northwest and southern slopes of La Palma, in pine forest and xerophytic communities in the south of the island. *P.

Argyranthemum webbii MARGARITA PALMERA Erect shrub to 90 cm, branched towards the base. Leaves glabrous, twice divided, the primary lobes sometimes broad (up to 15 mm), with a distinct petiole. Flowerheads up to 14 mm wide. Ligules white, large, up to 22 x 4 mm. Outer fruits of heads with 3 indistinct wings and sometimes in clusters. Laurel forests of the east of La Palma. *P.

Argyranthemum sventenii MARGARITA DE SVENTENIUS Shrub to 60 cm tall, branched at the base. Leaves glabrous, twice divided to the midrib into narrow lobes (primarily lobes up to 3 mm wide) and with a distinct petiole. Flowerheads small, up to 11 mm across. Ligules white, small, up to 10 x 2 mm. Outer fruits of heads with 3–4 wings and often coalesced into groups. A rare species growing in xerophytic scrub in the south of El Hierro. *H.

Argyranthemum hierrense MARGARITA HERREÑA Shrub to 1 m tall, branching throughout. Leaves twice or rarely three times divided, more or less to the midrib and with a distinct petiole. Primary lobes variable in size but sometimes up to 13 mm wide. Flowerheads small, up to 11 mm wide. Ligules white, up to 17 x 4 mm. Outer fruits of heads with 2–4 wings and often coalesced into groups. Locally common in the El Golfo region of El Hierro, at 100–650 m and in both xerophytic and laurel forest habitats. *H.

Argyranthemum winteri MARGARITA DE JANDÍA Shrub to 80 cm tall, branching throughout. Leaves glabrous, twice divided (but typically only shallowly so) and lacking a distinct petiole with the lobes up to 5 mm wide, extending more or less to the base. Flowerheads large, up to 20 mm across. Ligules white, up to 15 x 4 mm. Outer fruits of heads with 3 wings. A narrow endemic restricted to the mountains of Jandía on Fuerteventura. *F.

Argyranthemum maderense MARGARITA DE FAMARA A small shrub up to 70 cm tall. Leaves glabrous, divided into serrate lobes up to 8 mm wide, the leaves shortly petiolate or the lobes extending to the base. Flowerheads up to 15 mm across. *Ligules yellow*, 18–25 mm long. Outer fruits of heads distinctly 3-winged. Restricted to Lanzarote, on coastal cliffs of the Famara region and occasionally rocky areas away from the immediate coast. *L.

1. *Chamaemelum fuscatum*
2. *Gonospermum fruticosum*
3. *Gonospermum ptarmiciflorum*
4. *Gonospermum revolutum*
5. *Gonospermum ferulaceum*

ASTERACEAE | 351

Chamaemelum CHAMOMILE

Annual or perennial herbs with deeply dissected leaves. Flowerheads with disk florets, and with the base of the corolla tube enlarged and inflated (pouch-like). Pappus absent.

Chamaemelum fuscatum A more or less *hairless*, rather lax annual 60 mm–40(60) cm with erect to ascending stems. Leaves 2–3-pinnately divided below, 1-pinnately divided above into linear segments. Flowerheads somewhat *nodding* in bud; flower bracts with *blackish brown* papery margins and tips. Achenes 1–1.4 mm, ovoid with about 30 ribs. Meadows; uncommon. T.

Gonospermum

Shrubs. Leaves lobed. Flowerheads borne in dense yellow corymbs; bracts linear.

(a) Flowerheads with yellow ray florets.

Gonospermum fruticosum CORONA DE LA REINA A small shrub to 1.5 m with brown stems. Leaves sparsely woolly-hairy, with toothed lobes. Inflorescence a variably lax to dense, flat to domed corymb; flowerheads golden yellow. Laurel forest cliffs; frequent in suitable habitats in the west. *T, G, H, P.

(b) Flowerheads with white ray florets.

Gonospermum ptarmiciflorum MARGARITA PLATEADA, SILVERY LACE BUSH A small, densely *silvery-white*-hairy shrub to 50 cm with 2–3-pinnate leaves with 8–22 elliptic, deeply dissected segments. Flowerheads with large white ray florets, borne in corymbs. Mountains; rare. *C (centre).

1. *Pericallis appendiculata*
2. *Pericallis lanata*
3. *Pericallis steetzii* (habit) (RACHEL GRAHAM)
4. *Pericallis steetzii* (RACHEL GRAHAM)
5. *Pericallis echinata*
6. *Pericallis cruenta*
7. *Pericallis webbii*
8. *Pericallis webbii* (OLIVER WHITE)
9. *Pericallis webbii* (pink form) (OLIVER WHITE)

Gonospermum revolutum MARGARITA DE LUGO A perennial to 50 cm with deeply pinnately lobed green leaves with fleshy lobes and *downturned margins*, arranged in basal rosettes. Flowerheads to 20 mm with white ray florets; rather few. *T.

Gonospermum ferulaceum Similar to the species above but with green 1- to 2-pinnatisect leaves with linear segments. Flowerheads with 7–8 ray florets, borne in dense heads. C.

Pericallis FLOR DE MAYO, FLORIST'S CINERARIA

Perennial herbs or rhizomatous shrubs. Leaves entire, petioles often auriculate. Flowerheads usually corymbose, with ray and disk florets. Involucral bracts in a single row. Cypsela with longitudinal ridges. Pappus white or yellow, sometimes absent from the ray florets. Hybrids occur where species overlap, complicating identification. Endemic to Macaronesia.

(a) Shrubs with *woody stems*.

Pericallis hansenii ALAMILLO GOMERO Up to 4 m, often with creeping subaerial rhizomes. Leaves subcordate to deltoid with few, scattered thick hairs on the lower surface. Phyllary bracts usually glabrous. Ray floret ligules and *disk florets pink*. *G.

Pericallis multiflora Up to 4 m, often with creeping subaerial rhizomes. Leaves ovate to cordate, with dense white indumentum below when young. Phyllary bracts glabrous. Ray floret ligules pink. *Disk florets purple*. *T. *P. hadrosoma* FLOR DE MAYO is similar but has oblong to elliptic leaves. *C.

Pericallis appendiculata ALAMILLO Up to 1.5 m. Leaves cordate, ovate or lanceolate with dense white indumentum below. Phyllary bracts sometimes with echinules. Ray floret ligules flesh-coloured, cream or white, *disk florets white to light yellow, sometimes light pink*. *C, T, G, P. This species is variable and warrants further work.

(b) Perennial herbs, *herbaceous* or woody at the base.

Pericallis lanata PALOMERA Woody at the base. Leaves with white indumentum. *Capitula solitary, in pairs, or rarely in small synflorescences* (‹5). Phyllary bracts, glabrous or rarely with few purple echinules; ray florets 12–15, much longer than wide; purple or very rarely white; disk florets deep violet to mauve. *T (south to northwest).

Pericallis papyraceus ENCIMBA Herbaceous at the base. Leaves more or less glabrous below or with a white indumentum, sometimes light purple towards the veins. Flowerheads ›10 in an inflorescence. Phyllary bracts sometimes with few purple echinules. *Ray florets 7–8*, around 2 x as long as wide, purple, rarely white. *Disk florets deep violet to mauve*. *P.

Pericallis steetzii ARCILA Herbaceous at the base. Leaves with a white-to-grey indumentum below, sometimes light purple towards the veins. Flowerheads ›10 in an inflorescence. Phyllary bracts often with echinules. Ray florets 12–13, around 2 x as long as wide, pure white, cream, pink or purple. *Disk florets cream to yellow or light-pink*. *H, G. Plants from H have traditionally been separated as *P. murrayi*.

Pericallis echinata Herbaceous at the base. Leaves with sparse white indumentum below. Flowerheads with ›10 capitulae. Phyllary bracts *covered in purple echinules*. Ray florets typically ›12, ligules less than 3 x longer than wide, pink or light purple. Disk florets cream to yellow. *T (north and west).

Pericallis cruenta Herbaceous at the base. *Leaves usually dark purple* with white to grey hairs below. Flowerheads with ›8 capitulae. Phyllary bracts glabrous or sometimes with a few echinules. Ray florets 13; ligules less than 3 x as long as wide; purple. Disk florets deep violet to mauve. *T (north).

Pericallis webbii FLOR DE MAYO Herbaceous at the base. Leaves sometimes light purple close to the veins, white to grey lanate below. Synflorescence with >10 capitulae. Phyllary bracts glabrous or echinulate. Ray florets 11–13, *ligules often recurved, up to 4 x as long as wide*, pink or purple, rarely white. Disk florets purple to deep violet. *C (introduced in T).

Pericallis tussilaginis FLOR DE MAYO Herbaceous at the base with a *basal rosette of leaves when in flower*. Leaves white to grey hairy below. Flowerheads 6–25 in an inflorescence. Phyllary bracts rarely with echinules. Ray florets 11–13; ligules 2 x as long as wide, pink to light purple. Disk florets yellow or light to dark purple to mauve. *T. Plants from the north tend to have a dense indumentum and yellow disk florets, whereas those from the south are glabrous or only sparsely hairy and have purple of mauve disk florets. *P. tirmensis* was recently described from northwest *C. It has denser basal leaf rosettes and shorter ligules than *P. tussilaginis* and the *petioles are typically winged*.

1. *Pericallis tussilaginis*
2. *Schizogyne sericea*
3. *Schizogyne glaberrima*
4. *Aster squamatus*
5. *Cichorium pumilum*

Schizogyne

Shrubs. Flowerheads without ray florets. Pappus with an entire, unlobed rung of scales and a row of simple hairs.

Schizogyne sericea SALADO A shrub to 1.5 m with silky-white stems. Leaves linear, 30–50 mm, flat, greyish-white. Inflorescences dense; flowerheads 5–6 mm; bracts papery, the outermost small and triangular; ray florets absent, disk florets yellow. Coastal rocks; locally common. *L, F, C, T, G, H, P. *S. glaberrima* is similar but dense, with *linear, succulent*, hairless leaves and compact, pale yellow inflorescences. Coastal rocks and dunes; very local. *C.

Aster

Annuals or perennials with oval to linear stem leaves and flowerheads typically with yellow disk flowers and white, blue, pink or purple ray florets; bracts in several rows. Pappus of 1–2 rows of hairs. North American species traditionally classed in the genus now transferred to other genera such as *Tripolium* and *Galatella*.

Aster squamatus (syn. *Symphyotrichum squamatum*) LLUVIA SILVESTRE An erect, slender, almost hairless annual 10 cm–1 m (tiny in bare habitats). Leaves 40 mm–18 cm, linear–lanceolate, sometimes obscurely toothed. Flowerheads 7–9 mm long, with greenish bracts with broad, translucent margins, often purplish towards the tips; florets *inconspicuous,* whitish, cream, mauve or pink. Native to South America, naturalised in saline areas and sand flats. L, F, C, T, G, H, P.

SUBFAMILY CICHORIOIDEAE

Biennials and perennials, often *with a white latex*, and stem leaves spirally alternate, or absent. Flowerheads usually yellow and ligulate (florets comprising 5-toothed ligules only).

Cichorium CHICORY

Annual or perennial herbs with a white latex when cut. Leaves toothed or lobed. Flowerheads numerous; florets all rayed and toothed at the tips. Achenes angular (not flattened); pappus a series of short scales.

Cichorium pumilum A short, erect, branched perennial. Basal leaves pinnately lobed, short-stalked and hairy below, lanceolate and clasping the stem above. Flowerheads bright sky-*blue*, 25–40 mm across, borne in narrow, leafy, branched spikes. Bare and grassy habitats. L, F, C, T, G, H, P.

Scolymus SPANISH OYSTER PLANT

Stout, spiny, thistle-like perennials. Flowerheads with outer bracts leaf-like and grading into the true upper leaves. Achenes flattened, not beaked, with pappus absent or composed of a few rigid hairs.

Scolymus hispanicus CARDILLO, TAGARDINA, SPANISH OYSTER PLANT A robust, spiny, biennial or perennial to 80 cm with interrupted *spiny-winged* stems. Lower leaves yellowish green, oblong, pinnately lobed with sparse spines; upper leaves smaller and spinier; leaves with paler veins and border. Flowerheads golden-yellow, 25–40 mm across, *borne in long, narrow, spike-like panicles*; *florets rayed* (not tubular); flower bracts slightly hairy or hairless, with membranous margins, and narrowed into a sharp point. Frequent on fallow ground, roadsides and sandy waste ground. C, T,

G, H, P. ***S. maculatus*** CARDO DE LECHE is similar but with *broader wings 2–5 mm at the narrowest, with prominent white margins*. L, F, C, T, G, H, P. ***S. grandiflorus*** is distinguished by its *markedly hairy* oval to linear flower bracts. T.

Tolpis LECHUGUILLA

Annual or perennial herbs. Flowerheads with *very long, narrow, curved-spreading bracts*. Pappus typically with long, rough hairs.

Tolpis umbellata TOLPIS A variably sized, somewhat hairy annual to 75 cm with slender, spreading, branched stems. Leaves linear–lanceolate, toothed, lobed, or untoothed. Flowerheads *pale yellow, often with purplish-brown inner florets,* to 20 mm across, borne on thickened stalks; florets all rayed; *thread-like bracts spread untidily beneath the flowerheads*. Pappus with rigid hairs. Disturbed places. L, F, C, T, G, H, P.

Tolpis webbii A woody-based perennial with persistent leaf bases, branching at the base. Basal leaves sub-erect, linear to linear–lanceolate, toothed to unequally lobed, hairy, becoming smaller above. Flowerheads yellow, 10 mm, borne in lax corymbs. Mountain rocks. *T (Las Cañadas del Teide).

Tolpis crassiuscula A perennial with flat rosettes of leaves; leaves lanceolate with toothed to lobed margins, fleshy and glaucous. Inflorescences lax and few-flowered; flowerheads large, to 20 mm, golden yellow with a darker centre. Sea cliffs and barrancos; very rare. *T. ***T. coronopifolia*** is a similar endemic of P.

Tolpis lagopoda A small shrub with a woody stock. Leaves small, more or less entire or with wavy margins, thin (not succulent), slightly hairy to hairless. Inflorescences lax and few-flowered; flowerheads small, 10 mm. Pine forest cliffs; locally frequent. *C, T.

Tolpis laciniata A woody-based perennial with divergently branched, weak stems. Rosette leaves ovate–lanceolate, twice lobed, 30–80 mm, hairless or minutely grey-hairy. Flowerheads to 25 mm. Locally common in dry forests. ***T. santosii*** is similar but with *stout, erect* stems up to 30 mm across when mature. Leaves not deeply divided and flowerheads 30 mm. Coastal rocky habitats. *P (north to northeast). ***T. proustii*** A woody-based perennial, similar to *T. laciniata* but with lanceolate–ovate, rather large leaves with very broad lobes; stems and leaf stalks *white-woolly*. Dry cliffs and rocks in the west. *H.

Hedypnois

Small annuals with rosette leaves. Flowerheads yellow, with *fleshy bracts in a single row, persisting and encircling the fruiting heads*. Achenes ribbed, not flattened, not beaked; pappus of scales.

Hedypnois rhagadioloides (syn. *H. cretica*) A variable, more or less hairy annual to 45 cm with mostly basal leaves, and *many, branched stems*. Leaves narrowly elliptic, entire to deeply lobed, with winged stalks below; stalkless above. Flowerheads 5–15 mm across, yellow and dandelion-like, borne on stalks thickened immediately below; involucre with narrow linear–lanceolate bracts *strongly incurved in fruit*. Disturbed habitats; common. L, F, C, T, G, H, P. ***H. arenaria*** is similar but with flowerstalks scarcely thickened, and involucral bracts *not*, or only slightly, incurved in fruit. Coastal sands. F, T. P.

1. *Scolymus hispanicus* (OLIVER WHITE)
2. *Tolpis umbellata*
3. *Tolpis umbellata*
4. *Tolpis webbii* (OLIVER WHITE)
5. *Tolpis crassiuscula*
6. *Tolpis coronopifolia*
7. *Tolpis lagopoda*
8. *Tolpis santosii*
9. *Tolpis proustii* (OLIVER WHITE)

1. *Hedypnois rhagadioloides*
2. *Hedypnois rhagadioloides*
3. *Hedypnois rhagadioloides* (involucre)
4. *Hedypnois rhagadioloides* (involucre)
5. *Hypochaeris oligocephala*
6. *Tragopogon porrifolius* (in fruit)
7. *Sonchus asper*
8. *Sonchus asper*
9. *Sonchus oleraceus*
10. *Sonchus bourgeaui*

Rhagadiolus

Hairy annuals. *Achenes few, lobe-like, borne in a star-like formation*; pappus absent.

Rhagadiolus stellatus STAR HAWKBIT A rather weedy, coarsely hairy annual with branched, spreading stems to 40 cm. Leaves 25 mm–14 cm, oblong, sparsely toothed to lobed and indistinctly stalked. Flowerheads yellow, small to 10 mm across, long-stalked in lax panicles. *Fruiting head a star-shaped series of (5)7–8 long*, slender, *incurved*, lobe-like achenes 10–16 mm long. Fallow land. L, F, C, T, G.

Hypochaeris LECHUGUILLA, CAT'S EAR

Annual or perennial herbs with rosettes of leaves and solitary or few, branched stems. Flowerheads yellow, with rayed florets only; bracts in several overlapping rows; *receptacle with scales between the florets*. Achenes finely ribbed, beaked or not; pappus 1–2 rows of brownish hairs.

(a) Plants in grassy or wooded habitats.

Hypochaeris radicata COMMON CAT'S-EAR A perennial to 60 cm with almost *hairless, branched stems*, and a basal rosette of bristly hairy, wavy-toothed, oblong, dark-tipped leaves to 25 cm. Flowerstalks *thickened* below the flowerheads; flowerheads yellow, 20–40 mm across, with a bell-shaped involucre; *bracts hairless*, dark-tipped. Central achenes 8–17 mm, beaked (marginal achenes usually also beaked). Grassy habitats at high elevation. F, T, G, H, P. **H. glabra** SMOOTH CAT'S-EAR is similar but with virtually hairless, glossy leaves, and *small, partially closed flowerheads* 10–15 mm across with ray florets scarcely exceeding the bracts. Central achenes 6–9(14) mm (*marginal achenes not beaked*). L, F, C, T, G, H, P.

(b) Plants on cliffs.

Hypochaeris oligocephala A rosette cliff plant with a short, woody stock. Leaves lanceolate, virtually hairless, glossy, with undulate margins, coarsely toothed. Flowerheads few, involucre 15 mm long. Pappus with 12–14 unequal bristles. Very rare on basalt cliffs. *T (northwest).

Tragopogon GOAT'S BEARD

Annual or perennial herbs with a white latex when cut. Stems usually solitary. Leaves linear, often rush-like. Flowerheads with only ray florets. Fruiting head a large 'dandelion clock'; achenes ribbed, beaked; pappus with feathery and simple hairs.

Tragopogon porrifolius BARBÓN, SALSIFÍ, SALSIFY A variable annual or perennial, similar to the previous species, with broadly linear-lanceolate leaves with broad, sheathing bases, and with *robust flowerheads of purplish florets borne on stalks markedly swollen beneath the heads*. Achene with an equally long beak. Grassy places. C, T, G, P.

Sonchus CERRAJA, CERRAJÓN, SOW-THISTLE

Annual or perennial herbs or woody shrubs with a sticky latex when cut. Leaves pinnately lobed or wavy with spiny margins. Flowerheads yellow, only with ray florets. Achenes flattened, ribbed, not beaked; pappus with 2 or more rows of white, simple hairs. A complex genus in the Canary Islands.

(a) Annuals and small perennials that are herbaceous or woody only at the base.

Sonchus asper CERRAJILLA PICONA, PRICKLY SOW-THISTLE An erect, greyish or reddish, hairless (glandular above) annual to 1 m with simple or branched stems. Lower leaves sometimes pinnately lobed, with sharp, triangular toothed lobes, the lowest 2 *rounded* at the base, and *glossy* green on the upper surface; clasping the stem above. Flowerheads golden yellow, to 25 mm across. Achene thin, winged. Very common in a wide range of habitats. L, F, C, T, G, H, P.

Sonchus oleraceus CERRAJILLA COMÚN, SMOOTH SOW-THISTLE An erect annual to 1.5 m similar to *S. asper*; greyish or reddish, hairless (glandular above). Lower leaves variously, often deeply *pinnately lobed* with triangular toothed lobes, the *end-lobe distinctly larger* than the next pair down, and the lowest 2 *pointed* (not rounded) at the base, *matt* (not glossy) green on the upper surface; leaves clasping the stem above. Flowerheads pale yellow, to 25 mm across. Very common in a range of habitats. L, F, C, T, G, H, P. *S. tenerrimus* is similar but annual or perennial with deeply pinnately divided leaves with all the leaf lobes strongly constricted or linear at the base, and the *terminal lobe equalling the laterals (not distinctly larger)*. Flowerstalks often white-hairy. Ligule longer than the tube. Disturbed habitats. L, F, C, T, G, H, P.

Sonchus bourgeaui CERRAJILLA MORUNA An erect annual to 70 cm, simple or branched. *Leaves slightly hairy towards the base*. Lobes more or less triangular and with irregularly dentate margins. Flowerheads numerous, *dense lanate at the base, 30–40 mm wide*. Occurring only in the eastern islands, in a range of habitats. L, F, C.

Sonchus webbii (syn. *Lactucosonchus webbii*) LECHUGUILLA DE WEBB A herbaceous, glabrous perennial, with hollow stems and leaves in a basal rosette that range from highly divided to almost entire. *Flowerheads small (<5 mm across)* with 12–16 yellow florets. A rare species of the pine forests in the north of La Palma at 400–600 m. *P. *S. esperanzae* (syn. *Lactucosonchus beltraniae*) LECHUGUILLA DE TIJARAFE is similar but has linear, undivided leaves and *larger heads comprising 18–25 florets*. Known only from pine forest at around 1,250 m in the Barranco de Briestas in the northwest of La Palma. *P

Sonchus tuberifer CERRAJILLA DE TENO Herbaceous perennial up to 30 cm tall with tuberous roots. *Leaves mostly basal*, divided or coarsely toothed, glabrous below and finely hairy above and clasping the stem. Heads in groups of 3–4. Crevices on shady cliffs in the Teno massif. *T.

(b) Pendulous shrub with many-headed inflorescences, the heads long, slender and narrow (2 mm wide) and with 5–8 florets.

Sonchus pendulus (syn. *Chrysoprenanthes pendula*) CERRAJA COLGANTE Leaves glabrous, lobes lanceolate to triangular, acute. Mountain cliffs from the south to the west of Gran Canaria at 500–1,000 m. *C.

(c) Shrubs with terminal leaf-rosettes on branches, leaves with few lobes and inflorescences with few heads (<6).

Sonchus brachylobus CERRAJA BRILLANTE Small shrub up to 80 cm tall. Leaves glabrous, the terminal lobe deltoid and larger than the rounded lateral lobes. Heads few, around 20 mm wide, in a tight cluster. Locally frequent along the north and west coast of Gran Canaria, often on cliffs. *C.

(d) Shrubs with terminal leaf-rosettes on branches, leaf lobes very slender, inflorescences with >6 heads, heads very narrow (<5 mm across)

Sonchus leptocephalus BALILLO ALPISPILLO Tall, slender shrub to 1.5 m. Leaves glabrous, pendant, flat, not stiff and with at least three pairs of narrow (<1 mm) linear–filiform lobes. Heads around 3 mm wide and each with 12–20 florets, grouped in large panicles. Frequent along the north coast

1. *Sonchus tuberifer*	4. *Sonchus leptocephalus*	7. *Sonchus arboreus*
2. *Sonchus pendulus*	5. *Sonchus microcarpus*	8. *Sonchus heterophyllus*
3. *Sonchus leptocephalus*	6. *Sonchus capillaris*	9. *Sonchus ortunoi*

of Tenerife, in habitats up to 600 m; also occurring in the north and northwest of Gran Canaria. *C, T. **S. sventenii** (syn. **S. filifolius**) BALILLO GOMERO is similar but the leaves are very long and the heads narrower, with only 8–10 florets. West and southwest of La Gomera, at 200–1,000 m. *G.

Sonchus capillaris BALILLO FINO Tall, slender shrub to 2 m. Leaves pendant, flat, not stiff and with at least 5–9 pairs of narrow (<1 mm) linear–filiform lobes. Heads up to 2 mm wide with 10–15 florets. Achene 2–2.2 mm long. Locally frequent in the west of Tenerife, at 200–700 m *T. **S. microcarpus** BALILLO TINERFEÑO DEL SUR is similar but the leaves are more or less ascending, leaf lobes stiff, few, linear–filiform, less than 1 mm wide. Heads 1–2 mm across. Achene less than 2 mm long. Fairly common in the south and southwest of Tenerife up to 600 m. *T.

Sonchus gandogeri CERRAJÓN DE EL GOLFO Shrub to 1.5 m. Leaves glossy, lobes linear, flat, 5–10 mm wide. Heads in groups of 20–60; each head 4–5 mm across. The El Golfo region of El Hierro and sporadically on cliffs in the southeast of the island. *H. **S. lidii** CERRAJÓN DE LID and **S. pitardii** CERRAJÓN DE PITARD have also been recognised as distinct by some authors, but may be hybrids between S. gandogeri and S. hierrensis.

Sonchus arboreus BALILLO ARBÓREO Tall shrub up to 2.5 m with thick stems. Leaf lobes flat, usually 2–3 mm wide. Inflorescences corymbiform with more than 50 heads. Heads 4–5 mm wide and with up to 30 florets. Occurring at 300–600 m on the Teno and Anaga massifs and Güímar in Tenerife, and La Palma. *T, P. **S. regis-jubae** BALILLO COPUDO is similar but a shrub to 1.5 m with slender stems. Leaf lobes up to 5 mm wide. Heads few (less than 40), 5–6 mm wide with more than 30 florets. Very rare around Roque Cano. *G. **S. wildpretii** BALILLO DE AGANDO is also similar but with very long leaves up to 40 cm, lobes linear–filiform, obtuse or rounded. Inflorescences up to 250 narrow heads with up to 50 florets. Rare plant of the lower limits of the laurel forests in the north of La Gomera. *G.

Sonchus heterophyllus (syn. **Taeckholmia heterophylla**) Shrub up to 1 m. Leaves pinnatisect, very variable. Inflorescences branched not corymbiform. Heads 3 mm wide. Roadsides and cliffs at 200–600 m in the north of La Gomera. *G.

(e) Shrubs with terminal leaf-rosettes on branches, leaf lobes broad, inflorescences with >6 heads, heads at least 1 cm across.

Sonchus platylepis (syn. **Babcockia platylepis**) Leaves with a distinct white covering, lobes triangular, acute with a rounded sinus. Heads few, large *up to 6 cm across. Involucral bracts large and broadly ovate*. Mountains of the central region of Gran Canaria at 800–1,600 m. *C.

Sonchus ortunoi CERRAJÓN GOMERO Robust shrub to 1 m. Leaves glabrous, green, lobes deltoid–rounded. Heads few, *4–5 cm across. Involucral bracts ovate–lanceolate*, margin wavy. Locally frequent in the south and southwest of La Gomera, at 200–1,000 m. *G.

Sonchus hierrensis CERRAJÓN HERREÑO Robust shrub to 1.5 m. *Leaves tomentose*, lobes up to 20 mm wide. Heads 5 cm across, hairy, in large, many-headed inflorescences. Widespread throughout El Hierro and La Palma; in La Gomera, restricted to the north of the island. *G, H, P.

Sonchus congestus PIPE, CERRAJÓN Shrub to 1.5 m. Leaves glabrous, the lobes with serrate margins and acuminate apices. *Heads 2–2.5 cm wide, hairy*, in large, many-headed inflorescences. Widespread in forest zones of the north coast of Tenerife and in the central and northern regions of Gran Canaria. *C, T.

1. Sonchus hierrensis (OLIVER WHITE)
2. Sonchus hierrensis
3. Sonchus hierrensis
4. Sonchus congestus
5. Sonchus congestus (flower heads)
6. Sonchus pinnatifidus (habit)
7. Sonchus pinnatifidus
8. Sonchus palmensis (OLIVER WHITE)

Sonchus pinnatifidus CERRAJÓN DE RISCO Shrub to 1.5 m. Leaves glabrous, often with *small bracts on the petiole*, hardly lobed to lobed; lobes (where present) broadly triangular (rarely narrowly triangular), acute. Heads glabrous or sparsely hairy, not congested. L, F.

Sonchus palmensis CERRAJÓN PALMERO Tall shrub to 2 m. Leaves lacking bracts on the petiole, with 10–15 pairs of flat, obtuse or rounded lobes 6–40 mm wide. *Heads narrow, <5 mm, with up to 50 florets*, in large inflorescences. Widespread in La Palma in the lower and forest zones up to 1,000 m. *P.

Sonchus canariensis Shrub to 3 m. Leaves glabrous, lobes flat, oblong with acute apices. *Head up to 1 cm wide, with 100 or more florets*. Frequent in the west of Tenerife and rare in the north and west of Gran Canaria, up to 800 m. *C, T.

(f) Shrubs with leaves in basal rosette(s) on a short, woody stem.

Sonchus bupleuroides (syn. *Sventenia bupleuroides*) LECHUGÓN DE SVENTENIUS, HIJA DE DON ENRIQUE, CERRAJA DE DON ENRIQUE A glabrous perennial up to 30 cm tall, with a rosette of oblanceolate *entire leaves* with obtuse apices and a whitish covering on top of a short stem. Flowering stem with reddish-brown glandular hairs in the upper part. Flower heads yellow, around 1 cm across. Rare, on shady, humid rocks on high cliffs in the northwest of Gran Canaria at 600–800 m. *C.

Sonchus bornmuelleri CERRAJÓN BRILLANTE Leaf rosettes up to 1 m across, leaves green, slightly hairy at the base, *lobes rounded*. Inflorescence stem up to 80 cm long, with a few small bracts. Heads around 2.5 cm wide, densely hairy. Rare plant of north and east coastal cliffs on La Palma, up to 200 m. *P.

Sonchus acaulis CERRAJÓN DE MONTE Leaf rosettes up to 1 m across, leaves green, tomentose, *lobes acute*. Inflorescence stem very long, up to 1.5 m, lacking bracts. Heads around 5 cm across, densely white-hairy. Widespread in forests and xerophytic zones of Tenerife; in Gran Canaria, more frequent in the mountains at 500–1,600 m. *C, T, G (rare).

Sonchus fauces-orci CERRAJA DE ADEJE Stems very short, rosettes flat, less than 50 cm across. Leaves with a white dust covering, lobes irregular with acute apices. Heads 30–50 mm across. *Involucral bracts with black margins, extending down the petiole*. West of Tenerife at 300–600 m. *T.

Sonchus radicatus ANGOJA DE RISCO Stems very short, rosettes flat, less than 50 cm across, leaves with a white dust covering, deeply divided and with a *terminal lobe larger than the lateral lobes*. Heads in groups of up to 10, up to 3 cm wide. Involucral bracts not merging into the bracts on the inflorescence stem. Achene smooth up to 2.5 mm. Cliffs up to 600 m on the north coast of Tenerife. *T. *S. gomeraensis* (syn. *S. gonzalezpadronii*) CERRAJA GOMERA is similar but leaves deeply divided into *backward-pointing lobes*. Heads numerous, tomentose, 3–5cm across. Achene wrinkled. Locally frequent in the north of La Gomera at 400–1,200 m. *G.

Sonchus gummifer CERRAJA DE GÜÍMAR Stems up to 50 cm long, rosettes not flat. *Leaf lobes triangular, not overlapping*. Bracts of the inflorescence lobed. Heads few, around 2.5 cm wide. Cliffs up to 500 m in the southeast of Tenerife. *T. *S. tectifolius* ANGOJA DE ANAGA is like *S. gummifer* but the leaf lobes are *rounded and often overlapping*. Restricted to cliffs on the southern slopes of the Anaga peninsula in Tenerife. *T.

---

1. *Sonchus canariensis*
2. *Sonchus canariensis*
3. *Sonchus acaulis*
4. *Sonchus acaulis*
5. *Sonchus radicatus*
6. *Sonchus radicatus*
7. *Sonchus gummifer*
8. *Sonchus gummifer* (habit)
9. *Sonchus tectifolius*

Lactuca LETTUCE

Annual, biennial and perennial herbs with a white latex when cut. Flowerheads with a cylindrical involucre only with ray florets. Achenes flattened and ribbed; pappus with 2 rows of white simple hairs.

Lactuca serriola LECHUGA ESPINOSA, PRICKLY LETTUCE A tall, greyish, stiffly erect annual or biennial to 1.5(2) m, unbranched below, and branched above with whitish stems. Leaves *held stiffly erect*, oblong–lanceolate, entire to pinnately divided with distant lobes below and sharply toothed, more or less hairless but spiny along the midrib beneath, and along the margins; all leaves waxy and greyish. Flowerheads pale yellow and small, to 13 mm across, borne in a long, lax inflorescence. Achenes olive-grey, 3–4 mm (excluding beak). Waste ground. L, F, C, T, G, H, P.

Lactuca viminea LECHUGA DE BURRO, PLIANT LETTUCE A hairless biennial without prickles, often with numerous, erect stems 15–80 cm (1 m). Leaves dark grey-green, the lowermost pinnately lobed with linear–lanceolate toothed lobes, the stem leaves above lanceolate and more or less unlobed except at the base. Flowerheads pale yellow, borne in much-branched panicles; florets (4)5–8 per head. Achenes 6.5–12 mm. Local. P. *L. palmensis* is similar but with a cushion habit and blue florets. High mountains. *P.

Chondrilla

Flowerheads cylindrical with 8–10 long, inner bracts and a row of very short, leafy outer bracts; disk florets absent. Achene ribbed, beaked or beakless, with scales at the base of the beak; pappus of simple hairs.

Chondrilla juncea A greyish biennial, hairy, especially below, to 1 m. Stems normally solitary, stiff and *broom-like with few leaves*. Leaves oblong, lobed, withering below and linear and entire above. Flowerheads unstalked, in small clusters; florets all rayed. C, T.

Launaea

Shrubs and perennials of dry and arid habitats, often intricately branched, with spineless leaves that exude a latex when cut. Flowerheads with ray florets only. Pappus of feathery bristles.

Launaea nudicaulis LECHUGUILLA A perennial herb 40–50 cm with basal leaf rosettes and *many, divergently branched*, spreading to ascending flowering stems. Basal leaves 12–28 cm, with wavy-toothed to lobed margins; stem leaves few, smaller and semi-clasping the stems. Flowerheads yellow; involucre 10–16 mm long. Dry, arid habitats, flowering after winter rain; common. L, F, C, T, G, H, P.

Launaea arborescens AHULAGA, AULAGA, JULAGA A slightly mounded to open, *spreading shrub* with woody, spiny, zig-zagging stems branching at right angles. Leaves greenish, linear and alternate, soon withering and sparse. Flowerheads terminal and yellow. Abundant on lower ground. L, F, C, T, G, H, P.

---

1. *Lactuca serriola*
2. *Launaea nudicaulis*
3. *Launaea nudicaulis* (leaves)
4. *Launaea nudicaulis* (involucrum)
5. *Launaea arborescens* (habit); (inset) flower
6. *Urospermum picroides*
7. *Reichardia tingitana* (habit)
8. *Reichardia tingitana* (fruiting head)
9. *Reichardia tingitana* (involucrum)

Urospermum

Flowerheads with 7–8 bracts *all in 1 row* and fused at the base; only ray florets present. Achene beaked; pappus of 2 rows of plumose hairs.

Urospermum picroides An *annual* with bristly stems 20–60 cm. Leaves bristly, rather large to 20 cm long, and like sow-thistle; toothed to lobed, oblong below and linear–lanceolate above. Flowerheads yellow and dandelion-like, borne on stalks thickened below; involucre 25 x 28 mm, cylindrical. Pappus white and fluffy (plumose). Waste ground; common. L, F, C, T, G, H, P.

Reichardia CERRAJA

Annual or perennial herbs. Involucre pitcher-shaped with bracts in several rows, each with a white margin. Achenes 4–5-angled (at least the outermost); pappus of numerous rows of soft, simple hairs.

(a) Annuals or biennials, not woody-based.

Reichardia tingitana A variable, hairless *annual or biennial* with oblong, toothed, pinnately lobed leaves, often with white pimples; basal leaves broadly winged at the base, the stem leaves few and linear, ½-clasping the stem at the base. Flowerheads yellow, to 25 mm across, the *ray florets purplish at the base*, the outermost with a red stripe on the reverse; flower bracts to 15 mm long, with a membranous margin and hairless. Pappus *long, to 14 mm*. Hot, dry areas. L, F, C, T, G, H, P.

(b) Woody-based perennials.

Reichardia famarae A small cliff plant with a *woody stock* and dense rosettes. Leaves obovate to spathulate. Flowerheads solitary on short stems. Cliff crevices; rare. *L, F. *R. tingitana* × *R. famarae* hybrids occur infrequently in Famara (L).

Reichardia ligulata A *woody-based perennial* with lobed, hairless leaves with more or less spiny margins. Flowerheads borne on long stems with bracts. Locally abundant in rocky habitats at low elevation. *C, T, G, H, P.

Andryala ESTORNUDERA, PEORRERA

Flowerheads many in a cluster; bracts in a single row with some additional bracts; only ray florets present. Pappus of soft hairs.

Andryala integrifolia A variable, short to tall, white-hairy annual or perennial with sparingly branched, erect, leafy stems to 50 cm (1 m). Leaves oval–lanceolate, unlobed to lobed, semi-clasping the stem above, and densely covered in yellowish glandular hairs. *Flowerheads pale lemon yellow* (often with a darker centre), 20 mm across, borne abundantly in rather flat-topped clusters; flower bracts linear–lanceolate, and hairy; involucre 6–10 mm x 5–9 mm. Achenes 1 mm. Locally common in dry, sandy and grassy habitats. C, H, P.

Andryala pinnatifida A variable, silvery perennial, sometimes woody below, with white-woolly stems and leaves. Leaves ovate to lanceolate, variably toothed. Flowerheads borne in dense, flat-topped inflorescences. Common in a range of habitats. *C, T, G, H, P. Various forms described, including the common **subsp. *pinnatifida*** (C, T, G, H, P), and the rarer **subsp. *preauxiana*** (*C). *A. perezii* is similar but has leaves with undulate-crispate margins; longer peduncles; and smaller cypselae with a ring of short teeth at the apex. *L, F.

1. *Reichardia tingitana*
2. *Reichardia tingitana* (leaf)
3. *Reichardia famarae*
4. *Reichardia ligulata*
5. *Andryala integrifolia*
6. *Andryala integrifolia* (left); *A. pinnatifida* (right)
7. *Andryala pinnatifida* subsp. *preauxiana*
8. *Andryala pinnatifida* subsp. *pinnatifida*

Taraxacum DIENTE DE LEÓN, DANDELIONS

Perennial herbs with tap roots, a basal rosette of leaves, and flowerheads borne on leafless, hollow stems that exude a milky latex. Achenes finely ribbed; pappus of white simple hairs. Only one, cosmopolitan species section is described here. However, *Taraxacum* in its wider sense comprises a complex aggregate of numerous microspecies.

Taraxacum sect. **Ruderalia** (syn. *T. officinale*) DIENTE DE LEÓN, COMMON DANDELION A variable (group of) perennial(s) with leaves in a basal rosette and leafless, hollow stems. Leaves deeply pinnately lobed, with long, winged stalks at the base. Flowerheads bright yellow, the rays striped with purple beneath, on stalks to 40 cm; outer flower bracts 9–16 mm, backwardly curved. Fruiting head a familiar 'dandelion clock'; achenes brown, 2.5–4 mm. Disturbed habitats. C, T, G, H, P.

Crepis HAWK'S-BEARD

Annual or perennial herbs with spirally arranged leaves with lobes pointing backwards and erect, branched stems. Flowerheads yellow with florets all rayed; flower bracts in 2 rows, the outer row often shorter and spreading. Achenes ribbed and flattened; pappus of white, brittle hairs.

Crepis canariensis LECHUGUILLA DE RISCO A perennial to 30 cm. Basal leaves in rosettes, to 18 cm long, obovate–lanceolate to elliptic, hairless, irregularly toothed, with winged stalks; upper stem leaves triangular. Flowerheads 10–15 mm, borne in corymbs; stalks finely glandular-hairy. Achenes light brown, rough. Coastal rocks and interior slopes; very locally frequent in the east. *L, F.

Crepis vesicaria BEAKED HAWK'S-BEARD A robust, *hairy perennial* that has leaves with broad lobes, downy all over. Flowerheads to 15 mm across, with *orange-yellow* florets, the outer striped reddish externally; outer involucral bracts spreading; heads erect in bud. Achenes with *long, slender beaks*, 5–9 mm; fruiting pappus much exceeding the involucre. Meadows and waste places. **Subsp. *taraxacifolia*** is the form on the Islands. C, T.

Crepis foetida A very variable, erect to spreading, *strong-smelling* annual or biennial to 50 cm with few branches. Leaves bristly, with saw-like divisions or 1–2-pinnately lobed. Flowerheads drooping in bud, 1–many, yellow, the ray florets purplish on the outer surface; involucre 7–16 mm long. Achenes of 2 kinds, the outermost robust, beaked or not, the inner *longer and slender*, (10)12–17 mm with a slender beak; pappus protruding from the fruiting head. Common. C, T, G, H, P.

Crepis neglecta (syn. *Crepis capillaris*) CHICORIA LOQUILLA A slender annual to 30 cm with basal leaves in a rosette, lyre-shaped, saw-toothed or simply toothed; stem leaves few, clasping, the uppermost bract-like. Flowerheads yellow, nodding in bud; involucre 4–7 mm. Achenes spindle-shaped, 10-ribbed with a variably stout beak; pappus as long as the achene. Grassy habitats, roadsides. C, T, P.

Helminthotheca

Annuals to biennials similar to and previously grouped with *Picris*. Flowerheads with *oval outer bracts, much wider than the inner*. Achenes long-beaked; pappus white and feathery.

Helminthotheca echioides (syn. *Picris echioides*) RAPASAYA, BRISTLY OXTONGUE A robust *very bristly* annual or biennial to 80 cm; *bristles arising from pimple-like bases*. Leaves elliptic to oblong, wavy-edged and pimply, the lower with winged stalks, the upper clasping the stem. Flowerheads yellow and numerous; *the outermost involucral bracts 3–5, large and oval–heart-shaped, resembling an epicalyx,* 15–20 mm x 7–10 mm. Pappus white. A common weed of waste places. L, F, C, T, G, H, P.

ASTERACEAE | 371

Lapsana NIPPLEWORT

Annuals to perennials with leafy stems. Flowerheads many. Achenes flattened, not beaked; pappus absent.

Lapsana communis HIERBA PEZONERA An erect annual to 1 m, hairy below. Lower leaves narrowly obovate to oblanceolate, more or less pinnate with lateral leaflets exceeded by the terminal. Flowerheads small, yellow, diffuse. Thickets and woods in the west. T, P.

1. *Crepis canariensis*
2. *Crepis canariensis* (leaf)
3. *Crepis vesicaria* subsp. *taraxacifolia*
4. *Helminthotheca echioides*
5. *Lapsana communis*

SUBFAMILY CARDUOIDEAE

Plants *without* white latex, often *spiny* (thistle-like), and stem leaves spirally alternate (or absent). Flowerheads discoid (with tubular flowers only); the outer flowers often longer with larger lobes (appearing ray-like), often *red to blue or white* (rarely yellow).

Carlina MALPICA, CARDO DE CRISTO, CARLINE THISTLE

Shrubs. *Flowerheads large, without ray florets, but bracts conspicuous and spreading*; long-persisting. Pappus of feathery hairs.

Carlina xeranthemoides A silvery shrub to 50 cm, branching below. Leaves linear, woolly on both sides, with spiny margins. Flowerheads pale sulphur-yellow, borne on branches with bracts. Volcanic slopes. *T (Las Cañadas del Teide – common).

Carlina canariensis A shrub rather like the former but rather less silvery with broader, linear–lanceolate, remotely spiny leaves. Flowerheads large, dull yellow, 13–15 mm across. Cliffs and slopes mainly in the South. *C.

Carlina salicifolia A shrub to 1 m. Leaves narrowly lanceolate, densely hairy beneath, virtually hairless above; margins spiny. Flowerheads cream to dull yellow, *large,* 20–35 mm across. Frequent on cliffs and around forests. L, F, C, T, G, H, P. The following forms are recognised by some: **subsp.** *lancerottensis* which has spineless, broader leaves (*L) and **subsp.** *salicifolia* (on all the other islands).

Atractylis

Similar to *Carlina*. Inflorescence with leafy bracts deeply cut into spiny teeth, the innermost bracts papery-tipped, not brightly coloured. Pappus silvery-haired.

(a) Small perennials.

Atractylis cancellata CARDO ENREJADO A slender, thistle-like perennial with leaves in a lax rosette and with stems 30 mm–30 cm. Leaves hairy, oblong in outline, toothed and with short spines. Flowerheads large, pinkish-purple without ray florets, with an involucre to 20 mm, surrounded by upper leaves forming a *cage-like structure*; middle bracts with all spines similar. Common in dry fields, on roadsides and fallow land. L, F, C, T, G, P.

(b) Shrubs or compact shrublets.

Atractylis preauxiana PIÑA MAR A compact, silvery shrub to just 10 cm. Leaves entire, borne in loose rosettes, linear, felted. Flowerheads with few ray florets with distinct teeth at the tips, pinkish-cream; bracts brownish-black with soft, spine-like tips. Rocky coasts; rare. *C, T. **A. arbuscula** is similar but with a taller, less compact habit and broader leaves. Spine-like tips longer. Coastal rocks; very rare. Forms include: **subsp.** *arbuscula* (*L – Famara) and **subsp.** *schizogynophylla* (*C – north coast).

1. *Carlina xeranthemoides*
2. *Carlina canariensis*
3. *Carlina salicifolia*
4. *Carlina salicifolia* subsp. *lancerottensis*
5. *Carlina salicifolia* subsp. *lancerottensis* (fruiting head)
6. *Carlina salicifolia* subsp. *salicifolia*
7. *Atractylis cancellata*
8. *Atractylis preauxiana*
9. *Atractylis arbuscula* subsp. *arbuscula*

Carduus CARDO, THISTLE

Annuals to biennials with spiny-winged stems (at least in part), and alternate, spine-toothed leaves. Flowerheads rounded or cylindrical, often shaving-brush-shaped; ray florets absent. *Pappus with many rows of simple hairs*.

Carduus tenuiflorus SLENDER THISTLE An erect, narrowly branched biennial to 60(80) cm with stems *broadly winged* up to the flowerheads, grey to white-cottony. Flowerheads slender, 12–18 mm long and 5–10 mm across, borne in dense terminal clusters of 3–10; florets pale pink-red; bracts at least 1.5 mm wide. Disturbed habitats; common. L, F, C, T, G, H, P.

Carduus pycnocephalus An erect, narrowly branched biennial to 1 m with stems winged up to the flowerheads, similar to *C. tenuiflorus* but with leaves more densely white-cottony and stems not leafy and with interrupted spiny wings below the flowerheads; *flowerheads solitary or in clusters of 1–3;* 14–20 mm long and 7–12(14) mm across. Disturbed habitats. F, C, T, G, P.

Carduus baeocephalus An unbranched, robust biennial. Leaves shallowly wavy-lobed, *cobweb-hairy* with spiny margins; those below lanceolate, those above linear. Inflorescences with few flowerheads, *whitish* to pale pink with *dark pink anthers and stigmas*; bracts with short spine-tips. Coastal malpaís and barrancos. *C, H. **Subsp. *microstigma*** is the widespread form. *C. volutarioides* is similar but with cylindrical flowerheads entirely white, with white (not pink) anthers and stigmas. Disturbed habitats on the north coast at 10–250 m. Local. *T.

Carduus clavulatus A small herb with stems *scarcely winged*. Leaves *sparsely hairy*, ovate, entire, the uppermost ovate–lanceolate. Flowerheads few, *deep pink to mauve*. Forest margins and roadsides; fairly frequent. *F, C, T, G, H, P.

Cirsium THISTLE

Biennials and perennials similar to *Carduus*, with or without spiny wings. Flowerheads purple (or yellow). Pappus of many rows of *branched, feathery hairs* (not rough, unbranched hairs).

Cirsium vulgare CARDO, SPEAR THISTLE An erect biennial to 1.5 m with cottony stems that have *interrupted spiny wings*. Basal leaves to 30 cm long, deeply pinnately lobed with segments forked and spiny, and a *single long, pointed end-lobe*; upper leaves smaller, all leaves prickly. Flowerheads ovoid, 25–40 mm long and 20–50 mm across, with cottony bracts; the outer bracts with long spine-tips; florets reddish-purple. Disturbed habitats at high elevation; expanding locally. T.

Notobasis

Similar to *Carduus* but with *flowerheads encircled by tough, spiny upper leaves*.

Notobasis syriaca CARDO DE SIRIA, SYRIAN THISTLE A thistle-like annual 20 cm–2 m with rigid stems *not* spiny-winged. Leaves alternate, dark green with paler veins; narrowly elliptic, pinnately lobed with spine-tipped, narrow triangular lobes, reduced and *clustered* around the stalkless flowerheads above. Flowerheads solitary, purple; bracts with spine-tips; involucre 18–23 mm long. Achenes 67 mm long. Field margins, fallow land and roadsides. L, C.

1. *Carduus tenuiflorus*
2. *Carduus pycnocephalus*
3. *Carduus baeocephalus*
4. *Carduus baeocephalus* subsp. *microstigma* (LEOPOLDO MORO)
5. *Carduus volutarioides*
6. *Carduus clavulatus*
7. *Cirsium vulgare*
8. *Galactites tomentosus* (involucrum)

Galactites GALACTITES

Thistle-like perennials with deeply dissected leaves. Flowerheads without ray florets, but with disk florets much spreading at the margins. Pappus long and feathery.

Galactites tomentosus CARDO BORRIQUERO, GALACTITES A stiffly erect perennial 30–50 cm with conspicuously *white-veined and variegated* dark green leaves; alternate, oblong, pinnately lobed with spiny lobes, *white-downy beneath*. Flowerheads pale purple, borne solitary or in clusters; the outer *ray florets long, even and spreading*; flower bracts tapered abruptly into spine tips; white-downy; 12–18 mm long. Achenes 3.5–5 mm long. Common in a range of dry and disturbed habitats. C, T, G, H, P.

Onopordum SCOTCH THISTLE

Stout perennials with *spiny-winged stems*, often with cobweb-like hairs. Leaves spiny-margined and toothed to lobed. Flowerheads large and purple or white; all florets tubular and deeply 5-lobed. Pappus of many rows of simple hairs.

Onopordum nogalesii A robust perennial. Leaves white-felted, wavy-lobed with spiny margins. Flowering stems winged; flowerheads to 50 mm across, purplish; bracts with long spine-tips. Achenes rough, yellowish. Very local; dry slopes. *F (Jandía). *O. carduelium* has larger flowerheads with pink florets. Very rare; mountains. *C (centre).

Cynara ARTICHOKE

Stout perennials with leaves in a basal rosette or alternate; deeply divided into spiny segments. Flowerheads borne solitary or sparingly; purplish, blue or white. Pappus of many rows of feathery hairs.

Cynara cardunculus ALCACHOFA SILVESTRE, CARDOON A large, bushy, *greyish or whitish* perennial to 1.8 m with numerous basal leaves; stems white-hairy and unwinged. Leaves thick, lanceolate in outline, deeply 1–2-pinnately lobed into *lanceolate* (not narrow and linear), *flat* segments; toothed, shortly and sparsely hairy and green above, white-hairy beneath; lower leaves short-stalked and upper leaves stalkless; leaves with *long spines to 35 mm clustered* at the base of each segment. Flowerheads large, 35–95 mm with bracts in 5–8 rows, narrowed into spreading spine-tips, basally brown; florets violet-purple. **Var. *ferocissima*** is the common form. Dry hillsides; frequent. L, F, C, T, G, H, P.

Cynara humilis A bushy, dark grey-green perennial to 80 cm with numerous basal leaves; stems white-hairy and unwinged. Leaves lanceolate in outline, *deeply 2-pinnately divided into linear segments with downturned margins*, hairless and somewhat shiny above, white-hairy beneath; lower leaves short-stalked and upper leaves stalkless. Flowerheads large; involucre to 60 mm long, brownish when mature with purplish *bracts that have prominent spine-tips*; florets purplish. Dry fallow ground; rare. C.

Silybum MILK THISTLE

Robust annuals or perennials with spineless stems. Flowerheads with *filaments fused at the base to form a tube*. Pappus with many rows of simple hairs.

Silybum marianum CARDO MARIANO, MILK THISTLE A robust, weakly spiny biennial to 1 m. Basal leaves oblong, pinnately lobed and prominently *white-veined and variegated*, virtually hairless and stalked beneath; stem leaves smaller, with fewer lobes and clasping the stem. Flowerheads purple, 25–40 mm x 50 mm (14 cm), borne solitary and terminally; bracts terminating in long, stout spines to 7 mm. Common and widespread on fallow, cultivated, waste ground and field margins. L, F, C, T, G, H, P.

---

1. *Onopordum nogalesii*
2. *Cynara cardunculus* (habit)
3. *Cynara cardunculus* (in fruit)
4. *Cynara cardunculus* var. *ferocissima*
5. *Silybum marianum*
6. *Centaurea melitensis*
7. *Centaurea calcitrapa*
8. *Volutaria bollei*
9. *Volutaria tubuliflora*
10. *Volutaria canariensis*

Centaurea CORNFLOWER, KNAPWEED

A large genus of summer-flowering annual or perennial herbs with alternate simple, entire to pinnate leaves. Flowerheads purple, pink, white or yellow; bracts with an apical appendage (bract characteristics important for identification). Pappus absent (or simple to toothed hairs or scales).

(a) Flowerheads yellow.

Centaurea melitensis ABREPUÑO A white-hairy, much-branched annual with winged stems to 60 cm, grey woolly-hairy or felted. Lower leaves pinnately lobed or toothed, often *withered* during flowering. Flowerheads yellow; bracts terminating in short spines, 5–8 mm. Pappus equalling (not exceeding) the achene. Locally frequent. L, F, C, T, G, H, P.

(b) Flowerheads pink.

Centaurea calcitrapa ABROJO, RED STAR-THISTLE A medium, much-branched, virtually hairless perennial 50 mm–50(80) cm with grooved stems. Leaves grey when young, glandular, remotely shallowly to deeply pinnately lobed with bristle-pointed lobes, often withered below when in flower; upper leaves smaller and narrower. Flowerheads *reddish-purple*, 12–17 mm with equal florets, surrounded by conspicuous *spreading, long and star-like spines 15–30 mm, ›3 x as long as the longest laterals*; involucre ovoid–cylindrical. Pappus absent. Waste places, roadsides, bare, sandy ground. L, F, C, T, G, H.

Volutaria CARDOMANSO

Small annuals like *Centaurea* but with bracts without apical appendages. Hairs in the corolla tube an important diagnostic.

Volutaria bollei An erect, leafy annual to 45 cm with leaves extending along the stems in wings (decurrent); leaves green, ovate, *coarsely toothed*. Flowerheads 21 mm across, few, *white*. Volcanic rocks. *L, F.

Volutaria tubuliflora An annual to 80 cm with *deeply pinnately lobed* leaves forming loose basal rosettes; those above decurrent; leaves to 90 mm. Flowerheads small (10–15 mm across), *pink*; central florets containing adpressed, silky hairs. Achenes brown, ribbed. Disturbed habitats and roadsides; frequent. L, F, C, T, H.

Volutaria canariensis A small, *white-woolly* annual. Leaves deeply pinnately lobed. Flowerheads small with white ray florets and purplish central florets with long hairs within the tubes. Volcanic rocks. *C, T, G, H, P.

Cheirolophus CABEZÓN

Perennials like *Centaurea* with leaves lobed or unlobed; spineless. Flowerheads borne on long stalks that are swollen beneath the involucre. Outer achenes without a pappus. At least 16 similar species with limited distributions, only a few are described here.

Cheirolophus canariensis A shrub to 1.5 m. Leaves pinnately lobed with 4–5 pairs of linear–lanceolate lobes, sometimes almost entire. Flowerheads whitish-mauve; bracts large. Basalt cliffs; rare. *T. *C. burchardii* is very similar but with lanceolate, pale green, *entire* leaves; margins sometimes slightly toothed. Rocky shelves and cliffs; very rare. *T (Teno massif).

Cheirolophus teydis A profusely branched shrub. Leaves narrowly oblanceolate, with serrated margins. Flowerheads small, dull cream, borne on long slender stems. High mountains, in volcanic rocky slopes. *T, P.

Carthamus

Very spiny, often glandular *annuals*. Leaves pinnately lobed with spiny margins. Flowerheads solitary, surrounded by spiny, leaf-like bracts, yellow to orange. Pappus absent or *a series of narrow, pointed, persistent scales*.

Carthamus lanatus ALAZOR A very spiny, thistle-like annual 15–60 cm (1 m) with straw-coloured stems unbranched below, branched above, covered in white-woolly hairs when young. Leaves lanceolate, pinnately lobed with a spiny margin, withered below during flowering; clasping the stem above. Flowerheads 30 mm across, *yellow*; bracts with a spine-toothed appendage. Pappus a series of scales as long as the achene; achene 4.5–6 mm. Common in bare, dry and sandy places. L, F, C, T, G, P.

Carduncellus

Closely related to *Carthamus* but *woody, many-stemmed perennials*. Flowerheads yellow or purple. *Pappus a series of persistent or shedding bristles*.

Carduncellus caeruleus (syn. *Carthamus caeruleus*) AZAFRÁN AZUL A greyish, hairy perennial 20–80 cm with normally unbranched, erect, unwinged stems. Leaves rather shiny, grey-green, lyre-shaped and toothed or untoothed to pinnately lobed with bristle-tips; upper leaves semi-clasping the stem. Flowerheads blue-purple, borne solitary, surrounded by leafy bracts; involucre 16–30 mm x 20–30 mm; florets 14–22 mm, tubular and *deeply 5-lobed*. Fallow land and roadsides. L, F, C, T, G, P.

ADOXACEAE

A small family of perennials with opposite, toothed leaves. Inflorescences cymose; stamens 4–5; style 0 or 1. Fruit a 1–several-seeded drupe (rarely an achene). This family now includes species that were previously described under Sambucaceae or Caprifoliaceae.

Viburnum

Deciduous or evergreen shrubs with simple leaves. Flowers numerous, in compound clusters; stamens 5; style 0. Fruit a succulent 1-seeded drupe.

Viburnum rugosum FOLLAO An evergreen shrub to 5 m. Branches brown to reddish below, erect, more or less hairless, Leaves dark green, ovate, short-stalked, sparsely to densely hairy on both sides. Flowers white, 7 mm, pinkish in bud, borne in dense, flattened heads 10–15 cm across. Fruit a berry 6–7 mm, brownish to blue-black when ripe. Forests and heaths; locally common. *C, T, G, H, P.

Sambucus ELDER

Deciduous shrubs or perennials with pinnately divided leaves. Flowers borne in compound clusters; stamens 5. Fruit succulent, with 3–5 seeds.

Sambucus palmensis SAUCO A shrub to 5 m with arching branches and corky bark. Leaves with 3 pairs of lateral leaflets and the terminal largest; margins toothed; silky-hairy beneath. Flowers white, 5 mm, borne in umbels 10 cm across. Fruit small, brownish-black, 6 mm. Laurel forests; very rare and local. *C, T, G, P.

1. *Cheirolophus burchardii*
2. *Cheirolophus teydis*
3. *Carthamus lanatus*
4. *Viburnum rugosum* (habit)
5. *Viburnum rugosum*
6. *Viburnum rugosum*
7. *Sambucus palmensis*
8. *Scabiosa atropurpurea*

CAPRIFOLIACEAE   HONEYSUCKLE FAMILY

Woody shrubs and climbers with opposite, simple leaves. Flowers solitary, paired or in showy panicles on shrubs; calyx small; corolla regular or 2-lipped, fused below to form a tube; stamens mostly 5; ovary inferior. Fruit a berry or nutlet.

Lonicera HONEYSUCKLE

Deciduous or evergreen shrubs or climbers with simple, sometimes lobed leaves. Flowers stalkless, with a zygomorphic, 5-lobed corolla; stamens 5; style long. Fruit a several-seeded berry.

Lonicera etrusca MADRESELVA A deciduous, shrubby and woody climber to 3 m with leathery leaves, usually hairy beneath, and inflorescences with clusters of flowers in groups of 3 *on stalks 30–40 mm long*. Fruit 5.2–6.2 mm, ovoid and red. Thickets; uncommon. T.

Scabiosa SCABIOUS

Annual or perennial herbs with simple or pinnately divided leaves, the lowermost often in a rosette. Flowerheads flat or domed and long-stalked; outer flowers longer than the inner; calyx with 5 long bristles; corolla 5-lobed.

Scabiosa atropurpurea FLOR DE VIUDA, MOURNFUL WIDOW A hairy, bushy annual or biennial to 70 cm with erect, branched stems. Leaves oblong, untoothed and long-stalked below, pinnately lobed above. Flowerheads lilac to reddish-purple (often mixed in populations), 15–35 mm across, becoming oblong in fruit, the outer florets 2 x the size of the centrals; involucral bracts *not* longer than the florets. Fruits with long calyx teeth. Disturbed ground. C, T, G, H, P.

Dipsacus

Robust, prickly biennials and biennials with entire or pinnately divided leaves. Flowerheads very dense; corolla 4-lobed; stamens 4; style 1.

Dipsacus fullonum WILD TEASEL A tall biennial to 2 m with robust, prickly, angled stems. Basal leaves forming a large rosette; oblong–elliptic, toothed or not *and prickly*; stem leaves in pairs and fused together on the stem forming a water-collecting cup. Flowerheads 40–80 mm, pink-purple, oblong-cylindrical with long, spiny bracts at the base. Local in damp places. T.

FORMER VALERIANACEAE

This family is now considered part of the Caprifoliaceae.

Centranthus

Rhizomatous annual or perennial herbs with entire to deeply pinnately lobed, stalked or unstalked leaves. Flowers small, borne in dense panicles, pink or white; corolla tubular; stamen 1; stigma 1.

Centranthus calcitrapa A small, hairless annual with simple or branched stems to 30 cm. Leaves rounded to oval, green, flushed with purple, toothed or not, stalked below and stalkless above. Flowers *small*, with corolla tube 1–4.6 mm, white or pink, long, and short-spurred, borne in small terminal clusters. Fruit a feathery, persistent calyx. Sandy and disturbed places. C, T, G, H, P.

Centranthus ruber MILAMORES, RED VALERIAN A tufted, somewhat fleshy and waxy perennial to 80 cm. Leaves bluish, oval, pointed or blunt, clasping the stem above. Flowers with tube 5–12 mm, usually dark pink-red (sometimes white), spurred, borne in large, slightly fragrant, showy panicles. Fruit 1-seeded with a feathery, persistent calyx. Commonly planted and naturalised along roadsides and in towns, malpaís, and fields. L, F, C, T, G, H, P.

Valerianella CORNSALAD

Small annuals with symmetrical branches. Flowers borne in the branch axils and in terminal clusters; stamens and stigmas 3. Species rather similar; ripe fruits and calyx necessary for identification.

Valerianella locusta CANÓNIGO, COMMON CORNSALAD A small, variable, hairless annual to 15(40) cm with symmetrical, spreading branches, spoon-shaped, blunt, sometimes toothed leaves to 70 mm and dense terminal heads of pale lilac flowers; corolla 5-lobed; *calyx very small and 1-toothed*. Fruit hairless, 1.8–2.5 mm long and 1–1.5 mm across (scarcely longer than thick), shallowly grooved. Grassy habitats. C, T.

Valerianella eriocarpa CANÓNIGO ROJO, HAIRY-FRUITED CORNSALAD A small annual to 15(40) cm similar in appearance to *V. locusta* but with denser, compact flower clusters *and a distinctive calyx nearly as broad as the fruit, strongly net-veined and deeply (2)5–6-toothed;* corolla 1–1.6 mm. Fruit 0.9–2 mm, with rigid hairs. Grassy habitats. C, T. *V. dentata* CANÓNIGO SIMPLE, NARROW-FRUITED CORNSALAD is similar to *V. eriocarpa* but with laxer flower clusters and calyx *just ½ as broad as the fruit and scarcely veined*. Fruit 1.5–2 mm. P.

Valerianella discoidea CANÓNIGO ROSADO A small annual similar to *V. locusta* with opposite-spreading stems and oval to narrowly spathulate leaves, blunt, toothed or untoothed. Flowers with corolla 1.5–2 mm, whitish-blue or mauve; *calyx well-developed, with many (6–12) spine-tipped lobes, hairy outside and in*. Fruit 1.5–2.1 mm. Fallow land. F.

Pterocephalus

Shrubs and woody-based perennials with hairy stems. Leaves entire, hairy. Flowerheads scabious-like, pink; calyx with long, feathery bristles; florets 5-lobed.

Pterocephalus lasiospermus ROSALITO DE CUMBRE A mounded shrub to 1 m with greyish stems. Leaves pale greyish-green, narrowly oblanceolate, hairy. Flowerheads pale pink, solitary, borne numerously, on long stalks. Rocky slopes at high elevation; very locally common. *T (Las Cañadas del Teide). *P. porphyranthus* ROSALITO PALMERO has broader leaves and *deep magenta* flowerheads. High mountain pine forests. *P. *P. virens* ROSALITO DE ANAGA is a *dwarf cliff plant*; habit compact, leaves green, spathulate; flowerheads pale pink. Coastal cliffs; rare. *T.

PITTOSPORACEAE

Trees, shrubs and lianas with simple, alternate leaves. Flowers with 5 sepals, petals and stamens; style 1. Fruit a 2–4-parted capsule.

Pittosporum

Trees and shrubs with simple (rarely lobed) leaves. Flowers with 5 sepals and petals; style 1. Fruit a woody capsule with numerous seeds.

Pittosporum undulatum AZARERO, PITOSPORO A shrub with broad, mid-green leaves 50 mm–15 cm, narrowed at the base, with *undulate (not inrolled)* margins. Fruit 10–14 mm, splitting 2 ways. Widely planted; locally naturalised. C, T.

ARALIACEAE | IVY FAMILY

Woody climbers, trees and shrubs with alternate, simple, often lobed leaves. Flowers small, often borne in umbels and 5-parted; style(s) 1(5). Fruit a drupe or berry with 2–5 seeds.

Hedera HIEDRA, IVY

Woody vines that climb by means of rootlets. Flowers borne in terminal umbels. Fruit berry-like.

Hedera helix IVY A vigorous evergreen climber or creeper with flexuous stems with adhesive roots. Leaves glossy, hairless and dark green, palmately lobed (>½ to the base), 13 cm across; leaves on flowering stems oval and unlobed. Flowers pale yellowish-green and borne in terminal, dense spherical umbels. Berry spherical and black when ripe, 7–8.3 mm across. Cultivated on walls and in gardens. F, C, T, G, P.

Hedera canariensis An evergreen climber, like *H. helix* but with minute, scale-like hairs with 10 rays, reddish brown (starry, large, white with <10 rays in the former species). Leaves broader than long, *unlobed or 3-lobed* on non-flowering shoots; those of flowering shoots sub-circular. Berries 8 mm across. Locally frequent in laurel forests. *C, T, G, H, P.

APIACEAE | CARROT FAMILY

A large and important family. Mostly herbs with alternate leaves, often pinnately divided with sheath-like bases. Flowers normally borne in very characteristic compound *umbels*; often green, yellow or white with 5 free petals and 0 or 5 sepals; stamens 5; styles 2, often arising from a swelling. Fruit a dry, 2-parted schizocarp (composed of 2 mericarps), often with a central carpophore. Fruit characteristics are an important diagnostic.

Smyrnium

Biennial herbs in which the 2–3 leaf lobes arise from a single point (ternate). Umbels yellow, the flowers lacking sepals. Fruit egg-shaped with slender ridges and oil glands, hairless.

Smyrnium olusatrum APIO CABALLUNO, ALEXANDERS A tall, pungent, hairless biennial to 1.5(2.2) m. Stem robust, hollow when mature, bearing 1–2-pinnately *divided leaves* with triangular, toothed lobes, borne on short, somewhat inflated stalks; shiny green below, yellow above. Umbels terminal, borne in the leaf axils with 7–15(18) unequal rays; flowers yellow, to 3 mm across without sepals. Fruit ovoid, 5–7 mm and *black* when ripe, conspicuously *ridged*. Damp habitats and thickets. C, T, P.

Scandix

Slender annual herbs with thread-like, feathery leaves. Flowers with white petals and with minute sepals. *Fruits conspicuous, long, needle-like* and with cylindrical beaks, hairless.

Scandix pecten-veneris PEINE DE VENUS, SHEPHERD'S NEEDLE A small, spreading, hairless annual to 40(50) cm with leaves 2–3(4)-pinnately divided, the segments widened towards the tips. Umbels simple or with 2–3 rays; *bracteoles divided into linear lobes*; flowers small and white; petals 2–4.5 mm. Fruits highly distinctive, borne in claw-like clusters: *elongated, 50–95 mm long*, at least ½ the length comprising the seedless beak, which is *flattened and distinct*. Common on disturbed and fallow ground. L, F, C, T, G, H, P.

Crithmum

Hairless, fleshy perennials with leaves 1–2-pinnately divided. Petals yellow-green. Fruit oblong, hairless, not compressed, and rather corky (spongy when fresh).

Crithmum maritimum PEREJIL DE MAR, ROCK SAMPHIRE A somewhat bushy, greyish, hairless perennial to 45 cm, woody at the base. Leaves 1–2-pinnately divided with almost cylindrical, untoothed, *fleshy segments*; the base membranous and clasping the stem; *smelling of furniture polish when crushed*. Flowers yellow-green in umbels to 60 mm across with 10–32(36) rays, sepals absent; bracts and bracteoles numerous (5–10), triangular. Fruit 3.5–5 mm, oval and corky, later purple. Sea cliffs. F, C, T, G, H, P.

Astydamia

Woody-based perennials. Umbels with up to 10 basal bracts; flowers yellowish, the *innermost opening first*. Seeds with minute crystals scattered upon their surface and with a *long, forked carpophore*.

Astydamia latifolia LECHUGA DE MAR A fleshy, aromatic, yellowish to greyish-green biennial to perennial 10–40 cm high with leaves coarsely, jaggedly toothed to pinnately lobed, succulent, shiny. Umbels 60 mm–12 cm across, typically 10–15 rays, and a similar number of broadly linear bracts at the umbel base and secondary bracts at the base of the flowers; flowers yellow. Fruits ovoid, 3-veined and corky when ripe. Coastal dunes and rocks; locally common. L, F, C, T, G, H, P.

Foeniculum FENNEL

Perennials with 3–4-pinnately divided leaves with slender lobes, *strongly aromatic*. Flowers with sepals absent, petals yellow. Fruit distinctly ridged and scarcely compressed.

Foeniculum vulgare HINOJO, FENNEL A late summer-flowering, robust, tall, hairless, *strongly aromatic* perennial to 2.5 m, *smelling of aniseed*. Stems *bluish*, hollow when mature. Leaves feathery, 3–4-pinnately divided with *thread-like* segments 5–40 mm, light green and with sheathing bases. Umbels yellow with 5–44 rays without any bracts; petals 1.3–1.6 mm. Fruit oblong, 3–6(9) mm long, ridged. Hillsides and roadsides; common. L, F, C, T, G, H, P.

---

1. *Centranthus ruber*
2. *Pterocephalus lasiospermus*
3. *Pterocephalus virens*
4. *Hedera canariensis*
5. *Scandix pecten-veneris*
6. *Crithmum maritimum*
7. *Astydamia latifolia*
8. *Foeniculum vulgare*

Conium

Tall, erect, hairless biennials with compound leaves. Flowers with white petals. Fruits broadly ovoid with 5 prominent ribs.

Conium maculatum CICUTA, HEMLOCK A tall, erect, almost hairless biennial with *purple-spotted stems* to 2 m; strong-smelling and poisonous. Lower leaves large, to 50 cm long, triangular in outline, 2–4-pinnately divided with slender leaflets. Umbels terminal and axillary, to 50 mm across with 8–10 rays and few, small, backwardly turned bracts and similar bracteoles; flowers white. Fruit 2–3.9 mm, almost spherical with wavy ridges. Forests and thickets; local. C, T, G, H, P.

Bupleurum

Hairless annuals or perennials with *simple leaves*. Flowers pale green or yellow, borne in small umbels surrounded by petal-like bracts. Sepals absent, petals not notched. Fruits prominently ridged. Observation of the bracts and bracteoles is important.

Bupleurum semicompositum NEGRILLA A small annual 50 mm–12 cm, branched from the base. Leaves linear. Umbels with 4–6 very *unequal* rays; bracts 4, lanceolate, 3-veined, finely toothed; bracteoles 5, similar, slightly exceeding the flowers. Fruits 1–1.5 mm, ellipsoid, covered in whitish protuberances (papillae). Grassy habitats on clays and sands. L, F, C, T, G, H.

Bupleurum handiense ANÍS DE JANDÍA A small shrub with glaucous leaves, *broadly* oblong–ovate with 7–11 veins. Flowerheads with 5–8 rays, each with 8–12 flowers. North-facing cliffs; very rare. *L (Famara), F (Jandía).

Bupleurum salicifolium ANÍS DE RISCO A shrub to 1.5 m. Leaves glaucous, *linear* to *narrowly* lanceolate with 5–9 parallel veins. Flowerheads lax with 5–20 rays, each with 10–20 flowers. Fruits dark brown or black. Barrancos, slopes, ravine beds and coastal cliffs; locally common. **Subsp.** *aciphyllum* is the most common form on the islands (leaves 5–7 veins). *C, T, H, P. **Subsp.** *salicifolium* only recorded for La Gomera and Madeira (leaves 7–9 veins). G.

Apium MARSHWORT

Aquatic, hairless perennials with pinnately divided leaves. Flowers white. Fruits laterally compressed.

Apium graveolens APIO A strong-smelling, robust biennial to 1 m with solid stems. Leaves 1–2-pinnately divided into segments 5–50 mm long, diamond-shaped to lanceolate, lobed and toothed. Flowers borne in umbels, short-stalked to stalkless, mostly opposite the leaves; rays 4–12; bracts and bracteoles absent. Fruit 1.5–2 mm long, broadly ovoid. Damp habitats near the coast. L, F, C, T, G, H, P.

Ridolfia

Tall, erect, aromatic, hairless annuals with 3–4(6)-pinnately divided leaves with thread-like segments. Flowers yellow. Fruits laterally compressed, longer than wide.

Ridolfia segetum CILANTRO SALVAJE, RIDOLFIA A tall, fennel-like, bluish-green annual to 2.1 m, unpleasant-smelling when crushed (not like aniseed). Leaves 3–4(6)-pinnately divided with thread-like lobes 5–15 mm long and inflated stalks. Umbels yellow with many (8–56) slender, curved rays and no bracts or sepals; petals 1.9–2 mm, yellow. Fruit small, 1.2–2.7 mm long with slender ridges. L, F, C, T.

Ammi

Hairless annual or biennial herbs with erect, striated stems and leaves divided into linear or lanceolate segments. Flowers with white petals, borne in dense umbels. Fruits hairless and slightly laterally compressed.

Ammi majus AMEO BASTARDO A carrot-like annual to 1 m. Lower leaves 1-pinnately divided, upper leaves 1–2-pinnately divided, all somewhat feathery with linear–lanceolate leaf segments with dentate margins. *Flowers white*, borne in *dense* umbels; rays numerous (20–55), slender and spreading in flower but not *bunching* in fruit. Bracts divided, equalling or exceeding the rays. Fruit 1.5–2 mm. Common, sometimes abundant on disturbed ground. L, F, C, T, G, H, P. *A. visnaga* BIZNAGA is a similar, erect annual or biennial with narrower *linear, thread-like lobes*. Rays numerous (45–125), but much *thickened* and *erect (bunching)* in fruit. Fruit 2–2.8 mm. Rocky slopes; local. C, P.

1. *Bupleurum semicompositum*
2. *Bupleurum salicifolium*
3. *Apium graveolens*
4. *Ammi majus*

Ferula

Tall, robust perennials with leaves 3–4-pinnately divided into linear lobes. Flowers with yellow petals. Fruits strongly compressed dorsally. Species similar.

Ferula linkii CAÑAHEJA A tall, robust perennial to 3 m, with persistent, pole-like stems. Leaves with conspicuous sheathing bases; leaf lobes very narrow, linear to *thread-like*. Flowers yellow, borne in large umbels. Fruits oblong–elliptic, compressed, with narrow wings and thin ridges. Locally common on rocky slopes in the west. *C, T, G, H, P. The following are similar and most easily identified by their location: *F. lancerotensis* TAJASNOYO has *broader, flatter* leaf-lobes (2 mm wide). Abundant in the Famara area; rare or absent elsewhere. *L (north). *F. arnoldiana* is smaller, to 1.7 m with thread-like leaf lobes (0.8–1 mm wide) and scapes with long lower branches (to 60 cm) nearly at right angles with the main stem. Jandía cliffs; very rare. *F (south). *F. latipinna* has *celery-like* leaves with *lobed* segments. Flowering scapes rather slender with small umbels. Coastal cliffs; very rare. *G, P.

Torilis HEDGE PARSLEY

Bristly annuals with adpressed hairs, solid stems, and pinnately divided leaves. Flowers white or purplish. Fruits with *prominent protuberances*.

Torilis nodosa RASPILLO, KNOTTED HEDGE PARSLEY A short, rough-hairy annual to 50 cm. Leaves 1–2-pinnately divided with deeply toothed segments. Stems branched beneath; flowers with pinkish-white petals, *clustered on stalkless or short-stalked* leaf-opposed umbels with 2–3 very short rays (<5 mm). Fruit egg-shaped, 2.5–3.5 mm long with warts and straight bristles. Common in open grassy areas and waste ground. L, F, C, T, G, H, P. *T. elongata* CUERNECILLO is similar but with stems branched along their whole length (not just from the base) and *long* umbel rays, 15–40 mm. Similar habitats; common. L, F, C, T, G, H, P.

Torilis arvensis BARDANILLA An erect, slender annual to 1 m with adpressed hairs. Leaves 1–2-pinnately divided, the ultimate segments coarsely toothed or lobed. Flowers white, borne in terminal umbels with (2)4–12 rays, usually without bracts; bracteoles numerous. Fruit 3–6 mm, ovoid, slightly compressed, covered in spines, and with persistent styles 0.4–1.5 mm. Common in grassy habitats. L, F, C, T, G, H, P.

Pseudorlaya

Small, prostrate, densely bristly annuals in maritime habitats. Flowers white or pinkish. Fruits with prominent bristles. Possibly better-placed within the genus *Daucus*.

Pseudorlaya pumila PSEUDORLAYA A very *short, densely hairy,* rather fleshy annual 50 mm–20 cm, branched from the base. Leaves 2–3-pinnately divided with oval segments 2–5 mm. Umbels white to pale purple with 2–5(7) unequal rays of 8–12 flowers; petals more or less equal, some larger at the perimeter; bracts 2–5, linear. Fruit elliptic, 7.5–12 mm long, ridged and with hooked spines. Coastal dunes; introduced. L.

---

1. *Ferula linkii*
2. *Ferula linkii*
3. *Ferula linkii* (leaf lobes)
4. *Ferula linkii* (fruits)
5. *Ferula lancerotensis*
6. *Ferula lancerotensis* (leaf lobes)
7. *Ferula arnoldiana* (leaf lobes)
8. *Torilis elongata*
9. *Torilis elongata* (fruits)

Daucus CARROT

Summer-flowering annuals or biennials with leaves 2–3-pinnately divided. Umbels white, with pinnately lobed bracts; petals unequal, those of the outer flowers often larger. Fruit elliptic and spiny.

(a) Fruit spiny.

Daucus carota ZANAHORIA SALVAJE, WILD CARROT A variable, hairy or hairless biennial or perennial to 2 m with solid, often ridged stems. Leaves feathery with linear–lanceolate segments. Umbels with white flowers, many-rayed (9–130), often with a single purple flower in the centre, or purplish throughout; bracts pinnately lobed. Fruit oblong, 1.5–4 mm, shortly spiny; spines not distinctly fused at the base; fruiting rays becoming conspicuously incurved when dry. Common in grassy places. L, F, C, T, G, H, P. **Subsp.** *maximus* is *robust*, to 2 m, with basal leaves 1–3-pinnatisect, and *large umbels*, 12–23(30) cm across. Fruit 1.5–2.5 mm. (Throughout the range). The following are similar: ***D. aureus*** is an annual with stems hairless below, umbels with subequal rays and without a central sterile floret and fruits with large yellowish bristles *dilated* and fused at the base. Similar habitats. C, T. ***D. durieua*** is much smaller, with flowers not in large umbels, but in *lateral, stalkless* (or virtually stalkless) clusters. T.

(b) Fruit spineless.

Daucus elegans (syn. *Cryptotaenia elegans*) PEREJIL DE MONTEVERDE A herb with ridged stems. Leaves dissected with lanceolate lobes, sparsely hairy. Flowers borne in *diffuse* clusters with up to 10 compound rays. Fruits black, weakly ridged. Laurel forests, fairly common. *T, G, H, P.

Drusa

Prostrate, slender herbs. Leaves opposite. Inflorescences lateral, few-flowered; petals white. Fruits winged.

Drusa glandulosa PEGAJOSA A small, prostrate, slender and sparsely branched herb; *bristles prominent*. Leaves usually 3-lobed with toothed margins. Flowers white, few, often borne among the developing fruits. Fruits with 4 spiny-margined wings. Locally common at low altitude in a range of habitats. L, F, C, T, G, H, P.

Pimpinella

(a) Basal leaves 1-pinnate.

Pimpinella dendrotragium PEREJIL CABRUNO A perennial with hairless branches. Basal leaves 1-pinnate with ovate, hairless segments, finely toothed; stem leaves pinnate to 3-lobed, reduced. Umbels with 7–9 rays. Fruit hairy, ovoid. In and around forests; locally frequent. *T, P.

Pimpinella cumbrae PEREJIL DE CUMBRE A perennial with strongly ridged branches. Basal leaves 1-pinnate with ovate to circular segments, coarsely toothed, white-hairy. Umbels with 7–10 rays. Fruit velvety-hairy, globular. Subalpine zones. *T, P.

1. *Daucus carota*
2. *Daucus elegans*
3. *Drusa glandulosa*
4. *Pimpinella dendrotragium* (habit)
5. *Pimpinella dendrotragium*
6. *Pimpinella cumbrae* (habit)
7. *Pimpinella cumbrae*
8. *Pimpinella anagodendron*
9. *Pimpinella junoniae*

(b) Basal leaves 2-pinnate.

Pimpinella anagodendron PEREJIL DE ANAGA A perennial with very finely hairy branches. Basal leaves 2-pinnate, with deeply dissected primary branches. Umbels with 7–12 rays. Fruit shortly hairy, *oblong to ovoid*. Laurel forests and barrancos; very local. *T.

Pimpinella junoniae PEREJIL GOMERO A perennial with sparsely hairy branches. Basal leaves 2-pinnate with segments with up to 3 pairs of toothed to finely toothed, finely hairy lobes. Umbels with 7–10 rays. Fruit velvety-hairy, *globose*. Barrancos and forest slopes; very locally common. *G.

Rutheopsis

Hairless perennial herbs. Leaves variably divided. Flowerheads yellow. Fruits hairless, ovoid to oblong, with thickened ribs.

Rutheopsis herbanica TÁJAME An erect perennial to 50 cm, leafy below. Leaves pinnate with *large, ovate–elliptic*, toothed lobes, flushed *bronze-red* with age. Flowerheads pale to golden yellow, borne on slender stems, sparsely- to much-branched above; umbels slightly domed, with numerous rays. Fruits prominently ridged. Hillsides in the east; rather local. *F (centre), L.

Rutheopsis tortuosa (syn. *Seseli webbii*) APIO MARINO A perennial to 50 cm, often procumbent. Leaves shiny green or slightly glaucous, hairless with *linear to ovate, small* lobes. Flowerheads yellow; umbels compound; rays 10–25. Locally frequent on sea cliffs. *C, T, G, H, P.

Todaroa

Robust perennial herbs with lobed to 4-pinnate leaves. Flowers yellowish. Fruits hairless, the ridges expanded into wings.

Todaroa aurea CAÑAHEJA A perennial with leaves up to 4-pinnate, sparsely hairy, long-stalked. Umbels *congested*, dull greenish- to yellowish-cream, many-rayed. Fruits splitting into 2 mericarps, each 5-winged. Coastal rocks and rocky pine forest slopes; locally common. Regional forms include: **subsp.** *aurea* (*T); **subsp.** *suaveolens* CAÑAHEJA DE OLOR (the common form throughout the range). *C, T, G, H, P.

Athamanta

Robust perennial herbs with lobed to 4-pinnate leaves. Flowers whitish. Fruits densely hairy, ridged and beaked.

Athamanta montana CAÑAHEJA BLANCA A hairy, robust and erect perennial herb with leaves 2–4-pinnate, densely hairy with small, lanceolate lobes. Umbels compound, dull whitish-yellow. Pine forest, cliffs; locally frequent. *C, T, P. ***A. cervariifolia*** APIO DE RISCO is similar but with long-stalked *almost hairless, glaucous* leaves with large, ovate, double-toothed lobes. Flowerheads *whitish*. Basalt cliffs; common locally. *T, G, H, P.

---

1. *Rutheopsis herbanica*
2. *Rutheopsis tortuosa*
3. *Todaroa aurea*
4. *Todaroa aurea* subsp. *suaveolens*
5. *Athamanta montana*
6. *Athamanta montana* (leaf)
7. *Athamanta cervariifolia*
8. *Athamanta cervariifolia* (flowers and immature fruits)

Echium wildpretii

Glossary

Achene, a dry, one-seeded, rather hard, indehiscent (non-splitting) fruit; there are often many achenes in a fruiting head, as in *Ranunculus*.

Actinomorphic, (of a flower) with a radially symmetrical (regular) shape that has multiple axes of symmetry, e.g. a lily flower; syn. **regular**.

Acuminate, tapering to a point.

Alien, not native, introduced to a region.

Alternate, arising at one axis per node; e.g. leaves arising at different heights along the stem (not opposite or whorled).

Angiosperm, a flowering plant.

Annual, completing its life cycle in one year; typically without woody parts (herbaceous).

Anther, fertile, pollen-producing part of the stamen, typically on a terminal stalk (filament).

Anthocarp, a fruit in which the perianth tissue remains attached and forms part of a dispersal unit.

Apex, uppermost part of a structure.

Aril, succulent covering around the seed.

Ascending, arising upwards at an angle (curving).

Auriculate, shaped like an ear or earlobe.

Awl-shaped, long, pointed spike.

Awn, a long, stiff bristle e.g. in the florets of grasses (Poaceae).

Axil, the point at which the leaf or leaf stalk joins the stem; adj. **axillary**.

Beak, an elongated projection, usually on a fruit; adj. **beaked**.

Berry, a succulent fruit, typically with more than one seed; seeds without stony coats.

Biennial, completing its life cycle in two years (often with leaves in a rosette in the first year and flowering in the second year).

Bifid, divided into two parts, typically deeply at the apex.

Bipinnatisect, twice pinnatisect (pinnatisect: having lobes with incisions that extend almost, or up to the midrib).

Blade, the main, often flattened, part of the leaf (or petal).

Bract, a small, leaf-like structure, often subtending (beneath) a flower or inflorescence.

Bracteole, a small secondary floral bract.

Bulb, an underground storage organ composed of a condensed stem and fleshy leaves.

Bulbil, a small reproductive bulb borne in the leaf axil (sometimes among flowers e.g. *Allium*).

Caducous, soon-falling or easily detached.

Calyx, all the sepals of the flower (the outer whorls of the perianth if different from the inner).

Campanulate, bell-shaped.

Capitulum, a dense flower-head (inflorescence) composed of small, stalkless flowers (often ray and disk florets) crowded together on a compound receptacle; typical of the daisy family (Asteraceae) and Dipsacaceae.

Carpel, the female sporophyll within a flower; the floral organ that bears ovules; unit of a compound pistil.

Carpophore, a thin, sterile stalk above the pistil (typical in fruits of some Apiaceae and Caryophyllaceae).

Caruncle, a white spongy outgrowth of the seed.

Casual, introduced to an area and sporadic in its appearance, not persisting.

Cauliflory, when flowers are produced directly from the primary branch or trunk.

Chasmophyte, typically grows out of crevices and fissures in rock, often on cliffs or in gorges.

Chromosome, thread-like structure in the cell's nucleus, carrying genetic information in the form of genes.

Ciliate, fringed along the margin with hairs.

Cladode, flattened organs arising from the stem, typically resembling leaves.

Claw, the narrower part, resembling a stalk, at the base of a segment (e.g. the narrower part of a petal).

Compound, made up of more than one similar part or segment (not simple), typically a leaf.

Cordate, heart-shaped (often the base of a leaf).

Corm, underground storage organ formed from a swollen stem base.

Cone, a compact body of scales or bracts that contains the reproductive structures in gymnosperms.

Corolla, all the petals of a flower (which may form a tube), the inner whorls of the perianth.

Corona, trumpet- or cup-shaped extension of the corolla in *Narcissus*, or fused filaments in some Apocynaceae.

Corymb, a raceme in which the lower flowers have longer stalks, producing a flat-topped inflorescence; adj. **corymbose**.

Cosexual, flowers with both male and female reproductive organs, stamens and carpels, respectively.

Culm, the aerial (above-ground) stem of a grass or sedge.

Cupule, a hardened, cup-like structure composed of bracts in Fagaceae.

Cyathium, a specialised inflorescence of *Euphorbia* (Euphorbiaceae) with a cup-like structure containing a single carpellate (female) flower and several staminate (male) flowers.

Cyme, an inflorescence in which each flower terminates a branch; adj. **cymose**.

Cypsela (plural cypselae), an achene derived from an inferior ovary.

Deciduous, not persistent, for example leaves in autumn or petals after flowering.

Deflexed, bent abruptly downwards.

Dehiscent, splitting.

Desiccation, drying out.

Dioecious, with male and female flowers on separate plants (individual plants of one sex).

Diploid, having two sets of chromosomes.

Disk floret, small actinomorphic flower, forming part or all of a capitulum (in Asteraceae).

Divided, not entire (typically a leaf); divided into teeth, lobes or leaflets.

Drupe, succulent or spongy fruit, typically containing one seed with a stony coat.

Echinule, small spine or prickle.

Elliptic, flat shape (typically a leaf), widest at the middle.

Endemic, restricted to a particular country, region or island.

Entire, whole; without distinct lobes, teeth or divisions.

Epicalyx, an additional whorl of sepal-like bracts beneath the true calyx (e.g. in flowers of Malvaceae).

Epichile, the distal part of the lip (labellum) in some orchid species (e.g. *Epipactis*), separated from the basal part (hypochile) by a joint.

Exserted, protruding (not included, such as anthers from a corolla tube).

Falcate, sickle-shaped.

Falls, the outer perianth segments (tepals) of *Iris* flowers (Iridaceae).

Family, a monophyletic group of related genera; the taxonomic group between the lower rank of genus and the higher rank of order.

Farinose, floury; covered in a fine, mealy powder.

Fascicle, structures bunched or crowded.

Flexuous, not rigid.

Floccose, covered in soft, woolly hairs.

Floret, a small, individual flower that makes up part of a dense inflorescence, e.g. a component of the capitulum (in Asteraceae) or in grasses (Poaceae).

Flowerhead, a group of flowers (an inflorescence), such as a capitulum (Asteraceae).

Follicle, a dry, many-seeded fruit, dehiscent along one side; usually formed from one carpel.

Fruit, the ripened ovary or ovaries of a flower, containing one or more seed(s).

Genus (pl. **genera**), monophyletic group of related species, the taxonomic group between the lower rank of species and the higher rank of family; the generic name is the first part of the scientific binomial.

Geophyte, a plant that possesses a dormant, underground bulb, corm or tuber.

Glabrescent, somewhat or almost hairless.

Glabrous, hairless, not hairy.

Gland, organ of secretion, often in sticky plants; adj. **glandular** (often referring to hairs).

Glaucous, covered in a bluish, whitish or greyish waxy bloom (rather than green).

Globose, spherical.

Glume, of grasses, the bract below a spikelet

Gymnosperm, non-flowering, seed-bearing vascular plants such as conifers.

Halophyte, a salt-tolerant plant.

Head, a group of flowers crowded together at the end of a stalk.

Hemiparasite, a parasitic plant that gains some of its nutrition from another plant (its host) but also has chlorophyll and a root system; some hemiparasites can survive without a host (facultative hemiparasitism).

Herb, a plant without woody parts; a soft and leafy annual, biennial or perennial in which aerial parts naturally die to ground level at the end of the growing season; adj. **herbaceous**.

Hirsute, covered thickly in rather stiff hairs.

Hispid, covered in stiff hairs or bristles.

Holoparasite, a parasitic plant that gains all of its nutrition from another plant (the host) and that lacks chlorophyll and a true root system.

Hybrid, the offspring of a cross between two different species, races, or varieties.

Hybridisation, the formation of hybrid offspring.

Hypochile, the basal part of the lip (labellum) of some orchid species (e.g. *Epipactis*); see also epichile.

Indusium, a thin, membranous covering, especially around the sorus of a fern frond.

Inferior ovary, an ovary situated beneath the point of insertion of other floral organs; syn. **epigynous**.

Indehiscent, non-splitting.

Indumentum, the hairy covering of a plant (of any kind).

Inflorescence, a group of flowers together with their floral stem (axis) and any associated bracts.

Internode, a part of the stem between two nodes.

Involucre, a collection of involucral bracts (e.g. in Asteraceae).

Involucral bract, bracts surrounding a head of flowers (e.g. in Asteraceae).

Keel, a boat-shaped structure, formed by two lower petals in the pea family (Leguminosae), or a longitudinal ridge (typically on a leaf or petal).

Irregular, flower in which one or more members of the whorl, or several floral whorls, differ in form from the other members.

Jable, sand flat.

Labellum, lowermost petal of an orchid flower, often highly specialised, e.g. in bee orchids (*Ophrys*) and *Serapias*.

Lanate, woolly; clothed with long and soft entangled hairs.

Lanceolate, narrowly ovate, spear- or lance-shaped.

Lax, not crowded.

Legume, either the fruit or a plant from the pea family (Leguminosae).

Lemma, (of grasses) the lower of the pair of bracts (lemma and palea) that subtends the floret.

Ligule, (of grasses) a small membranous projection or ring of hairs at the junction of the leaf sheath and stem.

Limb, expanded portion of the calyx or corolla (distinct from the tube or throat).

Linear, long, narrow and parallel (typically the leaf of a grass).

Lip, region of the calyx or corolla sharply differentiated from the rest (see also **Labellum**).

Lobe, substantial division of a leaf, calyx or corolla.

Malpaís, rough volcanic flats.

Membranous, paper-like or membrane-like in consistency.

Mericarp, one-seeded portion formed by the splitting of a multiple-seeded fruit.

Midrib, the central or main vein.

Monoecious, with separate male and female reproductive structures on the same individual plant, as opposed to dioecious.

Morphology, the appearance, form or structure.

Mucronate, with a short bristle-tip.

Mycoheterotrophy, the process by which a non-photosynthetic plant obtains nutrition from a fungal symbiont (or sometimes from another plant via a shared fungal symbiont) living in its root system.

Native, naturally occurring in the area.

Naturalised, not native but well established.

Node, the position on the stem from which the leaves, branches or flowers arise.

Nut, a dry, indehiscent, one-seeded fruit with a woody, hard wall; often large in size.

Nutlet, a small, woody-walled nut (see also *achene*).

Obconical, an inverted cone, attached to the stalk by its pointed end.

Obcordate, heart-shaped and attached at the pointed end.

Oblong, elongated but wide in shape, the middle part parallel-sided (usually describing a leaf).

Obovate, ovate (oval), and narrower at the base.

Obtuse, with a point >90°.

Opposite, of two organs arising from a common node (e.g. leaves from a stem).

Ovary, part of the carpel or pistil containing the ovules, and later, the seeds.

Ovate, oval in outline, or egg-shaped.

Ovoid, a solid shape that is oval/ovate in side view.

Ovule, the structure that contains the egg and becomes the seed after fertilisation.

Palea, (of grasses) the upper of the pair of bracts (lemma and palea) that subtends the floret.

Palmate, having lobes or segments that radiate from a common axis.

Panicle, a branched, compound inflorescence.

Pappus, a structure consisting of hairs, bristles or scales on the fruit (achene or cypsela) of Asteraceae.

Parasite, a plant that obtains nutrients from another plant (it may be hemi- or holo-parasitic).

Pedicel, the stalk of an individual flower in an inflorescence or the stalk of a grass spikelet.

Peltate, leaf with stalk attached to the centre of the blade.

Perennial, living for more than two years, generally flowering every year. Often woody at the base.

Perfoliate, (of a leaf or bract) with the base united (around the stem).

Perianth, all the non-sexual segments (i.e. the calyx and corolla together) of a flower (see also **tepals**).

Petal, one of the segments of the inner whorl(s) of the perianth; **petaloid** is petal-like.

Petiole, the stalk of the leaf; adj. **Petiolate** (with a petiole).

Phonolitic, a grey or green volcanic rock.

Phyllary, one of the involucral bracts subtending the flowerhead in Asteraceae.

Phyllode, a flattened, expanded petiole that resembles and functions as a leaf.

Phytogeographic, a geographic region defined by its flora.

Pinnae, a primary division of a pinnate leaf, especially a fern.

Pinnate, (of a compound leaf) composed of leaflets arranged on opposite sides of a common axis (rachis); adv. **pinnately** (e.g. pinnately divided into leaflets).

Pinnatifid, (of a leaf) pinnately divided with the lobes cut nearly (but not quite) to the mid-vein.

Pinnatisect, (of a leaf) pinnately divided to the mid-vein, but the lobes are not contracted at the base to form discrete leaflets.

Pistil, the female organs of a flower, comprising stigma, style and ovary (see also **carpel**).

Plumose, with many fine filaments or branches, giving a feathery or 'fluffy' appearance (e.g. the pappus of some Asteraceae).

Pollinium (pl. **pollinia**), a mass of adhering pollen grains (e.g. of an orchid) that is shed and transported as a unit by a pollinator.

Polyploid, with more than two sets of chromosomes.

Prickle, a spiny outgrowth, broadened at the base.

Procumbent, trailing on the ground.

Pseudocopulation, the process by which a male insect (usually a bee or wasp) attempts to mate with the flower of a bee orchid (*Ophrys*) and in so doing, brings about cross-fertilisation.

Pubescent, hairy

Raceme, a simple unbranched inflorescence with stalked flowers borne on a single axis, with the youngest flowers at the top; adj. **racemose**.

Rachis, the stalk of a compound leaf or the central axis bearing the flowers.

Ray, a radiating branch of an umbel or cyathium (e.g. in Apiaceae or Euphorbiaceae).

Ray floret, a small, zygomorphic flower often resembling a single petal in the inflorescence (flower-head) of Asteraceae. Contrast with disk floret.

Receptacle, the thickened or expanded part of the stem from which the flowers or inflorescence arise (e.g. in Asteraceae).

Reflexed, bent downwards or backwards.

Regular, (of a flower) with a radially symmetrical (regular) shape that has multiple axes of symmetry, (see also **actinomorphic**). e.g. a *Lilium* flower.

Revolute, rolled back (e.g. the inrolled margins of a petal).

Rhizome, a horizontal underground stem; adj. **rhizomatous**.

Rosette, (of leaves) radiating from a central point.

Samara, a dry, indehiscent (non-splitting), one-seeded fruit with a membranous, wing-like extension for dispersal.

Scape, a flowering stem without leaves (all the leaves may be basal).

Schizocarp, a dry fruit that splits into two, one-seeded portions (mericarps) e.g. fruits of Apiaceae (see also **mericarp**).

Sclerophyll, a hard, leathery leaf that contains a high proportion of thickened cells (sclereids); adj. **sclerophyllous**.

Sepal, typically leaf-like segments of the outer whorl(s) of the perianth (sometimes like a petal, in which case it is called a tepal).

Sericeous, covered with fine silky hair.

Sessile, not stalked.

Silicula, a short and broad capsule that splits into two valves, in Brassicaceae.

Siliqua, a long, narrow capsule that splits into two valves, in Brassicaceae.

Simple, a structure (e.g. a leaf) that is not divided into segments or lobes (not compound).

Sori, groups of sporangia that contain spores, often on the undersides of the fronds.

Spadix, a spike-like organ bearing tiny male and female flowers at its base and surrounded by a **spathe**; characteristic of Araceae.

Spathe, a large, leafy bract, sometimes brightly coloured; characteristic of Araceae.

Spike, an unbranched inflorescence (raceme) of stalkless flowers.

Spikelet, the basic unit of the inflorescence of grasses (Poaceae) and sedges (Cyperaceae); each flower cluster (e.g. Plumbaginaceae).

Spinescent, becoming spiny/having or resembling spines.

Sporangia, a receptacle of spores (spore-producing structure).

Spur, a hollow, tubular projection originating from the sepals or petals, often containing nectar.

Stamen, male reproductive organ of the flower consisting of a filament and anther.

Staminode, aborted or sterile stamens (not pollen-producing), sometimes resembling petals.

Standard, the upper petals of flowers of species in the pea family (Leguminosae) and iris family (Iridaceae).

Stellate, star-shaped, e.g. of a hair.

Stigma, the part of the carpel or pistil that receives pollen and upon which the pollen germinates.

Stipule, a small, leaf-like organ at the base of some leaf petioles (stalks), often in pairs; either simple or lobed.

Stolon, a spreading, above-ground shoot, or runner; adj. **stoloniferous**.

Style, the stalk on an ovary bearing the stigma(s); sometimes absent.

Subshrub, a small shrub or perennial with woody stems.

Subspecies, the taxonomic subdivision of a species; usually a geographically, morphologically and/or genetically distinct race.

Succulent, fleshy or juicy, e.g. the stems of cacti and succulents.

Superior ovary, an ovary positioned above the attachment points of other floral organs; syn. hypogynous.

Syconium, a fleshy, hollow receptacle that develops into a multiple fruit (a fig), typically associated with symbiotic, pollinating wasps.

Synflorescence, a number of separate inflorescences clustered such that they appear as a single inflorescence (e.g. in Asteraceae).

Taxon (pl. **taxa**), the taxonomic unit of any rank, for example species, genus, subspecies or variety.

Teeth, divisions of a leaf, calyx or corolla; adj. **toothed.**

Tepals, the segments of the perianth, often used to describe segments that are not clearly petals or sepals (particularly in Liliaceae and related families).

Ternate, a compound leaf with three leaflets.

Terminal, at the top/apex.

Throat, the opening where the tube joins the limb of the corolla or calyx.

Tomentose, densely covered with short matted woolly hairs.

Traquitic, a form of volcanic rock.

Tree, a woody plant typically >5 m with a single trunk.

Trifid, cleft into three teeth or points.

Trifoliate, a compound leaf with three leaflets.

Tube, the narrow, cylindrical part of the calyx or corolla, distinct from the limb, lobes or throat; adj. **tubular.**

Tufted, clustered together, e.g. a plant with numerous stems.

Umbel, a flat-topped or convex 'umbrella-shaped' inflorescence consisting of a cluster of flowers with spreading stalks (pedicels) that arise from the apex of the peduncle; typical of Apiaceae.

Undulate, wavy at the margin.

Unisexual, flowers with either male or female reproductive organs only. Contrast cosexual.

Variety, a form of a species that is geographically, morphologically and/or genetically distinct (but not distinct enough to warrant subspecies status).

Vascular, pertaining to the veins (conducting tissue) of an organ; all the plants in this book are vascular.

Viscid, sticky.

Whorl, a group of lateral organs with at least two borne at each node (e.g. the petals of a flower).

Woody, hard and wood-like, typically persistent.

Woolly, clothed with soft, shaggy hairs.

Xerophyte, a drought-tolerant plant.

Zygomorphic, (of a flower) with a bilaterally symmetrical, irregular shape, e.g. an *Antirrhinum* flower.

Index of Spanish names

Entries with a photograph are indicated by a **bold** page number.

A

abrepuño **377**, 378
abrojo 120, **121**
abrojo **377**, 378
acacia azul **121**, 122
acacia majorera **121**, 122
acacia negra 122
acacia plateada 122
acanto 303
acebén 89
acebiño **326**, 327
acebuche 293, **293**
acedera de Madeira **228**, 229
acederilla 226, **227**
aceitilla 86
aceitilla de calcetines **93**, 95
acelga 245
adelfa de monte 173, **174**
adelfilla 186
aderno 258, **259**
adormidera 99
agonal 205
aguacate 54, **55**
aguileña común 105
ahulaga 366, **367**
ajillo **60**, 61
ajinajo **276**, 278
ajinajo herreño 278, **280**
ajo gato 64, **65**
ajo negro 69
ajo porro 68
ajo salvaje 68
alacranillo 273, **273**
alamillo **352**, 353
alamillo gomero 353
álamo 172
alazor 379, **380**
albaricoque 156
alcachofa silvestre 376, **377**
alcaparra 204
alcarcán **213**, 214
alcornoque 165
alfabega 312
alfalfa 142, **143**
alfilerillo 184
alfinelejo 184
alfolga olorosa 142, **143**
algáfita **157**, 159
algaraba 323
algaritofe 322, **322**
algarrobo **123**, 124
algoaera **244**, 245
algodón 193
algodonera **332**, 333
alhelí 208, **209**
alhelí canario **206**, 207
alhelí del Teide **206**, 207
alhelí de mar 207
alhelí menudo 208, **209**
alhelí montuno **206**, 207
alicacán 72, **73**
alicaneja 281
almácigo **187**, 189
almendro 156, **157**
almirón 330
alpiste 87 86
alpoaera 343
altabaca 334, **335**
alverjana 135
amagante de pinar **198**, 199
amapola 99, **100**
amapola cornuda **100**, 101
amapola de California **100**, 101
amapola espinosa 101
amargosa 346, **347**
ameo bastardo 387, **387**
amorseco **345**, 346
amuley 336
angoja de Anaga **364**, 365
angoja de risco **364**, 365
anís de Jandía 386
anís de risco 386, **387**
apio 386, **387**
apio caballuno 383
apio de risco **392**, 393
apio marino **392**, 393
arañuela 103
araucaria **52**, 53
árbol botella 196, **197**
árbol de la seda **267**, 268
árbol del caucho 161, **161**
árbol del cielo 192
árbol del Paraíso 193, **194**
arcila **352**, 353
arete perro **152**, 153
aromo 122
aromo blanco **121**, 122
arrebol 274, **277**, 278
arrebol azul 278, **280**
arrebol de Guelguén **277**, 278
arroyuelo 185
artisco 167 168
aulaga 367 366
ave del paraíso 78, **79**
avena Francesa 88
azafrán azul 379
azaico 262, **263**
azarero 383
azucena de mar 68

B

bainena 350
balancón 251, **252**
balango 86, **87**
balango canario 86, **87**
balillo alpispillo 360, **361**
balillo arbóreo 362, **363**
balillo copudo 362
balillo de Agando 362
balillo fino 362, **363**
balillo gomero 362
balillo tinerfeño del sur 362, **363**
ballico 89
ballico petudo **90**, 91
balo 264, **265**
balsamina 258, **259**
barba de chivo 124
barbón **358**, 359
barbusano 54, **55**
barbusano negro 54
bardanilla 389
barrilla 252, **253**
batatera 285
batatilla 48, **50**
batatilla de Indias 285, **286**
batatilla de playa 285
bayón **218**, 219

bea 112, 113
bea de Güímar 112, 113
bea de Tenerife 113, 114
bejeque 106
bejeque puntero de Tenerife 107, 109
bejeque rojo 109
beleño 288, **289**
bicácaro **328** ,329
bignonia de fuego **304**, 305
biznaga 387
bobo 290, **291**
boca de dragón 296
borraja 281
borriza 331, 333
botonera 345 344
brezo 263 262
brotona 320, **321**
brusquilla **250**, 251
burladora 290, **291**

C

cabezón 378
cachimba 56, **57**
cachurrera 344, **345**
caíl 226, **227**
cala 56, **57**
calabacilla 236, **237**
calabacilla 296, **298**
camellera 272, **273**
camosilla 208, **209**
campanilla **328**, 329
campanilla bastarda 329
campanilla canaria **328**, 29
campanilla morada 285
campanilla palmeada 285, **286**
campanita de hoz **328**, 329
caña 93, 94
cañaheja **388**, 389
cañaheja **392**, 393
cañaheja blanca **392**, 393
cañaheja de olor **392**, 393
candil 56, **57**
canónigo 382
canónigo rojo 382
canónigo rosado 382
canónigo simple 382
cañota 89
cañotilla 88
canutillo de Berthelot **237**, 238
capitana 265, **265**
capitana pegajosa 265
cardillo 355, **356**
cardo 374, **375**
cardo borriquero 375, **375**
cardo de Cristo 372

cardo de leche 356
cardo de Siria 374
cardo enrejado 372, **373**
cardomanso 378
cardo mariano 376, **377**
cardón 173, **174**
cardoncillo **270**, 271
cardón de Jandía 173, **174**
carrizo **93**, 94
casamelos 339
casia **123**, 124
castaño **164**, 165
casuarina 166, **167**
cebadilla 92, **93**
cebolla almorrana 74, **75**
cebolla almorrana menor 74, **75**
cebolleta 74, **75**
cebollín 68
cebollín de gato **60**, 61
cedro 51, **52**
cenizo 246
centaura 266
cepillito dorado 89, **90**
cerraja 359, 368
cerraja brillante 360
cerraja colgante 361 360
cerraja de Adeje 365
cerraja de Don Enrique 365
cerraja de Güímar **364**, 365
cerraja gomera 365
cerrajilla común 360, **361**
cerrajilla de Teno 360, **361**
cerrajilla moruna 360, **361**
cerrajilla picona **358**, 359
cerrajón 359, 362, **363**
cerrajón brillante 365
cerrajón de El Golfo 362
cerrajón de Lid 362
cerrajón de monte **364**, 365
cerrajón de Pitard 362
cerrajón de risco **363**, 365
cerrajón gomero 362, **363**
cerrajón herreño 362, **363**
cerrajón palmero **363**, 365
cerrillo **96**, 97
cerrillo de risco 84, **85**
cerrillón fino 91
césped de riñón 287
césped inglés 89
chabusquera 130, **131**
chabusquillo 130, **131**
chahorra 308
chaparro 72, **73**
chaparro canario 282, **283**
chicharaca 137, **138**
chicharilla canaria 135

chicharilla gomera 135, **136**
chicharilla mayor 135, **136**
chícharo 137
chícharo alberjana **136**, 137
chícharo altramuz 137, **138**
chícharo amarillo **136**, 137
chícharo cuchillero **136**, 137
chícharo de olor 137
chícharo morado **138**,139
chicoria loquilla 370
chinipa **134**, 135
chinipa amarilla 135, **136**
chirate 92, **93**
chirrigüela postrada 179
choco 129
chopo 172
chufa 82
cicuta 386
cilantro salvaje 386
ciprés 53
ciprés de Monterey 53
ciruelo 155
clavel de sol 252, **253**
clavelito silvestre **237**, 238
cochinita 48, **49**
codeso 129, **138**, 139
cohombrillo 166, **167**
cola de caballo 44, **45**
colaperro **90**, 91
col de risco 211
colderrisco de Güímar **211**, 212
colino 212, **213**
collejón **206**, 207
colmillo de perro **270**, 271
conejera 236, **237**
conejito mayor 296
conservilla 320, **321**
conservilla majorera 318, **321**
corazoncillo 146, **147**
corazoncillo de Aringa **145**, 146
corazoncillo de costa **145**, 146
corazoncillo de Gran Canaria 146, **147**
corazoncillo de Masca 146
corazoncillo de pinar **145**, 146
corazoncillo gomero 146
corazoncillo plateado 146
corneta **100**, 101
cornical **270**, 271
corona de la Reina 152, **152**
corona de la reina 351, **351**
corregüela blanca **283**, 284
corregüela de playa 285
corregüela rosada **283**, 284
corregüela tricolor 284
corregüelita azul **283**, 284

corregüelón de Famara **283**, 284
corregüelón de monte **283**, 284
cosco 252, **253**
cosco macho 251, **253**
coyeso **138**, 139
cresta de gallo **298**, 299
crestagallo 65, **65**
cruzadilla **169**, 170
cucharilla **213**, 214
cuchillera **81**, 82
cuchillera ancha **81**, 82
cuchillera canaria 80, **81**
cuchillera de Las Cañadas **81**, 82
cuchillera densa 82
cuello de cisne 76, **77**
cuernecillo 146
cuernecillo **388**, **389**
cuernecillo comestible 144
cuernecillo fino 146
cuernecillo grande 144
cuernúa **270**, 271
culantrillo de pozo 44, **45**
culantrillo marino 46, **47**
culantrillo negro 46, **47**

D
dama 217, 334, **335**
damasco 156
delfino **259**, 261
diente de león 370
Don Diego de noche 254
doradilla canela 46, **47**
doradilla fina 44, **45**
doradilla medicinal 46, **47**
doradilla velluda 46, **47**
drago 70, **71**
drago de Gran Canaria 70, **71**

E
embeleso **220**, 221
encimba 353
enredadera **283**, 284
enredadera de papa 290
escobilla 217
escobón 127
esparcilla mayor **234**, 235
espárrago borriquero 72, **73**
esparragón 72, **73**
esparragón colgante 72, **73**
esparragón rabo de burro 72
esparraguera 70
esparraguera común 72, **73**
esparraguera de monteverde 72
esparraguera majorera 72, **73**

especiero **187**, 189
espiguilla 86
espina blanca 72, **73**
espinaca de Nueva Zelanda **253**, 254
espino **288**, **289**
espino blanco 288
espino negro **158**, 160
espuela 105
espuela matapiojos 103, **104**
esterilla 233, **234**
estornudera 368
estramonio 290, **291**
estrellada espinosa **342**, 343
estrelladera **164**, **164**
eucalipto 188

F
falsa corregüela 229
falsa verdolaga **244**, 245
falso lino de lagartija **259**, 261
farolito 191
farroba 110 ,111
faya **164**, 165
fistulera 300
flamboyán 122, **123**
flechilla 86, **87**
flor de gofio **123**, 124
flor de mayo 353, 354, **354**
flor de pato 54
flor de piedra **112**, 13
flor de viuda **380**, 381
follao 379, **380**

G
gacia 125
gallocresta 320, **321**, 323
gamona 66, **67**
gamonilla 66, **67**
garbancera canaria 133, **134**
gibalbera 72, **73**
giralda 339, **340**
gladiolo 65
gomereta **110**, 111
gongarillo majorero 113, **114**
góngaro 111, **112**
góngaro de Anaga **112**, 113
gordolobo 302
grama 94, **96**, 97
gramilla **90**, 91
granadillo 170
granado 186
greña 287
greña 84, **85**
guaidil **282**, **283**
gualda 202, **203**

gualdón 202

H
hediondo 212
hediondo **240**, 241
helechera 46, **47**
helecho macho 48, **50**
helecho peine 48, **49**
helecho penco dentado 48, **49**
henequén 76, **77**
hiedra 383
hierba barroca 233
hierba candil 46, **47**
hierba clín **306**, 307
hierba de la pampa 98
hierba del diablo 290, **291**
hierba de San Juan 170
hierba gato 312, **313**
hierba huerto 317
hierba jabonera 106, **107**
hierba muda 146, **147**
hierba pastel 205
hierba pezonera 371, **371**
hierba pulguera 334
hierba turmera 199
higuera 160, **161**
higuereta **194**, 196
higuerilla 177, **178**
higuerilla de playa **176**, 177
hija 156
hinojo **384**, 385
huevito de gallo **291**, 292

I
ija de Don Enrique 365
ilex encina 165
incienso 336, **337**
incienso morisco 336, **337**

J
jabonera 233, **234**
jabonera **260**, 261
jacinto silvestre 74, **75**
jaguarzo **198**, 199
jara **198**, 199
jaramago 212, **213**
jara pringosa 199
jarilla 200
jazmín canario 292, **293**
jébana **209**, 210
jocama **306**, 307
jopillo 89
jopo 325
jorado 344, **345**
julaga 366, **367**
juncia amorosa 83, **83**

juncia clara 82, **83**
juncia negra 82, **83**
junco azul 80
junco cabezón 80
junco común 80, **81**
junco fino 80
junco marino 80
junquillo 82, **83**

K
kohlrauschia 238

L
labasa 226, **227**
lágrimas de la virgen 68, **69**
lantana 305
lavanda 318
lechuga de burro 366
lechuga de mar **384**, 385
lechuga espinosa 366, **367**
lechugón de Sventenius 365
lechuguilla 356, 359, 366, **367**
lechuguilla de risco 370, **371**
lechuguilla de Tijarafe 360
lechuguilla de Webb 360
leña blanca 292, **293**
leña buena **190**, 192
leñanoel 282, **283**
lengua de oveja **295**, 296
lengua de pájaro 239, **240**
lengua de serpiente 44, **45**
lengua de vaca 274, **275**
lengua oveja 274, **275**
lenguaperro 281
lenguaza 281
lentisco 189
limpiatubos 99
linaria 297
linillo 182
lino 181
lino bravo 181
lino de lagartija 260
lino silvestre **180**, 181
llantén **295**, 296
llantén blanco 294
llantén de agua 58
lluvia silvestre **354**, 355
loro 54, **55**
lucerna 142, **143**
lúzula canaria 80, **81**
lúzula de Forster 80

M
madama de risco 346, **347**
madama pegajosa 346, **347**
madreselva 381

madroño 261
madroño canario **260**, 262
magarza 346, **347**
magarza azulada 348, **349**
magarza de La Aldea 350
magarza de Lems 346, **347**
magarza de Lid 350
magarza de Santiago 348
magarza de Sunding 346, **347**
magarza fina 350
magarza grácil 348, **349**
majapelo 299
malfurada 169, **169**
malfurada de monte 170
malpica 372
malpica **242**, 243
malva 195
malva real 196
malvarosa 185 185
malvarrisco 194, **196**
manzanilla 338
manzanilla estrellada 338
marañuela 202, **203**
maravilla 343
margarita 346, **347**
margarita de Famara **349**, 350
margarita de Jandía **349**, 350
margarita del Teide 348, **349**
margarita de Lugo 351, 353
margarita de monte 347, 348
margarita de Sventenius **349**, 350
margarita gomera amarilla 348
margarita gomera de monte 348
margarita herreña **349**, 350
margarita palmera 350
margarita plateada 351, **351**
marmohay 247, **247**
marmulán 258, **259**
marrubio **306**, 307
marrubio negro 312
mastuerzo 317
matacán 268
matagallos 312, **313**
mata parda 146, **147**
mataperros **270**, 271
mataprieta 303, **304**
mato **248**, 249
mato azul 246
matocosta milengrana 230, **231**
matomoro 249, **250**
mato salado **244**, 245
mazoquera 311
mazorrilla del teide 84, **85**
melera 111
melera de jaguarzo 193, **194**

melera de jara 193, **194**
mellorina 101
melocotonero 156, **157**
melosa aserrada 140, **141**
melosa de arenas 140, **141**
melosa de damas 140
melosa pálida 140
melosa reclinada 140
membrillo 156, **157**
menta poleo 317
milamores 382, **384**
mirto 188
mocán **260**, 261
moco guirre 273, **273**
moqueguirre de costa **340**, 341
moqueguirre hediondo **340**, 341
moral 160, **161**
moralito 160
morera 160
morgallana 103, **104**
morterillo 236, **237**
moruja 232
morujilla 230, 233
mosquera 300
mostaza 210
muraje **259**, 260

N
ñamera 58, **59**
naranjero salvaje **326**, 327
narciso 68, **69**
nauta 313
negrilla 386, **387**
neguillón 236
nevadilla canaria **231**, 232
nevadilla de plata 232
nevadilla de risco 232
níspero 156
nogal 165
nomeolvides 282
norsa 58, **59**

O
ojo de buey 339, **340**
ojo de perdiz **104**, 105
olivillo **293**, 293
olivo 292, **293**
olmo 160
orégano 317
oreja de burro 56, **57**
oreja de burro 320, **321**
orijama **190**, 192
orobal 288, **289**
oro de risco 125, **126**
orquídea abejona 62, **63**
orquídea canaria 62, **63**

INDEX OF SPANISH NAMES

orquídea de dos hojas 64, **65**
orquídea de Tenerife 62, **63**
orquídea de tres dedos 64, **65**
orquídea gallo menuda 64
orquídea macho 62
orquídea manchada 62
ortiga 162
ortiga mansa **180**, 181
ortigón 162, **163**
ortiguilla mansa 300, 312, **313**

P

pajito 338, 339, **340**
pajito colorado 339, **340**
pajonera de cumbre 214, **215**
pajuco 95, **96**
palillo 83
palillo 303, **304**
palmera canaria **77**, 78
palmera datilera 78
palmera de abanico 76, **77**
palo blanco 293
palo de sangre **158**, 159
palomera **352**, 353
palomilla **298**, 299
palomina 274, **275**
palomino 274, **276**
pamplina 101
paniqueso 208
panizo 98
paragüitas 82, **83**
parra 118, **119**
pascueta 336
pasote 246, **247**
pastel de risco 111, **112**
pata camello 196, **197**
pata conejo 239, **240**
pataconejo basta 239, **240**
pataconejo carnosa 239, **240**
pataconejo fina 239
pata de gallina **93**, 94
pata de gallina **237**, 238
pata gallina **213**, 214
patagallo 46, **47**
patagallo blando 182
patagallo canario 182, **183**
patagallo cortado 182
patagallo púrpura 182, **183**
patagallo redondo 182, **183**
patilla 251, **253**
pegajosa 390, **391**
pega-pega 98
peine de Venus **384**, 385
pelotilla 113, **114**, 142, **143**, 294, **295**
pelotilla de Lanzarote 106, **107**

pensamiento de la cumbre 171, 172
peorrera 368
pepinillo del diablo 166, **167**
peral 156
peralillo **167**, 168
perejil cabruno 390, **391**
perejil de Anaga **391**, 393
perejil de cumbre 390, **391**
perejil de mar **384**, 385
perejil de Monteverde 390, **391**
perejil gomero **391**, 393
periquito 258
pico cernícalo 144, **145**
pico de El Sauzal 144, **145**
pico de fuego 144, **145**
pico pájaro 297
pico paloma 144, **145**
píjara 48, **49**
pimentera 290
piña mar 372, **373**
pincho **248**, 249
pinillo **295**, 296
pinillo de Famara **295**, 296
pininana 274, **276**
pinito **104**, 106
pino canario 51, **52**
pino carrasco 51
pino de Monterey 51
pino marítimo 51
pino marítimo 166, **167**
pinus 51
pipe 362, **363**
pirgua 48, **49**
pitaya 255, **256**
pitera 76, **77**
planta de jade **104**, 106
platanera 78
plumero 98
poleo de monte 322
polipodio del país 48, **50**
pompón de Las Cañadas 334, **335**
pseudorlaya 389
puerro de viña 68
pulicaria purpuraria 334, **335**

Q

quebradizo 266
quemón 206 205
quemoncillo 205, **206**

R

rábano marino **209**, 210
rabillo de conejo 88
rabo cordero 294, **295**

rabo cordero **322**, 323
rabo de burro 92, **93**
rabo de gato **96**, 97
rabo de gato blanco 97
rama cría 200, **201**
rama negra 124
ranúnculo campestre 102
ranúnculo centella 102
ranúnculo de agua 103
ranúnculo de almorrana 103
ranúnculo sardo 103
rapasaya 120, **121**
rapasaya 370, **371**
rapaso 262, **263**
raspilla azul 264, **265**
raspilla cuajaleches 264
raspilla de París 264
raspilla de sombra 264, **265**
raspilla enana 265
raspilla menuda 264
raspillo 389
ratonera 162, **163**
ratonera mansa 162
reina del Monte 267, **267**
rejalgadera **289**, 290
relinchón **209**, 210
relinchón bastardo 210, **211**
retama amarilla 127, **128**
retama del teide **128**, 129
retama fina **128**, 129
retama negra 127
retamón 125, **126**
rilla **237**, 238
romanillo 318
romerillo 217
romerillo 230, **231**, 235
romerillo **234**, 235
romerillo pardo 203, **204**
romero 320
romero marino 303, **304**
rompesacos 92
rosalito de Anaga 382, **384**
rosalito de cumbre 382, **384**
rosalito palmero 382
rosal silvestre 153
rosquilla **152**, 153
rúcula 208, 212
ruda 191

S

sábila 66, **67**
sábila arborea 66
sabina 51, **52**
saladillo 303, **304**
saladillo blanco 239, **240**
salado 243, **248**, 249

INDEX OF SPANISH NAMES

salado **354**, 355
salitrosa 225
salsifí **358**, 359
salvia 320
salvia canaria 320, **321**
sanalotodo 179, **180**
sangradera **242**, 243
sanguinaria 330, **340**, 341
sanguinaria cenicienta **340**, 341
sanguino **158**, 160
sanjora 109
sao **171**, 172
saquitero 258, **259**
sauce canario **171**, 172
sauco 379, **380**
sayón **270**, 271
sedero **267**, 268
serradilla amarilla 148, **149**
serradilla rosa 148
servellina 208
servellina verrugosa 208
siempreverde **301**, 302
siempreviva 221
simonillo 336, **337**
simonillo florido 336, **337**
sogal **248**, 249
sombrerillo 106, **107**
sonajera 202, **203**
sorgo 95, **96**

T

tabaco **291**, 292
tabaco moro 290, **291**
tabaiba amarga 177, **178**
tabaiba dulce **176**, 177
Tabaiba majorera **176**, 177
tabaibilla 177, **178**
taboire amarillo 139, **141**
taboire de Jandía **138**, 139
taboire fino **138**, 139
taboire rosado 140, **141**
tagardina 355, **356**
tagasaste **126**, 127
taginaste **277**, 278
taginaste azul de Gran Canaria **280**, 281
taginaste azul de Tenerife 278, **279**
taginaste blanco 278, **279**
taginaste blanco oriental 278, **279**
taginaste de Adeje 278
taginaste de Anaga **277**, 278
taginaste de Jandía 281

taginaste gigante **276**, 278
taginaste negro **280**, 281
taginaste palmero de cumbre **277**, 278
taginaste picante 274, **275**
taginaste rojo 274, **276**
taginaste rosado 278, **279**
tájame **392**, 393
tajasnoyo **388**, 389
tajinaste herreño 278, **280**
tajose **316**, 317
tarabaste 69, **69**
tarabaste gato 70, **71**
taragontía 58, **59**
taraguntía 58
tarajal 221
tarrillo 236, **237**
tártago **180**, 181
tasaigo 262, **263**
tebete **247**, 247
té canario 344, **345**
tecoma amarilla **304**, 305
tedera 133, **134**
tejo 262, **263**
tembladera **87**, 88
tepopote fino **52**, 53
tepopote frágil **52**, 53
teresita 336, **337**
til 55
tilo 55
tojia 344, **345**
tojia blanca 344, **345**
tojo **126**, 127
tolda 173, **175**
tomatera de culebra **286**, 287
tomillo 317
tomillo marino 219
tomillo salvaje 313
tonática **310**, 311
toronjil 311
torvisco 197, **197**
transparente **301**, 302
trebina 168, **169**
trébol 148, 168
trébol de olor 140, **141**
treintanudos 226
treintanudos de mar 225, **227**
trigo guanche 92
triguerilla 84, **85**
trompetero **291**, 292
tulipero africano **304**, 305
tunera **256**, 257
tunera india **256**, 257
turbinto **187**, 189

turgayte 339, **340**
turmera 200, **201**

U

uña de gato 252
uñagato 148
uva de mar 118, **119**
uva de playa **228**, 229

V

vara de San José 74, **75**
venenillo **167**, 168
verdolaga 255
verode 106
verode **340**, 341
verol 106
verónica 299
viborina 274, **275**
viborina canaria 274, **275**
viborina de Lanzarote 274, **275**
viborina triste 274, **275**
viña 118, **119**
vinagrera **228**, 229
vinagrerilla 226, **227**
viñátigo 54, **55**
vinca **267**, 268
violeta 170
violeta de Anaga **171**, 172
violeta del Teide **171**, 172

Y

yedra alemana **342**, 343
yerbamora **288**, 289
yerbarubí 95, **96**
yerbavino **186**, 187
yesca 333
yesquera amarilla **332**, 333
yesquera roja **332**, 333
yesquerilla 330
yesquerilla amarilla **331**, 332
yesquerilla de la fiebre **331**, 332
yesquerilla rastrera 330, **332**

Z

zanahoria salvaje **390**, 391
zaragatona 294
zarzaparrilla **60**, 61
zarzaparrilla canaria **60**, 61
zuaja 274, **275**
zumaque **187**, 189
zumaque blanco 189, **190**
zumaquero 159
zurrón de pastor 205

Index of English names

Entries with a photograph are indicated by a **bold** page number.

A

acacia 120
adder's-tongue family 44
Aleppo pine 51
alexanders 383
alkanet 281
allseed 233
almond 156, **157**
aloe **65**, 66
amaranth family 241
annual bellflower **328**, 329
annual daisy 336
annual mercury **180**, 181
annual pearlwort 233
annual rock rose 199
annual sea heath **218**, 219
annual yellow vetchling 137
apple mint 318
apple of Peru **286**, 287
artichoke 376
arum family 56
arum lily 57 56
asparagus 70
asphodel 66
autumn crocus 61
avocado 54, **55**

B

balm 311
balsam family 258
banana 78
banana family 78
bastard cabbage 210, **211**
bay laurel 54
beaked hawk's-beard 370, **371**
bean broomrape **326**, 327
bearded oat 86, **87**
bear's breech 303
bedstraw 264
bee orchid 62
beet 245
bellflower 327
bellflower family 327
Bermuda buttercup 168
bermuda grass 94

Bermuda grass **93**, 94
bidens 344, **345**
bindweed 282, **283**, 284, 285
bindweed family 282
bird of paradise 78, **79**
bird's foot 148
bird's-foot trefoil 144
birthwort family 54
black bryony 58, **59**
black medick 142
black nightshade 288, **289**
blackwood acacia 122
bladder campion **237**, 238
bladder dock **228**, 229
bledo 241
blinks family 255
blue gum **187**, 188
blue hound's tongue 281
blue-leaved wattle **121**, 122
blue water-speedwell 299
bog orchid 64
borage 281
borage family 272
bottle tree 197 196
bougainvillea 254
bracken 46, **47**
bramble 155
branched broomrape 325
branched horsetail 44, **45**
bristle-fruited silkweed **267**, 268
bristle grass 98
bristly oxtongue 370, **371**
broad-leaved cudweed 330
brome 86
brookweed **260**, 261
broom 127
broomrape 325
broomrape family 323
brooms 19
bryony 168
bucker-fern family 48
buck's horn plantain 294, **295**
buckthorn 159
bugle 307

bugloss 274
bumblebee orchid 62, **63**
bunch-flowered daffodil 68, **69**
bur-marigold 344
buttercup family 102
buttercups 102

C

cabbage family 204
cactus family 255
calcosa 228 229
Californian pepper tree **187**, 189
California poppy **100**, 101
campion 236
Canadian fleabane 336, **337**
Canary Island holly **326**, 327
Canary Island juniper 51, **52**
Canary Island palm 12
Canary Island pine 12, 19, 30, 51, **52**
Canary Island willow 12
Canary laurel 54, **55**
Canary palm **77**, 78
candelabra tree 173, **174**
caper 204
caper family 204
cardoon 376, **377**
carline thistle 372
carob 124
carob tree **123**, 124
carrot 390
carrot family 383
castor oil plant **180**, 181
cat's ear 359
centaury 266
century plant 76, **77**
Century plants 20
chain fern 48, **49**
chamomile 338, 351
charlock **209**, 210
cherry 155
chickpea 133
chicory 355
Chilean needle grass 86, **87**

clover 148
cockleburs 344
cock's foot 89
columbine 105
common broomrape 325
common cat's-ear 359
common centaury 266
common cornsalad 382
common cudweed 330
common dandelion 370
common field-speedwell 300
common fumitory 102
common myrtle 188
common poppy 99, **100**
common rue 191
common self-heal 311
common smilax **60**, 61
common spike-rush 83
common stork's-bill 184, **185**
common vetch 135
compact brome 86
convolvulus 282
coral necklace 232
cork oak 165
corn chamomile 338
corncockle 236
corn crowfoot 102
cornflower 378
corn marigold 339
cornsalad 382
cotton 193
cotton bush 246
cottonweed 338
creeping cinquefoil 155
Cretan catchfly 236
crow garlic 68
crown daisy 339, **340**
cucumber family 166
cudweed 330, 331
cultivar banana 78
cultivated flax 181
curry plant 333
curved sea hard-grass **90**, 91
cut-leaved crane's-bill 182
cypress 53

D

daffodil 67
daisy 336
dandelions 370
date palms 12
dead-nettle 311
deer fern 48, **49**
dense-flowered orchid 62
desert gourd 166, **167**

dock 226
dock family 225
dodder 287
dog rose 155
dove's-foot crane's-bill 182
downy woundwort 312, **313**
dragon arum 56
dragon fruit 255, **256**
dwarf convolvulus 284
dwarf mallow 195
dwarf rush 80
dwarf spurge 179

E

early purple orchid 62
eastern rocket 205
edible cyperus 82
edible lotus 144
elder 379
elephant ear 58, **59**
elm 160
elm family 160
emex 226, **227**
euphorbia family 173
everlasting pea 137

F

fat hen 246
fence post cactus 255, **255**
fennel **384**, 385
fenugreek 142
fern-grass **90**, 91
field marigold **342**, 343
field woundwort 312, **313**
fig 160, **161**
fig family 160
figwort 300
figwort family 300
flax 181
flax family 181
fleabane 334
florist's cineraria 353
fodder vetch 135
forget-me-not 282
forking larkspur 105
fountain grass **96**, 97
four-leaved allseed 233, **234**
foxglove 299
foxgloves 40
French lavender 318
French oat-grass 88
French rocket 205, **206**
fringed rue **190**, 191
fumitory 101
furrowed melilot 140, **141**

G

galactites 375, **375**
gentian family 266
geranium family 182
germander 307
giant reed **93**, 94
goat's beard 359
golden dog's tail 89, **90**
gongarillo 113
goosefoot 246
goose-grass 264
Gran Canarian dragon tree 30
grape 118, **119**
grape family 118
grapefruit 192
grass family 84
grass poly 185, **185**
greater musk-mallow 195
greater periwinkle **267**, 268
greater plantain **295**, 296
greater sea-spurrey 235
great mullein **301**, 302
great willowherb 186
green bristle grass 98
grey field-speedwell 300
groundsel 339
gum rock rose 199

H

hairy-fruited cornsalad 382
hairy medick **143**, 144
hairy rocket 214
hairy tare 133
hairy trefoil 151
hard fern family 48
hard grass 91
hare's-foot clover **149**, 150
hare's-foot fern 48, **50**
hare's-foot fern family 48
hare's tail 88
Hawaiian lily **291**, 292
hawk's-beard 370
heather 262
heather family 261
hedge mustard 205, **206**
hedge parsley 389
heliotrope 272
hemlock 386
henbane 288
henbit dead-nettle 312, **313**
herb Robert 182
hoary mustard **209**, 210
hoary stock 208, **209**
hollow-leaved asphodel 66

INDEX OF ENGLISH NAMES 413

holly family 327
hollyhock 196
holm oak 165
honeysuckle 381
honeysuckle family 381
hop trefoil 148, **149**
horehound 307
horned dock 226, **227**
horn-wort 99
horn-wort family 99
horse mint 317
horsetail family 44
hottentot fig 252
Hottentot fig 252
hound's tongue 281

I

ice plant 252, **253**
Indian bead tree 193, **194**
iris family 64
Italian arum 56, **57**
ivy 383
ivy broomrape **326**, 327
ivy family 383
ivy-leaved speedwell 300
ivy-leaved toadflax **298**, 299

J

jade plant **104**, 106
Japanese loquat 156
jasmine 292
Jersey cudweed **332**, 333
Johnson grass 95, **96**
joint pine family 53
juniper 12, 51
juniper family 51

K

knapweed 378
knotted hedge parsley 389

L

large blue alkanet 281
large disk medick 142
large quaking grass **87**, 88
larkspur 105
laurel 55 54
laurel family 54
lavender 318
lemon 192
lesser calamint 313
lesser celandine 103
lesser hop trefoil 148, **149**
lesser sea spurrey 235
lesser snapdragon 296, **298**

lesser swinecress 208
lettuce 366
lime 192
limoniastrum 225
linaria 297
little robin 182, **183**
loeflingia 239
London rocket 205
long-headed poppy 99
long-styled feather grass 97
loosestrife 185
loosetrife family 185
loquat 156
love grass 95
love-in-a-mist 103
lucerne 142, **143**
lupin 129

M

madder family 262
maidenhair fern 44, **45**
mallow 195
mallow family 193
mallow-leaved stork's-bill 184, **185**
Maltese fungus 118, **119**
marigold 343
maritime pine 51
marjoram 317
marsh willowherb 188
marshwort 386
marvel of Peru 254
mastic 189
medick 142
Mediterranean buckthorn 159
Mediterranean hair grass 88
Mediterranean stork's-bill 184
melilot 140
membranous nettle 162
mercury 181
mignonette family 202
milk thistle 376, **377**
mint 317
mint family 306
monkey-puzzle family 53
Monterey cypress 53
Monterey pine 51
mossy stonecrop 106
mountain aloe 67 66
mournful widow **380**, 381
mouse-ear 232
musk stork's-bill 184, **185**
mustard 210
myrtle 188
myrtle family 188

N

narrow-fruited cornsalad 382
narrow-leaved crimson clover 148, **149**
narrow-leaved cudweed 331, **332**
narrow-leaved red vetchling 137
narrow-leaved rock rose **198**, 199
nasturtium 202, **203**
nasturtium family 202
navelwort 106
nettle family 162
nettles 162
New Zealand spinach **253**, 254
nightshade 288
nipplewort 371
nit-grass 88, 89
Norfolk Island pine **52**, 53

O

oak 165
oak family 164
oat 86
olive 12
olive 292, **293**
olive family 292
opium poppy 99
orache 243
orange bird's foot 148, **149**
orchid family 61
oxalis family 168

P

pale flax 181
palm family 76
pampas grass 98
papaya **203**, 204
peach 156, **157**
pea family 120
pear 156
pearlwort 233
peas 137
pellitory-of-the-wall 162, **163**
pelotilla 116
pennyroyal 317
pepperwort 208
perfoliate St John's-wort 170
periwinkle 268
periwinkle family 268
petty spurge 179
pigweed 243
pine 51
pine family 51

pink family 230
plantain 78, 294
plantain family 294
pliant lettuce 366
plum 155
polypody family 48
pomegranate 186
pond water-crowfoot 103
poppy 99
poppy family 99
Portugal laurel 156
potato family 287
potato vine 290
prasium 307
prickly lettuce 366, **367**
prickly pear 20
prickly pear **256**, 257
prickly poppy 101
prickly sow-thistle **358**, 359
primula family 258
procumbent pearlwort 233
procumbent yellow sorrel 168
proliferous pink **237**, 238
purple spurge 179
purslane 255

Q
quince 156, **157**

R
radish 212
red clover 151
red dead-nettle 312
red-horned poppy **100**, 101
red star-thistle **377**, 378
red stonecrop 106, **107**
red-tipped cudweed 331, **332**
red valerian 382, **384**
red vetchling 137
restharrow 139
resurrection plant **342**, 343
retama **128**, 129
reversed clover 151
ribbon fern family 44
ribwort plantain 296
ridolfia 386
river red gum 188
rocket 205
rock rose 199
rock rose family 199
rockroses 12
rock samphire **384**, 385
rose 153
rose family 153
rosemary 320

rough clover **149**, 150
rough dog's-tail **90**, 91
rough-fruited buttercup 102
round-headed leek 68
rubber plant 161, **161**
rue 191
rupturewort 233
rye grass 89
ryegrass 89

S
sage 318, 320
salsify **358**, 359
saltwort 249
sandbur 97
sandwort 230
scabious 381
scarlet pimpernel **259**, 260
scilla 74
Scotch thistle 376
sea beet 245
sea bindweed 285
seablite 249
sea daffodil 68
sea heath family 219
seakale 211
sea knotgrass 225, **227**
sea lavender 24, 34, 221
sea purslane 245
sea rocket **209**, 210
sea rush 80
sea spleenwort 46, **47**
sea spurge 177
sea-spurrey 235
sea squill 74, **75**
sedge 80
self-heal 311
Seville orange 192
sharp rush 80, **81**
sheep's sorrel 226, **227**
shepherd's needle **384**, 385
shepherd's purse 205
shrubby orache **244**, 245
shrubby seablite 249, **250**
shrub tobacco 290, **291**
silkweed 268
silver wattle 122
silvery lace bush 351, **351**
silvery plantain 294
sisal 76, **77**
slender bird's foot trefoil 146
slender centaury 266, **267**
slender thistle 374, **375**
small caltrops 120, **121**
small cudweed 331

small-flowered catchfly 236, 237
small-flowered hairy willowherb 186
small medick **143**, 144
small melilot 140, **141**
small nettle 162, **163**
small tree mallow **194**, 195
smooth cat's-ear 359
smooth sow-thistle 360, **361**
snapdragon 296
soft rush 80
southern bugle **306**, 307
southern red bartsia **322**, 323
sow-thistle 359
spanish broom 127, **128**
Spanish oyster plant 355, **356**
spear-leaved orache 245
spearmint 317
spear thistle 374, **375**
speedwell 299
spicate Venus's looking glass **328**, 329
spiked centaury 266
spindle family 168
spiny cocklebur 344, **345**
spleenwort family 46
spotted medick 143
spurge 173
square-stalked willowherb 188
squirting cucumber 166, **167**
stalked bur grass **93**, 95
star clover 150
star hawkbit 359
sticky mouse-ear 232
stinking goosefoot 246
stinking mayweed 338
stock 207
stonecrop 106
stork's bill 184
stranglewort 268
strawberry clover 151
strawberry tree 261
suffocated clover 150
sumach **187**, 189
sun spurge 179
sweet alison 208
sweet chestnut 164, **164**, 165
sweet orange **190**, 192
sweet violet 170
Syrian thistle 374

T
tall ramping-fumitory **100**, 102
tamarisk 221

tamarisk family 221
tara **123**, 124
tassel hyacinth 74, **75**
tea tree 288
thistle 374
three-lobed sage 320
three-lobed stork's-bill 184
thrift family 221
thyme 317
thyme-leaved sandwort 230
thyme-leaved speedwell 299
tobacco 290
tolpis 356, **357**
tongue orchid 64
torpedo grass 98
tree aloe 66
tree heaths 12
tree houseleek 106
tree mallow 196
tree medick 142, **143**
tree of heaven 192
twiggy mullein **301**, 302
two-leaved gennaria 64, **65**

U

upright yellow flax **180**, 181

V

Venus's looking glass 329
verbena 306
vervain 306
vetch 133
violet cabbage **206**, 207
violet family 170
violets 170

W

wall barley 92, **93**
wallflower 207
wall rocket 208
wall speedwell 299
walnut 165
water plantain 58
weld 202, **203**
white broom 12
white clover **149**, 150
white henbane 288, **289**
white horehound **306**, 307
white lupin 130, **131**
white melilot 142
white mustard 210
white ramping-fumitory 101
wild carrot 390, **391**

wild clary 320, **321**
wild leek 68
wild madder 262, **263**
wild mignonette 202
wild radish 212, **213**
wild teasel 381
wild thyme 21
willow family 172
willowherb 186
willowherb family 186
willow-leaved rock rose 200
winged sea lavender 222
winged vetchling **136**, 137
woad 205
woolly trefoil **149**, 150
wormwood 336

Y

yellow bartsia **322**, 323
yellow centaury 266
yellow gromwell **273**, **273**
yellow horned-poppy **100**, 101
yellow lupin 129
yellow vetch 135, **136**
yellow vetchling **136**, 137
yellow-wort 266

Caralluma burchardii

Index of scientific names

Species entries in roman text are synonyms; entries with a photograph are indicated by a **bold** page number.

A

Acacia 120
  cyanophylla 122
  cyclops **121**, 122
  dealbata 122
  farnesiana 122
  melanoxylon 122
  saligna **121**, 122
ACANTHACEAE 303
Acanthus 303
  mollis 303
Achillea 338
  maritima 338
Achyranthes 243
  aspera **242**, 243
Adenocarpus 19, 129
  foliolosus 129, **131**
    var. *foliolosus* 129
  ombriosus 38
  viscosus 19, 33, 129, **131**
Adiantum **44**
  capillus-veneris 44, **45**
  reniforme 27, 44, **45**
    var. *pusillum* 44
    var. *reniforme* 44
Adonis 105
  microcarpa **104**, 105
ADOXACEAE 379
Aegilops 92
  geniculata 92
  neglecta 92
Aeonium 12, 40, 106
  aizoon 112 **113**
  appendiculatum 106
  arboreum **109**
    subsp. *arboreum* **108**, 109
    subsp. *holochrysum* 109
  aureum 30, **112**, 113
  balsamiferum 21, **110**, 111
  canariense 111, **112**
    subsp. *canariense* 111
    subsp. *christii* 36, 111
    subsp. *latifolium* 111
    subsp. *virgineum* 111
  castello-paivae 34, **108**, 109
  ciliatum 109, **110**
  cuneatum **112**, 113
  davidbramwellii 111
  decorum 34, **108**, 109
  diplocyclum 34, 38, 113
  dodrantale 113, **114**
  gomerense 111
  goochiae 36, 111
  haworthii 111
  hierrense 109
  holochrysum **108**, 109
  lancerottense 21, 22, 109, **110**
  lindleyi **110**, 111
    subsp. *lindleyi* 111
    subsp. *viscatum* 111, **112**
  manriqueorum 109
  nobile 36, 109
  palmense 36, 111
  percarneum **108**, 109
  pseudourbicum 33, 109
  rubrolineatum 34, **108**, 109
  saundersii 111
  sedifolium **110**, 111
  simsii 27, 30, 112, **113**
  smithii 111
  spathulatum 111
    var. *cruentum* 111
    var. *spathulatum* 111
  subplanum 111
  tabulaeforme **6**, 30, 33, 111, **112**
  undulatum **108**, 109
  urbicum **107**, 109
    var. *meridionale* **107**, 109
    var. *urbicum* 109
  valverdense 38, 111
  vestitum 109
  viscatum 34
  volkeri
    subsp. *paucifolium* **110**, 111
Agave 20, 76
  americana 20, 76, **77**
  attenuata 20, 76, **77**
  fourcroydes 20, 76, **77**

sisalana 20, 76, **77**
Agrimonia 153
   eupatoria 153
Agrostemma 236
   githago 236
Agrostis 84
   castellana 84, **85**
Aichryson 12, 33, 40, 113
   bethencourtianum 113, **114**
   bollei 113
   brevipetalum 36, 113
   diplocyclum **114**
   laxum 113, **114**
   pachycaulon 116
      subsp. *gonzalezhernandezii* **115**, 116
      subsp. *immaculatum* **114**, 116
      subsp. *pachycaulon* 116
      subsp. *parviflorum* 116
      subsp. *praetermissum* 116
   parlatorei 113
   porphyrogennetos 113
   punctatum **115**, 116
   tortuosum 21, 113, **114**
Ailanthus 192
   altissima 192
AIZOACEAE 251
Aizoanthemopsis hispanica 251
Aizoon 251
   canariense **11**, 251
   canariense 253 **252**
   hispanicum 251, **253**
Ajuga 307
   iva **306**, 307
      var. *pseudoiva* 307
Alcea 196
   rosea 196
Alisma 58
   lanceolatum 58
ALISMATACEAE 58
Allagopappus 346
   canariensis 346, **347**
   viscosissimus 346, **347**
Allium 68
   ampeloprasum 68
   canariense **69**
   nigrum 69
   pallens 68, **69**
   sphaerocephalum 68
   subhirsutum **69**
   vineale 68
Aloe 66
   arborescens 66
   marlothii 66, **67**
   vera 20, 66, **67**
Alyssum 204

simplex 205
AMARANTHACEAE 9, 241
Amaranthus 241
   albus 243
   blitoides 243
   blitum 242
   cruentus 242
   deflexus **240**, 241
   hybridus 242
   hypochondriacus 242
   lividus 242
   muricatus **240**, 241
   retroflexus 243
   viridis 241, **242**
AMARYLLIDACEAE 67
Ammi 387
   majus 387
   visnaga 387
ANACARDIACEAE 189
Anacyclus 338
   clavatus 338
   radiatus 339
      subsp. *coronatus* 339, **340**
      subsp. *radiatus* 339
Anagalis foemina 260
Anagallis arvensis 260
Anagyris 125
   latifolia 33, 125, **126**
Anchusa 281
   azurea 281
Androcymbium
   hierrense 61
   psammophilum 61
Andropogon 97
   distachyos 97
Andryala 368
   integrifolia 368, **369**
   perezii 368
   pinnatifida 368, **369**
      subsp. *pinnatifida* 368, **369**
      subsp. *preauxiana* 368, **369**
      subsp. *teydensis* 33
Anthemis 338
   arvensis 338
   cotula 338
Antirrhinum 296
   majus 296
APIACEAE 383
Apium 386
   graveolens 386, **387**
APOCYNACEAE 268
Apollonias barbujana 54
Aptenia cordifolia 252
Apteranthes 271
   burchardii 24, **25**, 269, **270**, 271

INDEX OF SCIENTIFIC NAMES | 419

subsp. *burchardii* 271
AQUIFOLIACEAE 327
Aquilegia vulgaris 105
ARACEAE 56
ARALIACEAE 383
Araucaria 53
  heterophylla **52**, 53
ARAUCARIACEAE 53
Arbutus 12, 261
  canariensis **17**, 33, **260**, 262
  unedo 261
ARECACEAE 76
Arenaria 230
  leptoclados 230
  serpyllifolia 230
Argyranthemum 40, 41, 346
  adauctum 346
  broussonetii 41, **347**, 348
    subsp. gomerensis 348
  callichrysum 348, **349**
    subsp. *callichrysum* 348
    subsp. *gomerensis* 348
  coronopifolium 33, 348, **349**
  escarrei 350
  filifolium 350
  foeniculaceum 348
  frutescens 41, 346, 348
    subsp. *canariae* 346, **347**
    subsp. *pumilum* 346, **347**
    subsp. *succulentum* 346, **347**
  gracile 348, **349**
  haouarytheum 350
  hierrense 38, **349**, 350
  lemsii 41, 346, **347**, 348
  lidii 350
  maderense 21, **349**, 350
  sundingii 41, 346, **347**
  sventenii **349**, 350
  tenerifae 19, 20, 33, 348, **349**
  vicentii 348
  webbii 350
  winteri 26, **349**, 350
Arisarum 56
  simorrhinum 56, **57**
Aristida 92
  adscensionis 92, **93**
Aristolochia 54
  paucinervis 54
ARISTOLOCHIACEAE 54
Arnebia 272
  decumbens 272
Arrhenatherum 84
  calderae 19, 84, **85**
Artemisia 336

ramosa 336, **337**
reptans 336
thuscula 336, **337**
Arthrocaulon 21, 249
  macrostachyum 9, **248**, 249
Arthrocnemum
  macrostachyum 249
  perenne 249
Arum 56
  italicum
    subsp. *canariense* 56, **57**
Arundo 94
  donax 20, **92**, **93**, 94
ASPARAGACEAE 70
Asparagus 70
  arborescens 72, **73**
  fallax 72
  horridus 72, **73**
  nesiotes **72**
    subsp. *purpuriensis* 72, **73**
  pastorianus 24, **71**, 72
  plocamoides 72, **73**
  scoparius 72
  umbellatus 72, **73**
ASPHODELACEAE 66
Asphodelus 66
  ayardii 66
  fistulosus 66
  ramosus 24, 66, **67**
  tenuifolius 9, **11**, 66, **67**
ASPLENIACEAE 46
Asplenium
  aureum 46, **47**
    var. *aureum* 46
  hemionitis 46, **47**
  marinum 46, **47**
  onopteris 12, **13**, 46, **47**
Aster 355
  squamatus **354**, 355
ASTERACEAE 330
  subfamily ASTEROIDEAE 330
  subfamily CARDUOIDEAE 372
  subfamily CICHORIOIDEAE 355
Asteriscus 344
  aquaticus 344
  graveolens 344
    subsp. *odorus* 344
    subsp. *stenophyllus* 344, **345**
  intermedius **10**, 344, **345**
  pygmaeus 343
  schultzii 21, 24, **345**
  sericeus 24, 344, **345**
Astragalus 130
  boeticus 130, **131**, 132
  edulis 130, 132

hamosus 130
mareoticus 130, **131**, 132
mareoticus 131
sinaicus 130, **131**, 132
solandri 130, **131**, 132
stella **132**, 133
Astydamia 385
latifolia **8**, 24, **384**, 385
Athamanta 393
cervariifolia **392**, 393
montana **392**, 393
Atractylis 372
arbuscula 372
subsp. *arbuscula* 372, **373**
subsp. *schizogynophylla* 372
cancellata 372, **373**
preauxiana 30, 372, **373**
Atriplex 243
glauca 245
subsp. *ifniensis* 9, **244**, 245
halimus 7, **8**, **244**, 245
hastata 245
lindleyi **242**, 243
portulacoides **244**, 245
prostrata 245
semibaccata **242**, 243
semilunaris **242**, 245
suberecta **242**, 243
Austrocylindropuntia 257
cylindrica **256**, 257
exaltata 257
subulata **256**, 257
Avena 86
barbata 86, **87**
canariensis 86, **87**

B

Babcockia platylepis 362
Ballota 312
nigra 312
BALSAMINACEAE 258
Bartsia 323
trixago 323
Bassia 245
tomentosa **10**, **244**, 245
Bellardia trixago 323
Bellis 336
annua 336
perennis 336
Bencomia 159
brachystachya **158**, 159
caudata 33, **157**, 159
exstipulata **158**, 159
sphaerocarpa 38, 159
Beta 245

macrocarpa **244**, 245
vulgaris 245
Bethencourtia 339
hermosae 34
palmensis 33, 339, **340**
rupicola 34
Bidens 344
aurea 344, **345**
pilosa **345**, 346
BIGNONIACEAE 305
Biserrula 133
pelecinus 133, **134**
Bismarckia 78
nobilis **77**, 78
Bituminaria 133
bituminosa 133, **134**, 325
var. *albomarginata* 133, **134**
var. *crassiuscula* 133, **134**
Blackstonia 266
perfoliata 266
BLECHNACEAE 48
Blechnum **48**
spicant **48, 49**
Boerhavia helenae 254
BORAGINACEAE 272
Borago 281
officinalis 281
Bosea 241
yervamora **240**, 241
Bougainvillea 254
glabra 254
Brachychiton 196
populneus **196**, **197**
Brachypodium
arbuscula 34
BRASSICACEAE 204
Briza 88
maxima **87**, 88
minor **87**, 88
Bromus 86
diandrus 86, **87**
hordeaceus 86
subsp. *hordeaceus* 86
subsp. *molliformis* 86, **87**
madritensis 86
squarrosus 86
Bryonia 168
verrucosa **167**, 168
Bryophyllum daigremontianum 116
Buglossoides 272
arvensis 272
Bupleurum 386
handiense 386
salicifolium 386, **387**
subsp. *aciphyllum* 33, 386

subsp. *salicifolium* 386
semicompositum 386, **387**
Bystropogon 322
  canariensis **321,** 322
  origanifolius 322, **322**
  plumosus 322

C

CACTACEAE 255
Caesalpinia 124
  gilliesii 124
  spinosa **123,** 124
Cakile 210
  maritima **209,** 210
Calamintha nepeta 313
Calendula 343
  arvensis **9, 342,** 343
  officinalis 343
  suffruticosa 343
  sventenii **342,** 343
Calotropis 268
  procera **267,** 268, 269
Calystegia 285
  soldanella 285
Campanula 327
  erinus **328,** 329
  occidentalis **328,** 329
CAMPANULACEAE 327
Camptoloma 303
  canariense 27, 303, **304**
Campylanthus 303
  salsoloides 22, 303, **304**
Canarina 329
  canariensis 2, **29, 328,** 329
Canna indica 79, **79**
CANNACEAE 79
CAPPARACEAE 204
Capparis 204
  spinosa 204
CAPRIFOLIACEAE 381
  former VALERIANACEAE 381
Capsella 205
  bursa-pastoris 205
Caralluma burchardii 271
Cardiospermum 191
  grandiflorum 191
Carduncellus 379
  caeruleus 379
Carduus 374
  baeocephalus 374, **375**
    subsp. *microstigma* 374, **375**
  clavulatus 374, **375**
  pycnocephalus 374, **375**
  tenuiflorus 374, **375**

volutarioides 374, **375**
Carex 80
  canariensis 80, **81**
  divulsa **81,** 82
  otrubae 82
  paniculata 82
    subsp. *calderae* **81,** 82
  perraudieriana **81,** 82
Carica papaya **203,** 204
CARICACEAE 204
Carlina 372
  canariensis 372, **373**
  salicifolia 372, **373**
    subsp. *lancerottensis* 372, **373**
    subsp. *salicifolia* 372, **373**
  xeranthemoides 372, **373**
Caroxylon 9, 22, 249
  tetrandrum **248,** 249
  vermiculata **10**
  vermiculatum 9, **248,** 249
Carpobrotus 252
  edulis 252
Carrichtera 214
  annua **10, 213,** 214
Carthamus 379
  caeruleus 379
  lanatus 379, **380**
CARYOPHYLLACEAE 230
Cascabela 271
  thevetia **270,** 271
Cassia
  corymbosa 124
  didymobotrya 124
Castanea 164
  sativa **164,** 165
Casuarina 166
  equisetifolia 166, **167**
CASUARINACEAE 166
Catapodium 91
  rigidum **90,** 91
Ceballosia fruticosa 273
Cedronella 322
  canariensis 322, **322**
CELASTRACEAE 168
Cenchrus 97
  ciliaris **10, 96,** 97
  longisetus 97
  setaceus 20, 97
Centaurea 378
  calcitrapa 378
  melitensis 378
Centaurium 266
  erythraea 266
  maritimum 266
  spicatum 266

tenuiflorum 266, **267**
Centranthus 381
  calcitrapa 381
  ruber 382, **384**
Cerastium 232
  fontanum 232
  glomeratum 232
Ceratonia 124
  siliqua **123**, 124
CERATOPHYLLACEAE 99
Ceratophyllum 99
  demersum 99
Ceropegia 271
  ceratophora 271
  chrysantha 271
  dichotoma 33, **270**, 271
    subsp. *dichotoma* 271
    subsp. fusca 271
    subsp. *krainzii* 271
  fusca 27, **29**, **270**, 271
Ceterach aureum 46
Chamaecytisus 127
  proliferus 19, 127
    subsp. *angustifolius* **126**, 127
    subsp. *meridionalis* 127
    subsp. *proliferus* 127
      var. palmensis 127, **128**
Chamaemelum 351
  fuscatum 351, **351**
  mixtum 338
Chamaesyce
  canescens 179
  maculata 179
  peplis 179
  prostrata 179
  serpens 179
Chasmanthe 65
  floribunda **65**
Cheilanthes **44**
  maderensis 44, **45**
  pulchella 44, **45**
Cheirolophus 378
  burchardii 378, **380**
  canariensis 378
  duranii 38
  ghomerythus 34
  junonianus 36
  santos-abreui 36
  satarataensis 34
  sventenii 36
  teydis 378, **380**
Chenoleoides tomentosa 245
Chenopodiastrum 246
  murale **244**, 246
Chenopodium 246

album 246
murale 246
nutans **244**, 246
vulvaria 246
Chondrilla 366
  juncea 366
Chrysanthemum 339
  coronarium 339
  segetum 339
  pendula 360
Cicer 133
  canariense 133, **134**
Cichorium 355
  pumilum **354**, 355
Cirsium 374
  vulgare 374, **375**
CISTACEAE 199
Cistanche 323
  phelypaea 7, 22, **322**, 323, 324
Cistus 12, 199
  asper 38
  chinamadensis **198**, 199
    subsp. *chinamadensis* 199
    subsp. *gomerae* 34, 199
    subsp. *ombriosus* 38, 199
  grancanariae **198**, 199
  horrens **198**, 199
  ladanifer 199
  monspeliensis 19, **198**, 199
  ocreatus 30, **198**, 199
  osbeckiifolius 33, **198**, 199
  symphytifolius 19, **198**, 199
Citrullus 166
  colocynthis 166, **167**
Citrus 192
  × *aurantiifolia* 192
  × *aurantium* 192
  × *limon* 192
  × *paradisi* 192
  × *sinensis* **190**, 192
Cladanthus 338
  mixtus 338
Cladonia
  foliacea 9
Clinopodium 313
  nepeta 313
Cneorum 192
  pulverulentum 192
Coccoloba 229
  uvifera **228**, 229
COLCHICACEAE 61
Colchicum 61
  hierrense **59**, 61
  psammophilum 24, **59**, 61
Coleostephus 339

myconis 339, **340**
Colocasia 58
   esculenta 58, **59**
Commicarpus 254
   helenae **253**, 254
Conium 386
   maculatum 386
Consolida 105
   ajacis 105
   regalis 105
CONVOLVULACEAE 282
Convolvulus 282
   althaeoides **283**, 284
      subsp. *althaeoides* 284
   arvensis **283**, 284
   canariensis 27, **283**, 284
   caput-medusae 30, 282, **283**
   farinosus **283**, 284
   floridus 282, **283**
   fruticulosus **283**, 284
      subsp. glandulosus 284
   glandulosus 27, 284
   lopezsocasi **283**, 284
   perraudieri 284
   scoparius 30, 282, **283**
   siculus **283**, 284
   subauriculatus 34, 284
   tricolor 284
Conyza
   bonariensis 336
   canadensis 336
   sumatrensis 336
Coronilla 152
   valentina 152
   viminalis **152**
Coronopus
   didymus 208
   squamatus 208
Cortaderia 98
   selloana 98
Cosentinia **46**
   vellea 46, **47**
Crambe 211
   arborea 212
      var. *arborea* 212
      var. *indivisa* 33, **211**, 212
   feuilleei 38, 211
   gomerae 212
   laevigata 212
   pritzelii 211
   santosii 212
   scaberrima **211**
   scoparia 30, 212
   strigosa 33, **211**

sventenii 212, **213**
tamadabensis 212
Crassula 105
   lycopodioides **104**, 106
   ovata **104**, 106
   tillaea 106
CRASSULACEAE 12, 105
Crepis 370
   canariensis 370, **371**
   capillaris 370
   foetida 370
   neglecta 370
   vesicaria 370
      subsp. *taraxacifolia* 370, **371**
Cressa 287
   cretica **286**, 287
Crinum 70
   × *amabile* 70
   × *amabile* 71 **70**
   asiaticum × *C. zeylanicum* 70
Crithmum 385
   maritimum **384**, 385
Cryptotaenia elegans 390
CUCURBITACEAE 166
CUPRESSACEAE 51
Cupressus 53
   macrocarpa 53
   sempervirens 53
Cuscuta 287
   approximata 9, **22**, **286**, 287
   planiflora **286**, 287
Cydonia 156
   oblonga 156, **157**
Cylindropuntia 257
   imbricata 257
Cymbalaria 299
   muralis **298**, 299
Cynanchum 268
   acutum 268
Cynara 376
   cardunculus 24, 376, **377**
      var. *ferocissima* 376, **377**
   humilis 376
Cynodon 94
   dactylon **93**, 94
Cynoglossum 281
   creticum 281
CYNOMORIACEAE 118
Cynomorium
   coccineum 22, 118, **119**
Cynosurus 89
   echinatus **90**, 91
CYPERACEAE 80
Cyperus 82
   alternifolius

subsp. *flabelliformis* 82, **83**
capitatus 82, **83**
eragrostis **83**
esculentus 82
involucratus 82
juncia clara **82**
laevigatus 82, **83**
longus 82
teneriffae 82, **83**
Cystanche
phelypaea **7**
CYTINACEAE 193
Cytinus 193
hypocistis
subsp. *subexsertum* 193, **194**
ruber 193, **194**
Cytisus 127
scoparius 127

D

Dactylis 89
glomerata 89
metlesicsii 89, **90**
smithii 89, **90**
Daphne 197
gnidium 12, **197**
Datura 290
innoxia 290, **291**
metel 290, **291**
stramonium 290, **291**
Daucus 390
aureus 390
carota 390, **391**
subsp. *maximus* 390
durieua 390
elegans 390, **391**
Davallia
canariensis 12, **13**, 48, **50**
DAVALLIACEAE 48
Delairea 343
odorata **342**, 343
Delonix 122
regia 122, **123**
Delphinium 103
staphisagria 103, **104**
Dendriopoterium
pulidoi 30
DENNSTAEDTIACEAE 46
Descurainia 214
artemisioides 214
bourgaeana 33, 214, **215**
gilva **216**, 217
gonzalesii 214, **215**
lemsii 214, **215**
millefolia **216**, 217

preauxiana 27, 214, **215**
Desmazeria rigida 91
Dicheranthus 238
plocamoides **237**, 238
Dichondra 287
micrantha 287
Digitalis 40, 299
canariensis **3**, 33, **298**, 299
chalcantha 27, 299
isabelliana 30, 299
Dioscorea 58
edulis 58, **59**
DIOSCOREACEAE 58
Dipcadi 70
serotinum 9, 70, **71**
Diplazium
caudatum 12
Diplotaxis 208
tenuifolia 208
Dipsacus 381
fullonum 381
Dittrichia 334
viscosa 334, **335**
Dorycnium
broussonetii 146
eriophthalmum 148
spectabile 148
Dracaena 70
draco 36, 70, **71**
tamaranae 30, 70, **71**
Dracunculus 56
canariensis **30**, 33, **57**, 58, **59**
Drimia 74
hespera **75**
hesperia 74
maritima 74, **75**
Drosanthemum 252
floribundum 252
Drusa 390
glandulosa 390, **391**
DRYOPTERIDACEAE 48
Dryopteris
guanchica 48, **49**
oligodonta 12, **13**, 48, **50**
Dutra 290
Dysphania 246
ambrosioides 246, **247**

E

Ecballium 166
elaterium 166, **167**
Echium **3**, 12, 40, 274
acanthocarpum **280**, 281
aculeatum **276**, 278
auberianum 30, 274, **275**

bethencourtii 36, **277**, 278
bonnetii 274, **275**
brevirame 36, **277**, 278
callithyrsum **280**, 281
decaisnei 27, 278
   subsp. *decaisnei* 278, **279**
   subsp. *purpuriense* 21, 278
   subsp. *purpuriense* **279**
famarae 278
gentianoides 36, **277**, 278
giganteum **276**, 278
handiense 26, 281
hierrense 278, **280**
horridum 274
lancerottense 274, **275**
leucophaeum **277**, 278
onosmifolium 27, **280**, 281
   subsp. *spectabile* 281
perezii 36, **37**, 40, 274, **276**
pininana 36, 40, 274, **276**
plantagineum 274, **275**
simplex 33, 40, 274, **276**
strictum 278, **279**
sventenii 278
triste 34, 274, **275**
virescens 278, **279**
webbii 36, 278, **280**
wildpretii 3, 19, 30, 36, 40, 274, **276**
   subsp. trichosiphon 274
Einadia nutans 246
Eleocharis 83
   multicaulis 83
   palustris 83
Eleusine 94
   indica **93**, 94
Emex spinosa 226
Ephedra 53
   fragilis **52**, 53
   nebrodensis **52**, 53
EPHEDRACEAE 53
Epilobium 186
   hirsutum 186
   palustre 188
   parviflorum 186
   tetragonum 188
EQUISETACEAE 44
Equisetum
   ramosissimum 44, **45**
Eragrostis 95
   barrelieri 95, **96**
Erica 262
   canariensis 12, **14**, 33, 36, 262, **263**
   platycodon **14**, 262, **263**
ERICACEAE 261

Erigeron 334
   bonariensis 336, **337**
   calderae 334, **335**
   canadensis 336, **337**
   karvinskianus 336, **337**
   sumatrensis 336, **337**
Eriobotrya 156
   japonica 156
Erodium 184
   botrys 184
   chium 184
   cicutarium **183**, 184
   hesperium **183**, 184
   laciniatum 184
   malacoides **183**, 184
   moschatum 184, **185**
   neuradifolium 184
   salzmannii 184
   touchyanum 184, **185**
Eruca 212
   sativa 212
   vesicaria 212
Erucastrum 214
   canariense **213**, 214
   cardaminoides 214
Erysimum 207
   albescens 27, 207
   scoparium 33, **206**, 207
   virescens 38, **206**, 207
Erythrostemon gilliesii 124
Eschscholzia 101
   californica **100**, 101
Eucalyptus 188
   camaldulensis 188
   globulus **187**, 188
Euphorbia 24, 173
   aphylla 27, 173, **175**
   atropurpurea 33, **176**, 177
   balsamifera 3, 9, 21, 30, 33, **176**, 177
   berthelotii 177, **178**
   bourgaeana 177
   bravoana 34, 177, **178**
   canariensis 9, 24, **29**, 33, 173, **174**
   chamaesyce 179
   exigua 179
   grandicornis 173, **175**
   handiensis 3, 26, 173, **174**
   helioscopia 179
   ingens 173, **174**
   lamarcki 38
   lamarckii 177, **178**
   lambii 177, **178**
   maculata 179
   mellifera **18**, 173, **174**

neriifolia 173, **175**
paralias 7, **8**, 22, 177, **180**
peplis 179
peplus 179
prostrata 179
regis-jubae 21, **176**, 177
segetalis 177, **178**
serpens 179, **180**
serrata 177
terracina 179, **180**
tirucalli 173, **175**
trigona 173, 175
EUPHORBIACEAE 173
Evax pygmaea 331

F

FAGACEAE 164
Fagonia 120
  cretica **11**, 120, **121**
Fallopia 229
  convolvulus 229
Ferula 389
  arnoldiana **388**, 389
  lancerotensis **22**, **388**, 389
  latipinna 389
  linkii **388**, 389
Festuca 84
  agustinii 84, **85**
Ficaria 103
  verna 103
Ficus 160
  benjamina **161**
  carica 160, **161**
  elastica **161**
Filago 330
  desertorum 330, **332**
  gallica 331
  germanica 330
  lutescens 331, **332**
  minima 331
  pygmaea 331
  pyramidata 330
  vulgaris 330
Foeniculum 385
  vulgare **384**, 385
Forsskaolea 162
  angustifolia 162, **163**
Frankenia 219
  boissieri 219
  capitata 219, **220**
  ericifolia 219, **220**
  pulverulenta **10**, **218**, 219
FRANKENIACEAE 219
Fumaria 101
  bastardii **100**, 102

capreolata 101
coccinea **100**, 102
montana **100**, 102
muralis **100**, 102
officinalis 102
parviflora **100**, 102
vaillantii 102

G

Galactites
  tomentosus 375, **375**
Galium 264
  aparine 264
  intricatum 264
  murale 264
  parisiense 264
  scabrum 264, **265**
  spurium 264
  tricornutum 264
  verrucosum 264
Gamochaeta antillana 331
Gastridium 88
  phleoides 89, **90**,
  ventricosum 89
Gaudinia 88
Gennaria 64
Genista 125
  benehoavensis 36, 125, **126**
  canariensis 125, **126**
  microphylla 27, 125, **126**
  nervosa 125, **126**
  osyrioides 125
  pallida 125
    subsp. *gomerae* 125, **126**
    subsp. *pallida* 125, **126**
    subsp. *silensis* 125
  rosmarinifolia 125, **126**
  salsoloides 33
  stenopetala 125
Gennaria 64
  diphylla 64, **65**
GENTIANACEAE 266
GERANIACEAE 182
Geranium 182
  canariense 182
  dissectum 182
  molle 182
  purpureum 182, **183**
  reuteri 182, **183**
  robertianum 182
  rotundifolium 182, **183**
Gesnouinia 164
  arborea **164**
Gladiolus 64
  italicus 65
Glaucium 101

corniculatum **100,** 101
flavum **100,** 101
Glebionis 339
   coronaria 339, **340**
   segetum 339
Globularia 300
   ascanii 300
   salicina 300, **301**
   sarcophylla 300
Gnaphalium 331
   antillanum 331, **332**
   luteoalbum 333
Gomphocarpus 268
   fruticosus **267,** 268, 269
Gonospermum 351
   ferulaceum **351,** 353
   fruticosum 351, **351**
   ptarmiciflorum 27, 351, **351**
   revolutum **351,** 353
Gossypium 193
   herbaceum **194,** 195
Greenovia 12
   aizoon 113
   aurea 113
   diplocycla 113
   dodrentalis 113
Gymnocarpos 230
   decandrus 230, **231**
Gymnosporia 168
   cassinoides **167,** 168
   cryptopetala **167,** 168

H

Habenaria 64
   tridactylites 64, **65**
Halimione portulacoides 245
Heberdenia 258
   excelsa **16,** 33, 258, **259**
Hedera 383
   canariensis 383, **384**
   helix 383
Hedypnois 356
   arenaria 356
   cretica 356
   rhagadioloides 356, **358**
Helianthemum 200
   bramwelliorum 200, **201**
   broussonetii 200, **201**
   bystropogophyllum 200
   canariense 12, 200, **201**
   cirae 200
   gonzalezferreri 200, **201**
   henriquezii 200
   juliae 200, **201**

ledifolium 200, **201**
linii 200
salicifolium 200
teneriffae 200
tholiforme 27, 200
thymiphyllum 200, 201
Helichrysum 333
   gossypinum **22, 332,** 333
   monogynum **22, 332,** 333
   orientale 333
Heliotropium 272
   bacciferum 7, **8,** 272, **273**
   curassavicum 273
   europaeum 272
   messerschmidioides 273
   supinum 272
Helminthotheca 370
   echiodes 370, **371**
Herniaria 233
   canariensis 233
   cinerea 233, **234**
   fontanesii 233
   hartungii 233
Himantoglossum 62
   metlesicsianum 62, **63**
Hippocrepis 153
   multisiliquosa **152,** 153
Hirschfeldia 210
   incana **209,** 210
Hordeum 92
   murinum 92, **93**
Hylocereus 255
   undatus 255, **256**
Hyophorbe 76
   verschaffeltii 76, **77**
Hyoscyamus 288
   albus 288, **289**
Hyparrhenia 97
   sinaica **96,** 97
HYPERICACEAE 169
Hypericum 169
   canariense 12, 170
   coadunatum 27
   glandulosum **169,** 170
   grandifolium 12, **169**
   perfoliatum 170
   perforatum 170
   reflexum **169,** 170
Hypochaeris 359
   glabra 359
   oligocephala **358,** 359
   radicata 359

I

Ifloga 331

spicata 331, **332**
Ilex
   canariensis 12, **18**, 33, 34, 327
      subsp. *lopezlilloi* 327
      subsp. *platyphylla* 327
   canariensis 328 **329**
   perado 327
      subsp. *platyphylla* 12, **17**, 33, 327, **328**
Illecebrum 232
   verticillatum 232
Impatiens 258
   sodenii 258, **259**
   walleriana 258
Inula viscosa 334
Ipomoea 285
   batatas 285
   cairica 285, **286**
   corymbosa 285
   hederacea 285
   imperati 287
   indica 285, **286**
   nil 285
   pes-caprae 285
   purpurea 285
IRIDACEAE 64
Isatis 205
   tinctoria 205
Isoplexis canariensis 299
Ixanthus 267
   viscosus 267

J

Jasminum 292
   odoratissimum 292, **293**
JUGLANDACEAE 165
Juglans 165
   regia 165
JUNCACEAE 80
Juncus 80
   acutus 80, **81**
      subsp. *leopoldii* 80
   capitatus 80
   effusus 80
   inflexus 80
   maritimus 80
Juniperus 51
   canariensis 12, 34, 38, 51, **52**
   cedrus **30**, 33, 36, 51, **52**
   turbinata subsp. canariensis 51
Justicia 303
   hyssopifolia 303, **304**

K

Kalanchoe 116
   beharensis **115**, 116
   daigremontiana **115**, 116
Kickxia 297
   commutata 297, **298**
   elatine 297
   heterophylla 297, **298**
   pendula 27
   scoparia 297, **298**
Kleinia 341
   neriifolia **11**, 12, 24, 341, **342**
Kohlrauschia
   prolifera 238
   velutina 238
Kunkeliella
   canariensis 27

L

Lactuca 366
   palmensis 366
   serriola 366, **367**
   viminea 366
Lactucosonchus
   beltraniae 360
   webbii 360
Lagurus 88
   ovatus 88
Lamarckia 89
   aurea 89, **90**
LAMIACEAE 306
Lamium 311
   amplexicaule **311**, 312
   purpureum 312
Lantana
   camara 305
Laphangium luteoalbum 333
Lapsana 371
   communis 371
Lathyrus 137
   angulatus 139
   annuus **136**, 137
   aphaca **136**, 137
   articulatus **138**, 139
   cicera 137
   clymenum **138**, 139
   ochrus **136**, 137
   odoratus 137
   setifolius 137
   sphaericus 137, **138**
   tingitanus 137, **138**
Launaea 366
   arborescens 9, 366, **367**
   nudicaulis 366, **367**
Launaea
   arborescens **11**

LAURACEAE 54
Laurus 54
   novocanariensis 12, **15**, 33, 34, 54, **55**
Lavandula 318
   bramwellii 318
   buchii 318, **319**
   canariensis 318, **319**
      subsp. *canariae* 318
      subsp. *canariensis* 318
      subsp. *fuerteventurae* 318
      subsp. *gomerensis* 318
      subsp. *hierrensis* 318
      subsp. *lancerottensis* 318
      subsp. *palmensis* 318
   minutolii 318, **319**
   pinnata 21, 318, **319**
   stoechas 318
Lavatera
   acerifolia 27
   arborea 196
   cretica 195
Legousia 329
   falcata **328**, 329
   hybrida 329
LEGUMINOSAE/FABACEAE 120
   subfamily CAESALPINIOIDEAE 120
   subfamily PAPILIONOIDEAE 125
LEMNACEAE 56
Leopoldia 74
   comosa 74, **75**
Lepidium 208
   coronopus 208
   didymum 208
Leucaena 122
   leucocephala **121**, 122
Limoniastrum 225
   monopetalum 225
Limonium 221
   arboreum 222, **223**
   bollei 24, **224**, 225
   bourgaei 222, **223**
   bourgeaui 21
   brassicifolium 222
      subsp. *brassicifolium* 222
      subsp. *macropterum* 222
   dendroides 34, **224**, 225
   imbricatum 36, **220**, 221
   lobatum 222, **223**
   macrophyllum **220**, 222
   papillatum 22, 24, **224**, 225
      var. *papillatum* 225
   pectinatum 222
      var. *solandri* 222, **224**
   preauxii 27, 222, **223**
   puberulum 222, **223**
   sinuatum 222
   spectabile 33, 222
   sventenii 222
   tuberculatum **224**, 225
LINACEAE 181
Linaria 297
   simplex 297
   spartea 297
Linum 181
   bienne 181
   strictum **180**, 181
   usitatissimum 181
Lithospermum arvense 272
Lobularia 208
   canariensis 208
      subsp. *intermedia* 208, **209**
      subsp. *marginata* 208, **209**
   libyca 208, **209**
   maritima 208
Loeflingia 239
   hispanica 239
Logfia 331
   gallica 331, **332**
   minima 331
Lolium 89
   canariense 89, **90**
   multiflorum 89
   perenne 89
Lonicera 381
   etrusca 381
Lophocereus 255
   marginatus 255
Lotus 144
   angustissimus 146
   arinagensis 30, **145**, 146
   berthelotii 144, **145**
   broussonetii 146
   campylocladus **144**, 146
   edulis 144
   emeroides 34, 146
   eremeticus 144, **145**
   eriophthalmus 148, **149**
   glinoides 146, **147**
   holosericeus 146
   lancerottensis 146, **147**
   leptophyllus **145**, 146
   maculatus 144, **145**
   mascaensis 146
   ornithopodioides 146
   pedunculatus 144
   pyranthus 144, **145**
   sessilifolius 146, **147**
      subsp. *villosissimus* **147**

spartioides 146, **147**
tenellus 146
tenellus 145 **144**
uliginosus 144
Lupinus 129
   albus 130
   albus 131 **130**
   angustifolius 130
   luteus 129
   pilosus 129
Luzula 80
   canariensis 80, **81**
   forsteri 80
Lycium 288
   europaeum 288
   intricatum 9, **11**, 12, 288, **289**
Lysimachia 260
   arvensis **259**, 260
   foemina 260
   linum-stellatum 260
   wildpretii **259**, 261
LYTHRACEAE 185
Lythrum 185
   hyssopifolia **185**
   junceum 185

M

Maireana 246
   brevifolia 246
   brevifolia 244 **245**
Mairetis 273
   microsperma 273
Malcolmia 207
   littorea 207
   maritima 207
Malephora 252
   crocea 20, 252
   crocea 253 **252**
Malva 195
   acerifolia **30**, 33, 196
   acerifolia 194 **195**
   alcea 195
   arborea 196
   multiflora 195
   multiflora 194 **195**
   neglecta 195
   nicaeensis 195
   parviflora 195
   parviflora 194 **195**
   phoenicea 196
   phoenicea 194 **195**
   pseudolavatera 195
MALVACEAE 193
Marcetella 159
   moquiniana **158**, 159

Marrubium 307
   vulgare **306**, 307
Matthiola 207
   bolleana 207
      subsp. *bolleana* **206**, 208
      subsp. *morocera* **206**, 208
      subsp. *viridis* **206**, 208
   incana 208, **209**
   parviflora 208, **209**
Medicago 142
   arabica 143
   arborea 142, **143**
   ciliaris 143
   intertexta 143
   italica 144
   laciniata 142, **143**
   littoralis 143
   lupulina 142
   minima **143**, 144
   orbicularis 142
   polymorpha **143**, 144
   sativa 142, **143**
   soleirolii 143
   tornata 144
   truncatula 144
   turbinata 144
Melia 193
   azedarach 193, **194**
MELIACEAE 193
Melica 84
   canariensis 84, **85**
Melilotus 140
   albus 142
   indicus 140, **141**
   segetalis 140
   sulcatus 140, **141**
Melinis 95
   repens **95**, **96**
Melissa 311
   officinalis 311
Mentha 317
   longifolia 317
   pulegium 317
   spicata 317
   suaveolens 318
Mercurialis 181
   annua **180**, 181
Mesembryanthemum 252
   cordifolium 252
   crystallinum 7, **8**, 20, 252, **253**
   nodiflorum 7, 252, **253**
Micromeria 313
   benthamii **314**, 315
   densiflora 315
   ericifolia **314**, 315

glomerata **314,** 315
helianthemifolia 315
herpyllomorpha 315
lachnophylla 315, **316**
lanata **314,** 315
lasiophylla 315
mahanensis 315, **316**
pedro-luisi 315, **316**
pineolens 30, **314,** 315
rivas-martinezii **314,** 315
tenensis 315
teneriffae **311,** 313
tenuis 315
   subsp. *linkii* **314,** 315
tragothymus 315, **316**
Minuartia geniculata 230
Mirabilis 254
   jalapa 254
Misopates 296
   calycinum 297, **298**
   orontium 296, **298**
   salvagense 297, **298**
Monanthes 12, 40, 116
   anagensis 116, **117**
   brachycaulos **115,** 116
   icterica 116, **117**
   laxiflora 116, **117**
      var. *chlorotica* 116
      var. *laxiflora* 116, **117**
      var. *microbotrys* 116, **117**
   minima 116
   muralis 116
   pallens 33, 116, **117**
   polyphylla 33, 116, **117**
   wildpretii **115,** 116
MORACEAE 160
Morella faya 165
Moricandia 207
   arvensis **206,** 207
Morus 160
   alba 160
   nigra 160, **161**
Musa 78
   acuminata 78
MUSACEAE 78
Muscari comosum 74
Myoporum 302
   laetum **301,** 302
   tenuifolium **301,** 302
Myosotis 282
   discolor 282
      subsp. *canariensis* 282
   latifolia 282, **283**
   ramosissima 282

Myrica 165
   faya 12, **17,** 34, 36, **164,** 165
MYRICACEAE 165
MYRTACEAE 188
Myrtus 188
   communis 188

N

Narcissus 67
   tazetta 68, **69**
Nassella 86
   neesiana **85,** 86
Neatostema 273
   apulum 273
Neochamaelea pulverulenta 192
Neotinea 62
   maculata 62
Nepeta 311
   teydea 19, 33, **310,** 311
Neurada 196
   procumbens 196, **197**
NEURADACEAE 196
Nicandra 287
   physalodes **286,** 287
Nicotiana 290
   glauca 9, 290, **291**
   glutinosa **291,** 292
   paniculata 290
   tabacum **291,** 292
Nigella 103
   damascena 103
Notholaena **46**
   marantae
      subsp. *subcordata* 46, **47**
   marantae 47 **46**
Notobasis 374
   syriaca 374
Notoceras 214
   bicorne **213,** 214
NYCTAGINACEAE 254

O

Ocotea 55
   foetens 12, **15,** 36, **55**
Oenothera 186
   parodiana 186, **187**
   rosea 186, **187**
Olea 292
   cerasiformis 12, 293
   europaea 292, **293**
OLEACEAE 292
Oligomeris 204
   linifolia **203,** 204
ONAGRACEAE 186
Ononis 139

angustissima **138,** 139
    subsp. *angustissima* 139
    subsp. *longifolia* 139
catalinae 139, **141**
christii **138,** 139
diffusa 140
hebecarpa 139, **141**
hebecarpa 141
hesperia 7, **138,** 139
mitissima 140
pendula 140, **141**
reclinata 140
serrata 140, **141**
sicula 139
tournefortii **139,** 140
Onopordum 376
    nogalesii 376, **377**
OPHIOGLOSSACEAE 44
Ophioglossum
    lusitanicum 30
    polyphyllum 44, **45**
Ophrys 62
    bombyliflora 62, **63**
Opuntia 20, 257
    ficus-indica **256,** 257
    imbricata 257
    maxima 257
    stricta 257
    tuna **256,** 257
ORCHIDACEAE 61
Orchis 62
    canariensis 38, 62, **63**
    intacta 62
    mascula 62
    patens subsp. *canariensis* 62
Origanum 317
    vulgare 317
Ornithogalum 74
    narbonense 74, **75**
Ornithopus 148
    compressus 148
    perpusillus 148
    pinnatus 148, **149**
OROBANCHACEAE 323
Orobanche 325
    amethystea 325
    calendulae 325
    castellana affin. 325, **326**
    cernua 324, **326,** 327
    cernua 326
    crenata **326,** 327
    hederae **326,** 327
    minor 324, 325
Oryzopsis 91
    coerulescens 91

miliacea 91
Osyris 219
    lanceolata **218,** 219
Otanthus maritimus 338
OXALIDACEAE 168
Oxalis 168
    corniculata 168
    latifolia 168
    pes-caprae 168, **169**

P

Pallenis 343, 344
    hierichuntica 9, **342,** 343
    spinosa 343, **343**
Pancratium 68
    canariense 68, **69**
    maritimum 68
Panicum 98
    miliaceum 98
    repens 98
Papaver 99
    argemone 101
    dubium 99
    hybridum **100,** 101
    rhoeas 99, **100**
    somniferum 99
PAPAVERACEAE 99
Parapholis 91
    incurva **90,** 91
Paraserianthes 120
    lophantha 120, **121**
Parentucellia 323
    latifolia **322,** 323
    viscosa **322,** 323
Parietaria 162
    debilis 162
    filamentosa 162, **163**
    judaica 162, **163**
    mauritanica 162
Parolinia 217
    aridanae 217
    filifolia **216,** 217
    glabriuscula 27, 217
    intermedia 33, **216,** 217
    ornata **216,** 217
    platypetala 27, 217
    schizogynoides 34, 217
Paronychia 232
    argentea 232
    canariensis **231,** 232
    capitata 232
        subsp. *canariensis* 232
    echinulata 232
Patellifolia 247
    patellaris **11,** 247

procumbens 247
Pelargonium 185
  capitatum 185
Pelletiera wildpretii 261
Pennisetum
  setaceum **96,** 97
  villosum 97
Pericallis 353
  appendiculata **352,** 353
  cruenta **352,** 353
  echinata **352,** 353
  hadrosoma 30, 353
  hansenii 353
  lanata 33, **352,** 353
  multiflora 353
  murrayi 38, 353
  papyraceus 353
  steetzii 38, **352,** 353
  tirmensis 354
  tussilaginis 354
  webbii **29, 352,** 354
Periploca 271
  laevigata **22,** 27, 269, **270,** 271
Persea 54
  americana 54, **55**
  barbujana 12, **15,** 33, 54, **55**
    subsp. *barbujana* 54
    subsp. *ceballosi* 54
  indica 12, **18,** 33, 34, 54, **55**
Petrorhagia 238
  dubia 238
  nanteuilii 238
  prolifera **237,** 238
  velutina 238
Phagnalon 333
  purpurascens 333, **335**
  rupestre 333
  saxatile 333, **335**
  umbelliforme 334, **335**
Phalaris 86
  canariensis 86, **87**
Pharbitis preauxii 287
Phelipanche 325
  gratiosa 22, 324, 325, **326**
  lavandulacea 324, 325
  mutelii 325
  nana 325, **326**
  purpurea 325
  ramosa 325
Phillyrea 293
  angustifolia 293
Phoenix 78
  canariensis 12, 27, **77,** 78
  dactylifera 12, 78

Phragmites 94
  australis **93,** 94
  communis 94
Phyllis 265
  nobla 265
  viscosa 265
Picconia 12, 293
  excelsa **14,** 33, 293
Picris echiodes 370
Pimpinella 390
  anagodendron **391,** 393
  cumbrae 390, **391**
  dendrotragium 390, **391**
  junoniae **391,** 393
PINACEAE 51
Pinus 51
  canariensis 12, 19, 30, 51, **52**
  halepensis 51
  pinaster 51
  radiata 51
Piptatherum 91
  coerulescens **90,** 91
  miliaceum 91
Pistacia 189
  atlantica 12, **187,** 189
  lentiscus 12, 189
Pittosporum 382
  pitosporo 383
  undulatum 383
PITTSPORACEAE 382
PLANTAGINACEAE 294
Plantago 294
  afra 294
  albicans 294
  amplexicaulis 294, **295**
  arborescens **295,** 296
  aschersonii 294
  bellardii 294
  coronopus 294, **295**
  famarae **295,** 296
  lagopus **295,** 296
  lanceolata 296
  major **295,** 296
  ovata 294, **295**
  phaeostoma 294, **295**
  webbii **295,** 296
Pleiomeris 261
  canariensis **16,** 33, **259,** 261
Pleudia
  aegyptiaca 320
  herbanica 318
Plocama 264
  pendula 264, **265**
PLUMBAGINACEAE 221
Plumbago 221

auriculata **220,** 221
POACEAE 84
　subfamily ARISTIDOIDEAE 92
　subfamily ARUNDINOIDEAE 94
　subfamily CHLORIDOIDEAE 94
　subfamily DANTHONOIDEAE 98
　subfamily PANICOIDEAE 95
　subfamily POOIDEAE 84
Polycarpaea 239
　aristata 239, **240**
　carnosa 239, **240**
　divaricata 239, **240**
　filifolia 27, 239
　latifolia 239, **241**
　nivea 22, 24, 26, 239, **240**
　smithii 239, **240**
　tenuis 239
Polycarpon 233
　tetraphyllum 233, **234**
POLYGONACEAE 225
Polygonum 225
　aviculare 226
　equisetiforme 226
　maritimum **8,** 225, **227**
POLYPODIACEAE 48
Polypodium
　cambricum subsp. *macaronesicum* 48
　macaronesicum **13,** 48, **50**
Populus 172
　alba 172
　nigra 172
Portulaca 255
　canariensis **253,** 255
　oleracea 255
PORTULACACEAE 255
Potentilla 155
　indica **154,** 155
　reptans 155
Prasium 307
　majus 307
Prenanthes
　pendula 30
PRIMULACEAE 258
Prunella 311
　vulgaris 311
Prunus 155
　armeniaca 156
　domestica 155
　dulcis 156, **157**
　lusitanica
　　subsp. *hixa* 12, **17,** 33, 156, **157**
　persica 156, **157**
　　var. *nucipersica* 156
Pseudognaphalium 333
　luteoalbum **332,** 333

Pseudorlaya 389
　pumila 389
Pteranthus 233
　dichotomus 233
PTERIDACEAE 44
Pteridium
　aquilinum **13,** 46, **47**
Pterocephalus 382
　lasiospermus 19, 30, 33, 382, **384**
　porphyranthus 382
　virens 382, **384**
Pulicaria 334
　arabica
　　subsp. *hispanica* 334
　burchardii 26, 334, **335**
　canariensis **25,** 334, **335**
　　subsp. *lanata* 334,**335**
　odora 334
Punica 186
　granatum 186
Pyrostegia 305
　venusta **304,** 305
Pyrus 156
　communis 156

Q

Quercus 165
　ilex 165
　suber 165

R

Radiola 182
　linoides 182
Ramalina 9
RANUNCULACEAE 102
Ranunculus 102
　arvensis 102
　cortusifolius 103, **104**
　ficaria 103
　muricatus 102
　peltatus 103
　sardous 103
Raphanus 212
　raphanistrum 212, **213**
Rapistrum 210
　rugosum 210, **211**
Reichardia 368
　famarae 26, 368, **369**
　ligulata 368, **369**
　tingitana 9, **10,** 367, 368, **369**
Reseda 202
　crystallina 202, **203**
　lutea 202
　luteola 202, **203**
　scoparia 202

RESEDACEAE 202
Retama 129
   rhodorhizoides 12, 36, 38, **128,** 129
Rhagadiolus 359
   stellatus 359
RHAMNACEAE 159
Rhamnus 159
   alaternus 159
   crenulata 12, **158,** 160
   glandulosa **15, 158,** 160
   integrifolia 160
Rhodalsine 230
   geniculata 230, **231**
   platyphylla 26, 230, **231**
   webbii 230
Rhus 189
   albida 189
   coriaria **187,** 189
Ricinus 181
   communis **180,** 181
Ridolfia 386
   segetum 386
Rivasgodaya 125
Romulea 64
   grandiscapa 64, **65**
   hartungii 64
Rosa 153
   canina 155
   micrantha **154,** 155
ROSACEAE 153
Rosmarinus officinalis 320
Rostraria 88
   cristata 88
Rubia 262
   fruticosa 262, **263**
      subsp. *melanocarpa* 262, **263**
      subsp. *periclymenum* 262
   occidens 262, **263**
   peregrina 262
RUBIACEAE 262
Rubus 155
   bollei 12, **154,** 155
   palmensis 155
   ulmifolius **154,** 155
Rumex 226
   acetosella 226, **227**
   bipinnatus **228,** 229
   bucephalophorus 226, **227**
      subsp. *canariensis* 226
   conglomeratus 226, **227**
   crispus 226, **227**
   lunaria 12, 36, **228,** 229
   maderensis **228,** 229
   pulcher 226, **227**

   scutatus 229
   spinosus 226, **227**
   vesicarius **10, 228,** 229
Ruta 191
   chalepensis **190,** 191
   graveolens 191
   microcarpa 191
   oreojasme 192
   pinnata 33, **190,** 191
RUTACEAE 191
Rutheopsis 393
   herbanica **392,** 393
   tortuosa **392,** 393

S

Sagina 233
   apetala 233
   procumbens 233
SALICACEAE 172
Salix 172
   canariensis 12, **16, 27, 34, 171,** 172
Salpichroa 292
   origanifolia **291,** 292
Salsola 249
   divaricata 7, **248,** 249
   tetrandra 249
   tragus **247,** 249
   vermiculata 249
Salvia 318
   aegyptiaca 320, **321**
   broussonetii 320, **321**
   canariensis 27, 33, 320, **321**
   fruticosa 320
   herbanica 318, **321**
   officinalis 320
   rosmarinus 320
   triloba 320
   verbenaca 320, **321**
Sambucus 379
   palmensis 379, **380**
Samolus 261
   valerandi **260,** 261
Sanguisorba 159
   megacarpa **157,** 159
SANTALACEAE 217
SAPINDACEAE 191
SAPOTACEAE 258
Sarcocornia 249
   perennis 9, 249, **250**
Scabiosa 381
   atropurpurea **380,** 381
Scandix 383
   pecten-veneris **384,** 385
Schenkia 266
   spicata 266

Schinus 189
  molle **187**, 189
  terebinthifolia **187**, 189
Schizogyne 355
  glaberrima **354**, 355
  sericea 38, **354**, 355
Scilla 74
  haemorrhoidalis 74, **75**
  latifolia **22**, 74, **75**
Scolymus 355
  grandiflorus 356
  hispanicus 355, **357**
  maculatus 356
Scorpiurus 153
  muricatus **152**, 153
  sulcatus **152**, 153
  vermiculatus 153
Scrophularia 300
  arguta 300
  calliantha **301**, 302
  glabrata 19, 33, 300, **301**
  smithii 302
    subsp. *smithii* **301**, 302
SCROPHULARIACEAE 300
Searsia 189
  albida 189, **190**
Sedum 106
  nudum 21
    subsp. *lancerottense* 21, 106, **107**
  rubens 106, **107**
Seirophora 9
Semele 72
  androgyna 72, **73**
  gayae 72
Senecio 339
  angulatus 341
  flavus 341
  flavus 342 **343**
  glaucus 341
    subsp. *coronopifolius* 341
    subsp. *coronopifolius* 340 **341**
  incrassatus 341
  incrassatus 342 **343**
  leucanthemifolius 341
    var. *falcifolius* **340**, 341
  leucanthemifolius 340 **341**
  massaicus **340**, 341
  vulgaris 339
Senna 124
  bicapsularis **123**, 124
  corymbosa 124
  didymobotrya **123**, 124
Serapias 64
  parviflora 64
Seseli webbii 393

Setaria 98
  adhaerens **96**, 98
  parviflora **96**, 98
  viridis 98
Sherardia 264
  arvensis 264, **265**
Sideritis 19, 308
  canariensis 308, **309**
  cystosiphon 308
  dasygnaphala 27, 308
  dendrochahorra 308, **309**
  discolor 27, 308, **310**
  gomeraea **310**, 311
  infernalis 33, 308
  kuegleriana 308, **309**
  lotsyi 308
  macrostachys **306**, 308
  nervosa 308
  nutans 34, 311
  oroteneriffae 308, **309**
  pumila 21, 308, **310**
  soluta 311
    subsp. *gueimaris* **310**, 311
    subsp. *soluta* **310**, 311
Sideroxylon 258
  canariense **18**, 258, **259**
Silene 236
  apetala 236, **237**
  behen 237, 238
  berthelotiana **237**, 238
  conica 236
  cretica 236
  gallica 236, **237**
  gracilis 236, **237**
  inaperta 236
  lagunensis 238
  nocteolens 238
  nutans 238
  rubella 236
  sabinosae 38, 238
  tamaranae 238
  tridentata **237**, 238
  vulgaris **237**, 238
Silibum 376
  marianum 376, **377**
SIMAROUBACEAE 192
Sinapis 210
  alba 210
  arvensis **209**, 210
  mostaza **210**
Sisymbrium 205
  erysimoides 205, **206**
  irio 205
  officinale 205, **206**
  orientale 205

SMILACACEAE 61
Smilax 61
  aspera 61
    subsp. mauritanica **60**, 61
  canariensis **60, 61**
Smyrnium 383
  olusatrum 383
SOLANACEAE 287
Solandra 292
  maxima **291**, 292
Solanum 288
  jasminoides 290
  laxum 290
  lidii 290
  mauritianum 289, 290
  nigrum 288, **289**
  vespertilio 290
    subsp. vespertilio **289**, 290
  villosum 290
Solenopsis 329
  laurentia 329
Sonchus 40, 359
  acaulis **29, 364**, 365
  arboreus **361**, 362
  asper **358**, 359
  bornmuelleri 36, 365
  bourgeaui **358**, 360
  brachylobus 360
  bupleuroides 365
  canariensis **364**, 365
  capillaris **361**, 362
  congestus 362, **363**
  esperanzae 360
  fauces-orci 365
  filifolius 362
  gandogeri 362
  gomerensis 365
  gonzalezpadronii 365
  gummifer **364**, 365
  heterophyllus **361**, 362
  hierrensis 38, 362, **363**
  leptocephalus 360, **361**
  lidii 362
  microcarpus **361**, 362
  oleraceus **358**, 360
  ortunoi **361**, 362
  palmensis 36, **363**, 365
  pendulus 360, **361**
  pinnatifidus **363**, 365
  pitardii 362
  platylepis 362
  radicatus **364**, 365
  regis-jubae 362
  sventenii 362
  tectifolius 365
  tectifolius 364 **365**
  tenerrimus 360
  tuberifer 360, **361**
  webbii 360
  wildpretii 362
Sorghum 95
  bicolor 95
  halepense 95, **96**
Spartium 127
  junceum 127, **128**
Spartocytisus 19
  filipes **128**, 129
  supranubius 19, **128**, 129
Spathodea 305
  campanulata **304**, 305
Spergula 235
  arvensis 235
  fallax **234**, 235
  pentandra 235
Spergularia 235
  bocconei **234**, 235
  diandra 235
  fimbriata **234**, 235
  marina 235
  media 235
Squamarina
  cartilaginea 9
Stachys 312
  arvensis **311**, 312
  germanica 312
    subsp. cordigera **311**, 312
  ocymastrum **311**, 312
Stereocaulon 9
Stipa 92
  capensis 92, **93**
Strelitzia 78
  reginae 78, **79**
STRELITZIACEAE 78
Suaeda 7, 9, 21, 249
  ifniensis **9**, 22, **250**, 251
  mollis 251
  spicata **250**, 251
  vera **9**, 249, **250**
Sventenia bupleuroides 365
Symphyotrichum squamatum 355

T

Taeckholmia heterophylla 362
TAMARICACEAE 221
Tamarix 221
  africana 221
  canariensis 9, **220**, 221
Tamus edulis 58
Tara spinosa 124

Taraxacum 370
  officinale 370
    sect. *Ruderalia* 370
Tecoma 305
  stans **304**, 305
Telene microphylla 125
Teline
  canariensis 125
  nervosa 125
  osyrioides 125
  pallida 125
  rosmarinifolia 27, 125
  splendens 36
  stenopetala
    subsp. *microphylla* 38
  stenopetala 125
Tetraena 22, 118
  fontanesii 7, **8**, 118, **119**
  gaetula **10**, 118, **119**
    subsp. *gaetula* 118, **119**
    subsp. *waterlotii* 118
Tetragonia 254
  tetragonoides **253**, 254
Teucrium 307
  heterophyllum 307
    subsp. *brevipilosum* **306**, 307
    subsp. *hierrense* 307
  spinosum 307
THEACEAE 261
Theligonum 266
  cynocrambe 266
Thesium 217
  canariense 219
  humile 217
  retamoides 33, **218**, 219
  subsucculentum **218**, 219
THYMELAEACEAE 197
Thymus 317
  origanoides 21, **316**, 317
  vulgaris 317
Todaroa 393
  aurea **392**, 393
    subsp. *aurea* 393
    subsp. *suaveolens* **392**, 393
Tolpis 356
  calderae 36
  coronopifolia 356, **357**
  crassiuscula 356, **357**
  laciniata 356
  lagopoda 356, **357**
  proustii 38, 356
  santosii 36, 356, **357**
  umbellata 356, **357**
  webbii 33, 356, **357**
Toninia

tristis 9
Torilis 389
  arvensis 389
  elongata 389
  nodosa 389
Toris
  elongata 388 **389**
Traganum 251
  moquinii 7, **8**, 22, 24, 251, **253**
Tragopogon 359
  porrifolius **358**, 359
Tragus 95
  racemosus **93**, 95
Tribulus 120
  terrestris 120, **121**
Trifolium 148
  angustifolium 148, **149**
  arvense **149**, 150
  campestre 148, **149**
  cherleri 150
  dubium 148, **149**
  fragiferum 151
  glomeratum 150
  hirtum 151
  lappaceum 150
  pratense 151
  repens **149**, 150
  resupinatum 151
  scabrum **149**, 150
  spumosum 151
  squamosum 151
  squarrosum 151
  stellatum 150
  striatum 151
  subterraneum 151
  suffocatum 150
  tomentosum **149**, 150
Trigonella 142
  stellata 142, **143**
TROPAEOLACEAE 202
Tropaeolum 202
  majus 202, **203**
Tuberaria 199
  guttata 199
  lignosa 200
Turbina corymbosa 285

U

Ulex 127
  europaeus **126**, 127
ULMACEAE 160
Ulmus 160
  minor 160
Umbilicus 106
  gaditanus 106, **107**

Urginea maritima 74
Urospermum 368
   picroides **367**, 368
Urtica 162
   dubia 162
   membranacea 162
   morifolia 162, **163**
   stachyoides 162, **163**
   urens 162, **163**
URTICACEAE 162

V

Valantia 265
   hispida 265
Valerianella 382
   dentata 382
   discoidea 382
   eriocarpa 382
   locusta 382
Verbascum 302
   sinuatum 302
   thapsus **301**, 302
   virgatum **301**, 302
Verbena 306
   officinalis 306
VERBENACEAE 305
Veronica 299
   anagallis-aquatica 299
   arvensis 299
   hederifolia 300
   persica 300
   polita 300
   serpyllifolia 299
Viburnum 379
   rugosum **14**, 379, **380**
Vicia 133
   aphylla 135
   benghalensis **10**, 135, **136**
   chaetocalyx 135
   disperma 133
   ervilia 135
   filicaulis 135
   hirsuta 133
   lutea 135, **136**
   nataliae 135, **136**
   parviflora 133, **136**
   pubescens 135
   sativa 135
   scandens 135, **136**
   tetrasperma 135
   villosa 135
   voggenreiteriana 135
   vulcanorum 135
Vieria 346

   laevigata 346, **347**
Vinca 268
   major **267**, 268
Viola 170
   anagae **171**, 172
   arvensis 172
   cheiranthifolia 20, 33, **171**, 172
   kitaibeliana 172
   odorata 170
   palmensis 36, **171**, 172
   riviniana 170, **171**
VIOLACEAE 170
Visnea 12
   mocanera **16**, 33, **260**, 261
VITACEAE 118
Vitis 118
   vinifera 118, **119**
Volutaria 378
   bollei 378
   canariensis 378
   tubuliflora 378

W

Wahlenbergia 330
   lobeliodes 330
Washingtonia 76
   filifera 76
   robusta 76, **77**
Withania 288
   aristata 288, **289**
   frutescens 288
   somnifera 288
Woodwardia 48
   radicans 12, 36, **48**, **49**

X

Xanthium 344
   spinosum 344, **345**
XANTHORRHOEACEAE 66
Xolantha 199

Z

Zantedeschia 56
   aethiopica 56, **57**
ZYGOPHYLLACEAE 118